Linear Partial Differential and Difference Equations and Simultaneous Systems with Constant or Homogeneous Coefficients

Linear Partial Differential and Difference Equations and Simultaneous Systems: With Constant or Homogeneous Coefficients is part of the series "Mathematics and Physics for Science and Technology," which combines rigorous mathematics with general physical principles to model practical engineering systems with a detailed derivation and interpretation of results. Volume V presents the mathematical theory of partial differential equations and methods of solution satisfying initial and boundary conditions, and includes applications to: acoustic, elastic, water, electromagnetic, and other waves; the diffusion of heat, mass, and electricity; and their interactions. This is the third book of the volume.

The book starts with five different methods of solution of linear partial differential equations (p.d.e.) with constant coefficients. One of the methods, namely characteristic polynomial or exponential of several variables, is then extended to a further five classes, including linear p.d.e. with homogeneous power coefficients and finite difference equations and simultaneous systems of both (simultaneous partial differential equations [s.p.d.e.] and simultaneous finite difference equations [s.f.d.e.]). The applications include detailed solutions of the most important p.d.e. in physics and engineering, including the Laplace, heat, diffusion, telegraph, bar, and beam equations. The free and forced solutions are considered together with boundary, initial, asymptotic, starting, and other conditions.

The book is intended for graduate students and engineers working with mathematical models and can be applied to problems in mechanical, aerospace, electrical, and other branches of engineering dealing with advanced technology, in the physical sciences, and in applied mathematics.

Mathematics and Physics for Science and Technology

Series Editor: L.M.B.C. Campos
former Director of the Center for Aeronautical
and Space Science and Technology
Lisbon University

Volumes in the series:

For more information about this series, please visit: http://www.crcpress.com/Mathematics-and-Physics-for-Science-and-Technology/book-series/CRCMATPHYSCI

Linear Partial Differential and Difference Equations and Simultaneous Systems with Constant or Homogeneous Coefficients

Luiz Manuel Braga da Costa Campos
and
Luís António Raio Vilela

CRC Press
Taylor & Francis Group
Boca Raton London New York

CRC Press is an imprint of the
Taylor & Francis Group, an **informa** business

Cover image: Luis Manuel Braga da Costa Campos and Luís António Raio Vilela

First edition published 2024
by CRC Press
2385 NW Executive Center Drive, Suite 320, Boca Raton, FL 33431

and by CRC Press
4 Park Square, Milton Park, Abingdon, Oxon, OX14 4RN

CRC Press is an imprint of Taylor & Francis Group, LLC

© 2024 Luis Manuel Braga da Costa Campos and Luís António Raio Vilela

Library of Congress Cataloging-in-Publication Data

Names: Campos, Luis Manuel Braga da Costa, author. | Vilela, Luís António Raio, author.
Title: Linear partial differential and difference equations and simultaneous systems : with constant or homogeneous coefficients / Luis Manuel Braga da Costa Campos and Luís António Raio Vilela.
Description: First edition. | Boca Raton, FL : CRC Press, 2024. | Series: Mathematics and physics for science and technology | Includes bibliographical references and index.
Identifiers: LCCN 2023048341 | ISBN 9781032624730 (hardback) | ISBN 9781032635187 (paperback) | ISBN 9781032635200 (ebook)
Subjects: LCSH: Differential equations, Partial--Numerical solutions. | Differential equations, Linear--Numerical solutions. | Continuation methods. | Equations, Simultaneous.
Classification: LCC QA377 .C275 2024 | DDC 515/.353--dc23/eng/20240201
LC record available at https://lccn.loc.gov/2023048341

ISBN: 978-1-032-62473-0 (hbk)
ISBN: 978-1-032-63518-7 (pbk)
ISBN: 978-1-032-63520-0 (ebk)

DOI: 10.1201/9781032635200

Typeset in Times
by KnowledgeWorks Global Ltd.

to Francelina

Contents

Sections / subsections

Volume V – Partial Differential Equations with Applications to Waves and
       Diffusion
Book 12 – Linear Partial Differential and Finite Difference Equations and
       Simultaneous Systems

List of Tables

Notes: Generalised Equation of Mathematical Physics and Engineering (GEMPE)

List of Figures and Diagrams

FIGURES

DIAGRAMS

Preface

The preface to the series "*Mathematics and Physics Applied to Science and Technology*" can be found in book 10 and is not repeated in this book 12 of the same series. Likewise, the preface to volume V of the series on *Partial Differential Equations with Applications to Waves and Diffusion* is found in book 10, which is the first book of volume V, and is concerned with *Vector Fields with Application to Thermodynamics and Irreversibility*. Among the major applications of reversible and irreversible thermodynamics is *Compressible Flow with Applications to Engines, Shocks and Nozzles*, which is book 11 of the series and is the second book of volume V. The third book of volume V is the present book 12 with title *Linear Partial Differential and Finite Difference Equations and Simultaneous Systems*. It starts with five methods of solution of a linear partial differential equation with constant coefficients: (i) separation of variables (Section 3.1); (ii) similarity functions (Section 3.2); (iii) symbolic differential operators (Section 3.3); (iv) exponentials of several variables leading to a characteristic polynomial (Subsections 3.4.1–3.4.39); and (v) factorisation considered before (Sections 1.6–1.9 of book 10) and extended here (Subsections 3.4.40–3.4.44). The most general of the five methods is (iv) exponentials of several variables, which is both the simplest to apply and the most readily extended from one to six cases when characteristic polynomials exist, namely, single (simultaneous system of) linear: (a/b) partial differential equations (p.d.e.) with constant coefficients [Section 3.4 (3.5)]; (c/d) p.d.e. with power coefficients and homogeneous derivatives [Section 3.6 (3.7)]; and (e/f) multiple finite difference equations with constant coefficients [Section 3.8 (3.9)]. The examples of solution include the generalised equation of mathematical physics and engineering (notes 3.1–3.12), and particular cases such as the harmonic, wave, diffusion, telegraph, elastic bar, and beam equations.

ORGANISATION AND PRESENTATION OF THE SUBJECT MATTER

Volume V (*Partial Differential Equation with Applications to Waves and Diffusion*) is organised like the preceding four volumes of the series *Mathematics and Physics applied to Science and Technology*: (volume IV) *Ordinary Differential Equations with Applications to Trajectories and Oscillations*; (volume III) *Generalized Calculus with Applications to Matter and Forces*; (volume II) *Elementary Transcendentals with Applications to Solids and Fluids*; and (volume I) *Complex Analysis with Applications to Flows and Fields*. Volume V consists of ten chapters: (i) the odd-numbered chapters present mathematical developments; (ii) the even-numbered chapters contain physical and engineering applications; and (iii) the last chapter is a set of 20 detailed examples of (i) and (ii). The chapters are divided into sections and subsections, for example, Chapter 1, Section 1.1, and Subsection 1.1.1. The formulas are numbered by chapters in curved brackets, for example (1.2) is Equation 2 of Chapter 1. When referring to volume I, the symbol I is inserted at the beginning, for example: (i) chapter I.36.1, Subsection I.36.1.2; and (ii) equation (I.36.33a). The final part of each chapter includes: (i) a conclusion referring to the figures as a kind of visual summary; and (ii) the notes, lists, tables, diagrams, and classifications as additional support. The latter (ii) appear at the end of each chapter and are numbered within the chapter (for example, diagram–D2.1, note–N1.1, table–T2.1); if there is more than one diagram, note, or table, they are numbered sequentially (for example, notes–N2.1 to N2.20). The chapter starts with an introductory preview, and related topics may be mentioned in the notes at the end. The lists of mathematical symbols and physical quantities appear before the main text, and the index of subjects and bibliography are found at the end of the book.

About the Authors

L.M.B.C. Campos was born on March 28, 1950, in Lisbon, Portugal. He graduated in 1972 as a mechanical engineer from the Instituto Superior Técnico (IST) of Lisbon Technical University. The tutorials as a student (1970) were followed by a career at the same institution (IST) through all levels: assistant (1972), assistant with tenure (1974), assistant professor (1978), associate professor (1982), chair of Applied Mathematics and Mechanics (1985). He has served as the coordinator of undergraduate and postgraduate degrees in Aerospace Engineering since the creation of the programs in 1991. He was the coordinator of the Scientific Area of Applied and Aerospace Mechanics in the Department of Mechanical Engineering and also the director (and founder) of the Center for Aeronautical and Space Science and Technology until retirement in 2020. He is currently Emeritus Professor.

In 1977, Campos received his doctorate on "waves in fluids" from the Engineering Department of Cambridge University, England. Afterwards, he received a Senior Rouse Ball Scholarship to study at Trinity College while on leave from IST. In 1984, his first sabbatical was as a Senior Visitor at the Department of Applied Mathematics and Theoretical Physics of Cambridge University, England. In 1991, he spent a second sabbatical as an Alexander von Humboldt scholar at the Max-Planck Institut fur Aeronomic in Katlenburg-Lindau, Germany. Further sabbaticals abroad were excluded by major commitments at the home institution. The latter were always compatible with extensive professional travel related to participation in scientific meetings, individual or national representation in international institutions, and collaborative research projects.

Campos received the von Karman medal from the Advisory Group for Aerospace Research and Development (AGARD) and Research and Technology Organization (RTO). Participation in AGARD/RTO included serving as a vice-chairman of the System Concepts and Integration Panel, and chairman of the Flight Mechanics Panel and of the Flight Vehicle Integration Panel. He was also a member of the Flight Test Techniques Working Group. There, he was involved in the creation of an independent flight test capability, active in Portugal during the last 40 years, which has been used in national and international projects, including Eurocontrol and the European Space Agency. The participation in the European Space Agency (ESA) has afforded Campos the opportunity to serve on various program boards at the levels of national representative in Program Boards and General Council.

His involvement in activities sponsored by the European Union (EU) has included the following: (i) participation in 25 research projects with industry, research, and academic institutions including coordination of two; (ii) membership of various Committees, including Vice-Chairman of the Aeronautical Science and Technology Advisory Committee; and (iii) participation on the Space Advisory Panel on the future role of EU in space. Campos has been a member of the Space Science Committee of the European Science Foundation, which liaises with the Space Science Board of the National Science Foundation of the United States. He has been a member of the Committee for Peaceful Uses of Outer Space (COPUOS) of the United Nations. He has served as a consultant and advisor on behalf of these organisations and other institutions. His participation in professional societies includes member and vice-chairman of the Portuguese Academy of Engineering, fellow of the Royal Aeronautical Society, Royal Astronomical Society, and Cambridge Philosophical Society, associate fellow of the American Institute of Aeronautics and Astronautics, and founding and life member of the European Astronomical Society.

Campos has published and worked on numerous books and articles. His publications include 20 books, of which 16 as a single author, 2 with co-authors, 1 as an editor, and 1 as a co-editor. He has published 159 papers (82 as the single author, including 12 reviews) in 60 journals, and 278 communications to symposia. He has served as reviewer for 43 different journals, in addition to 37 reviews published in *Mathematics Reviews*. He is or has been a member of the editorial boards of several journals, including *Progress in Aerospace Sciences, International Journal of Aeroacoustics, International Journal of Sound and Vibration*, and *Air & Space Europe*.

Campos' areas of research focus on four topics: acoustics, magnetohydrodynamics, special functions, and flight dynamics. His work on acoustics has concerned the generation, propagation, and refraction of sound in flows with mostly aeronautical applications. His work on magnetohydrodynamics has concerned magneto-acoustic-gravity-inertial waves in solar-terrestrial and stellar physics. His developments on special functions have used differintegration operators, generalising the ordinary derivative and primitive to complex order; they have led to the introduction of new special functions. His work on flight dynamics has concerned aircraft and rockets, including trajectory optimisation, performance, stability, control, and atmospheric disturbances. There are additional publications in solid mechanics including elasticity, also in general relativity and gravitation, in quantum mechanics, in number theory, and statistics and population models.

Campos' professional activities on the technical side are balanced by other cultural and humanistic interests. Complementary non-technical interests include classical music (mostly orchestral and choral), plastic arts (painting, sculpture, architecture), social sciences (psychology and biography), history (classical, renaissance, and overseas expansion), and technology (automotive, photo, audio). Campos is listed in various biographical publications, including *Who's Who in the World* since 1986, *Who's Who in Science and Technology* since 1994, and *Who's Who in America* since 2011.

L.A.R. Vilela was born on March 11, 1994, in Vila Real, Portugal. He has obtained a Bachelor's degree in Aerospace Engineering and is currently working toward a Master's degree, also in Aerospace Engineering, at Instituto Superior Técnico (IST) of Lisbon University. He has worked as co-author on the two previous books of this series. Besides physical sciences, Vilela has a special interest in philosophy and psychology, and, in general, an interest in an advanced understanding of the human nature and mind.

Acknowledgements

The fifth volume of the series justifies renewing some of the acknowledgments made in the first four volumes. Thanks are due to those who contributed more directly to the final form of this book: L. Sousa for help with manuscripts; Mr. J. Coelho for all the drawings; and at last, but not least, to my wife, my companion in preparing this work.

Abbreviations for Mathematical Equations

F.D.E.   Finite Difference Equation
O.D.E.   Ordinary Differential Equation
P.D.E.   Partial Differential Equation
P.F.D.E.   Partial Finite Difference Equation
S.F.D.E.   Simultaneous Finite Difference Equations or System of F.D.E.
S.O.D.E.   Simultaneous Ordinary Differential Equations or System of O.D.E.
S.P.D.E.   Simultaneous Partial Differential Equations or System of P.D.E.
S.P.F.D.E. Simultaneous Partial Finite Difference Equations or System of
                P.F.D.E.

List of Physical Quantities

The location of first appearance is indicated, for example, "2.7" means *Section 2.7*; "6.8.4" means *Subsection 6.8.4*; "N.8.8" means *note 8.8*; and "E10.13.1" means *example 10.13.1*.

c Wave speed: 3.1.2
k Wavenumber: 3.1.3
q Stiffness parameter: 3.2.6
r Radial distance in plane: 3.3.1
s Slope: 3.3.3
t Time: 3.1.2
v Velocity: 3.3.3
x Cartesian coordinate: 3.1.1
y Cartesian coordinate: 3.1.1
z Complex variable: 3.1.1
z^* Complex conjugate: 3.1.1
E Young's modulus: 3.2.6
I Moment of inertia: 3.2.6
ϕ Phase: 3.1.2
φ Polar angle: 3.3.1
ϑ Friction coefficient: N3.1
λ Wavelength: 3.1.3
μ Spatial damping: 3.3.10
ν Temporal damping: 3.3.10
ω Frequency: 3.2.2
ρ Mass density: 3.2.6
τ Period: 3.2.2
χ Diffusivity: 3.1.4
ζ Displacement: 3.3.3
Φ Dependent variable of partial differential equation: 3.1.1

3 Linear Partial Differential and Difference Equations and Simultaneous Systems with Constant or Homogeneous Coefficients

The three most important partial differential equations (p.d.e.) of mathematical physics are the: (i) Laplace or harmonic equation that describes potential fields (Volumes I–III); (ii) wave equation that describes propagation problems (notes IV.5.1–IV.5.52); and (iii) diffusion or heat equation that describes diffusion processes (notes V1.1–V.1.5). Their simplest forms are: (i) the harmonic equation in two-dimensional Cartesian coordinates; (ii/iii) the one-dimensional wave (diffusion) equation that involves time in addition to one Cartesian coordinate. The (i) harmonic and (ii) wave equations have been solved by method I of similarity solutions (Sections 1.6–1.9) that applies to linear p.d.e. with constant coefficients and all derivatives of the same order. The (iii) diffusion equation involves derivatives of different orders, and together with the (i) harmonic and (ii) wave equations, can be solved by four other methods II to V, that apply to linear p.d.e. with constant coefficients without requiring the derivatives to be all of the same orders.

Method II of separation of variables (Section 3.1) applies to linear p.d.e., including some with variable coefficients, but not all with constant coefficients. Method III of similarity functions (Section 3.2) applies to all linear p.d.e. with constant coefficients, and if the derivatives are (are not) all of the same order, it leads to similarity solutions (may lead to non-similarity solutions) as (unlike) the method I. Method IV of symbolic differential operators (Section 3.3) has no "a priori" restrictions, but it leads to series expansions that have closed-form solutions only in simple cases. Method V of exponential solutions (Section 3.4) applies to all linear p.d.e. with constant coefficients, and derivatives of equal or unequal order, and is the simplest since it depends only on the roots of a characteristic polynomial. It is taken as the baseline to extend to: (i) linear p.d.e. of the Euler or homogeneous type with power coefficients with the same exponent as the order of the derivatives

DOI: 10.1201/9781032635200-1

they multiply (Section 3.5); (ii/iii) simultaneous systems of linear p.d.e. with constant (homogeneous power) coefficients [Section 3.6 (3.7)]; and (iv/v) single (simultaneous systems of) linear finite difference equations with constant coefficients [Section 3.8 (3.9)].

The most important class of ordinary (partial) differential equations o.d.e. (p.d.e.) [Sections IV.1.3–IV.1.5 (V.3.4)] is the linear type with constant coefficients: (i) the general integral of the equation without forcing term is a combination of exponentials, and its determination reduces to the algebraic problem of finding the roots of the characteristic polynomial in one (several) variables [Sections IV.1.3 (V.3.4)]; (ii) the exponentials have power coefficients when the characteristic polynomial has multiple roots; (iii) the characteristic polynomial can also be used to obtain a particular integral, when the forcing term is a polynomial, exponential, circular, or hyperbolic sine or cosine, or a product of these functions [Sections IV.1.4–IV.1.5 (V.3.4)]; (iv) a particular integral can also be obtained by the method of inverse characteristic polynomial in one (several) variables if the forcing term is an exponential, a polynomial, or their product by a smooth or infinitely differentiable function [Section IV.1.5 (V.3.4)]; and (v) the complete integral of the equation with the forcing term is then found as the sum of the general integral of the equation without the forcing term plus a particular integral of the equation with the forcing term.

The methods used to solve a single linear o.d.e. (p.d.e.) with constant coefficients also apply in the case of power coefficients [Sections IV.1.6–IV.1.8 (V.3.5)]. These methods are based on the characteristic polynomial, which also exists in a modified form, replacing ordinary with homogeneous derivatives, thus leading to linear p.d.e. with variable coefficients of the Euler type; that is, each derivative of order r is multiplied by a power of exponent r. In this case, the general integral of the equation without the forcing term involves powers for single roots of the characteristic polynomial and also logarithms for multiple roots of the characteristic polynomial [Section IV.1.6 (V.3.5)]; also a particular integral can be found when the forcing term is a power, a polynomial, a circular or hyperbolic sine or cosine of logarithms, or a product of these functions [Sections IV.1.7–IV.1.8 (V.3.5)]. The method of the characteristic polynomial in one (several) variables for linear o.d.e. (p.d.e.) with constant [Sections IV.1.3–IV.1.5 (V.3.4)] or homogeneous power [Sections IV.1.6–IV.1.8 (V.3.5)] coefficients extends to simultaneous systems of o.d.e (p.d.e.) with constant [Sections IV.7.4–IV.7.6 (V.3.6)] or power [Sections IV.7.7–IV.7.9 (V.3.7)] coefficients using a matrix of polynomial operators. The characteristic polynomials in several variables in the matrices for simultaneous linear p.d.e use ordinary (homogeneous) derivatives in the case of constant (power) coefficients [Sections V.1.6 (V.1.7)]. The characteristic polynomial in one (several) variables also exists for linear single (multiple) finite difference equations with constant coefficients [Section IV.7.8 (V.3.8)] and generalises to matrices of polynomials for the extension to simultaneous systems [Section IV.7.9 (V.3.9)].

3.1 METHOD OF SEPARATION OF VARIABLES

The method of separation of variables applies (Subsection 3.1.7) to linear p.d.e. in some cases of constant (Section 3.1) or variable (notes IV.8.1–IV.8.15) coefficients, and is illustrated by the: (i) two-dimensional harmonic equation in Cartesian coordinates, leading to solutions that are arbitrary functions of a complex variable or its conjugate (Subsection 3.1.1); (ii) one-dimensional wave equation involving time and one Cartesian coordinate, leading to similarity solutions with a permanent shape propagating with the same speed in opposite directions (Subsection 3.1.2) that may form standing modes by superposition (Subsection 3.1.3); (iii) one-dimensional diffusion equation involving time and one Cartesian coordinate leading to solution(s) that are real and non-permanent and decay in space-time (Subsections 3.1.4–3.1.5). The cases (i/ii/iii) are examples (Subsection 3.1.6) of elliptic/hyperbolic/parabolic second-order p.d.e., respectively. A linear p.d.e. with constant coefficients may be non-separable (Subsection 3.1.7) whereas another with variable coefficients can be separable (Subsection 3.1.8).

3.1.1 Two-Dimensional Cartesian Harmonic Equation (Laplace 1820)

The two-dimensional **harmonic equation** (I.11.6) in the plane of Cartesian coordinates is the linear second-order p.d.e. (3.1a) with constant coefficients and all derivatives of the same order:

$$0 = \frac{\partial^2 \Phi}{\partial x^2} + \frac{\partial^2 \Phi}{\partial y^2}: \qquad \Phi(x,y) = X(x)\,Y(y), \qquad \text{(3.1a, b)}$$

a solution is sought by the **method of separation of variables** as the product of separate functions of each variable (3.1b). Substitution of (3.1b) in (3.1a) and division by (3.1b) lead to (3.2a) where the right and left sides depend on different variables and can be equal only if they have the same constant value (3.2b):

$$\frac{1}{X}\frac{d^2X}{dx^2} = -\frac{1}{Y}\frac{d^2Y}{dy^2} = a^2: \qquad \frac{d^2X}{dx^2} - a^2\,X = 0 = \frac{d^2Y}{dy^2} + a^2\,Y. \qquad \text{(3.2a–d)}$$

From (3.2a) [(3.2b)], follow the linear second-order o.d.e. (3.2c) [(3.2d)] that have two linearly independent solutions (3.3a) [(3.3b)]:

$$X_\pm(x) = e^{\pm ax}, \quad Y_\pm(y) = e^{\pm iay}; \qquad \text{(3.3a, b)}$$

the product of (3.3a) and (3.3b) specifies, by (3.1b) $\equiv$ (3.4a), the four solutions (3.4b):

$$\Phi_{\pm\pm}(x,y) = X_\pm(x)\,Y_\pm(y) = \exp\left[\pm a\,(x \pm i\,y)\right] \qquad \text{(3.4a, b)}$$

of the two-dimensional Cartesian harmonic equation (3.1a).

Since the constant a in (3.4a) is arbitrary, the four solutions may be multiplied by any function of a and integrated over a, leading to two arbitrary functions:

$$\int \Phi_{\pm\pm}(x,y)\,C_{\pm\pm}(a)\,da = \int C_{\pm\pm}(a)\,\exp\left[\pm a\,(x \pm i\,y)\right]da = f_\pm(x \pm i\,y); \qquad \text{(3.5a, b)}$$

there are only two distinct solutions in (3.5a, b) since replacing a with $-a$ makes no difference, because a is arbitrary. The two solutions (3.5a, b) are functions of a complex variable (3.6a) and its conjugate (3.6b):

$$z = x + i\,y, \quad z^* = x - i\,y: \quad f_+(x + i\,y) = f_+(z), \quad f_-(x - i\,y) = f_-(z^*). \tag{3.6a–d}$$

The *sum of two arbitrary differentiable functions (3.7a) of the complex variable (3.6a) and its conjugate (3.6b) is the general integral (3.7b, c)*

$$f_\pm \in \mathcal{D}(\mathbb{I}\,C): \quad \Phi(x, y) = f_+(x + i\,y) + f_-(x - i\,y) = f_+(z) + f_-(z^*) \tag{3.7a–c}$$

of the two-dimensional Cartesian harmonic equation (3.1a), because: (i) the two solutions (3.5a, b) are linearly independent; (ii) the sum of solutions (3.7b) is a solution because the p.d.e. (3.1a) is linear; and (iii) the general integral (3.7c) involves two arbitrary functions as it should. The result agrees with (3.7a–c) $\equiv$ (1.232a–c), showing that the same solution can be obtained by different methods, namely, similarity functions (separation of variables) in Subsection 1.7.1 (3.1.1).

As simple examples of solutions of the two-dimensional Cartesian Laplace equation (3.1a), consider the complex functions of a complex (3.6a) or conjugate (3.6b) variable corresponding to the square (3.8a, b) [cube (3.9a–c)], that is, power with exponent two (three):

$$z^2, (z^*)^2 = (x \pm i\,y)^2 = x^2 - y^2 \pm i\,2\,x\,y, \tag{3.8a, b}$$

$$z^3, (z^*)^3 = (x \pm i\,y)^3 = x^3 \pm 3\,i\,y\,x^2 + 3\,x\,(\pm i\,y)^2 + (\pm i\,y)^3$$
$$= x^3 - 3\,x\,y^2 \mp i\,(y^3 - 3\,y\,x^2). \tag{3.9a–c}$$

The real and imaginary parts of (3.8b) $\equiv$ (3.10a, b) and (3.9c) $\equiv$ (3.10c, d) specify four harmonic functions that are solutions of the two-dimensional Cartesian Laplace equation (3.1a):

$$\Phi(x, y) = x^2 - y^2, \ 2\,x\,y, \ x^3 - 3\,x\,y^2, \ y^3 - 3\,x^2\,y, \tag{3.10a–d}$$

as can be checked by direct substitution of (3.10a–d) in (3.1a). Changing one sign in the Laplace equation (3.1a) leads to the equivalent of the wave equation that is considered next (Subsection 3.1.2).

3.1.2   ONE-DIMENSIONAL CARTESIAN WAVE EQUATION (D'ALEMBERT 1747, LAGRANGE 1760)

The classical **wave equation** (note IV.7.17) in one dimension in Cartesian coordinates is the linear second-order p.d.e. with constant coefficients and all derivatives of the same order (3.11a), where t is time and c is the wave speed of propagation as will be shown in the sequel:

$$0 = \frac{\partial^2 \Phi}{\partial t^2} - c^2 \frac{\partial^2 \Phi}{\partial x^2}; \qquad \Phi(x,t) = X(x)\, T(t), \qquad (3.11\text{a, b})$$

seeking a solution by separation of variables (3.11b), substitution in (3.11a) and division by Φ leads to (3.12a) where the r.h.s. and l.h.s. are functions of different variables and, thus, must equal the same constant (3.12b):

$$\frac{1}{c^2\, T} \frac{d^2 T}{dt^2} = \frac{1}{X} \frac{d^2 X}{dx^2} = a^2; \qquad \frac{d^2 X}{dx^2} - a^2\, X = 0 = \frac{d^2 T}{dt^2} - a^2\, c^2\, T \qquad (3.12\text{a–d})$$

The linear second-order o.d.e. with constant coefficients (3.12b) $\equiv$ (3.12c) [(3.12a) $\equiv$ (3.12d)] have two linearly independent solutions (3.13a) [(3.13b)]:

$$X_{\pm}(x) = e^{\pm ax}, \qquad T_{\pm}(t) = e^{\mp act}:$$

$$\Phi_{\pm\pm}(x,t) = X_{\pm}(x)\, T_{\mp}(t) = \exp\left[\pm a\, (x \mp c\, t)\right]. \qquad (3.13\text{a–d})$$

The product (3.11b) $\equiv$ (3.13c) of (3.13a, b) specifies two solutions (3.13d) of the one-dimensional Cartesian wave equation (3.10a), since $\pm\, a$ is arbitrary.

The constant a in (3.13c, d) is arbitrary, and the solutions can be multiplied by arbitrary functions of a and integrated over a leading to two distinct solutions:

$$\int \Phi_{\pm\pm}(x,t)\, C_{\pm\pm}(a)\, da = \int C_{\pm\pm}(a) \exp\left[\pm a\, (x \mp c\, t)\right] = g_{\pm}(x \mp c\, t). \qquad (3.14\text{a, b})$$

The functions (3.14b) are constant (3.15a) for (3.15b) an observer moving (3.15c) at constant velocity c in positive (negative) x-direction, corresponding to (3.15d) with the upper + (lower −) sign:

$$g_{\pm}(x \mp c\, t) = const \quad \Rightarrow \quad x \mp c\, t = const \quad \Rightarrow \quad dx \mp c\, dt = 0 \quad \Rightarrow \quad \frac{dx}{dt} = \pm\, c;$$

$$(3.15\text{a–d})$$

The interpretation is that *the **similarity solutions** in space-time (3.14a, b) represent a **permanent waveform** propagating with unchanged shape (Figure 3.1) in the positive (negative) x-direction at the **wave speed** c, so that after a time t, the same point of the waveform is at a position changed by $\Delta x = +\, c\, t\, (\Delta x = -\, c\, t)$. The general integral of the one-dimensional Cartesian wave equation (3.11a) is a superposition of two permanent waveforms propagating at the same wave speed c in opposite directions (3.16c, d) with **phases** (3.16a, b):*

$$\phi_{\pm} = x \mp c\, t: \qquad \Phi(x,t) = g_{+}(x - c\, t) + g_{-}(x + c\, t) = g_{+}(\phi_{+}) + g_{-}(\phi_{-}), \qquad (3.16\text{a–d})$$

because: (i) both terms of (3.16c, d) are solutions of the p.d.e. (3.11a); (ii) the p.d.e. is linear, so the sum is also a solution; (iii) in (3.16c, d) two arbitrary functions appear as should be the case for the general integral of a second-order p.d.e.

3.1.3   PERMANENT PROPAGATING WAVES AND STANDING MODES

The wave equation (3.11a) is a particular case of (1.223a) that has been solved by the method of similarity functions (Subsection 1.6.8). Thus, the wave equation (3.11a) can be solved equivalently (3.16a–d) by the methods of similarity functions (separation of variables) in Subsection 1.6.8 (3.1.2). As simple examples of **propagating waves** in the positive + (negative −) direction with velocity, consider the sinusoidal waveforms in one-dimensional space-time (3.17) with amplitudes $A_\pm$:

$$\Phi_\pm(x,t) = A_\pm \sin[k(x \mp c\,t)]. \tag{3.17}$$

The **nodes** where the function vanishes (3.18a) are the space-time **events** (3.18b, c):

$$\Phi_\pm(x,t) = 0 \quad \Rightarrow \quad n \in |N, \qquad x \mp c\,t = \frac{\pi\,n}{k} \quad \Rightarrow \quad dx = \pm c\,dt, \tag{3.18a–d}$$

implying (3.18d) propagation at velocity c in the positive (negative) x-direction. The constant k in (3.17) is the **wavenumber**, related to the **wavelength** λ by (3.19a), so that the waveform takes the same value after travelling one wavelength (3.19b–e):

$$k = \frac{2\pi}{\lambda} : \Phi_\pm(x+\lambda,t) = A_\pm \sin[k(x+\lambda \mp c\,t)] = A_\pm \sin[k(x \mp c\,t) + 2\pi]$$

$$= A_\pm \sin[k(x \mp c\,t)] = \Phi_\pm(x,t). \tag{3.19a–e}$$

Adding two waves (3.17) with the same amplitude (3.20a, b) propagating in opposite directions (3.20c, d) leads to **standing modes** (3.20e):

$$A_+ = A_- \equiv A: \qquad \Phi(x,t) = \Phi_+(x,t) + \Phi_-(x,t)$$

$$= A\{\sin[k(x-c\,t)] + \sin[k(x+c\,t)]\}$$

$$= 2A\sin(k\,x)\cos(k\,c\,t); \tag{3.20a–e}$$

the **nodes** or zeros (3.21a) are (3.21b) at fixed positions (3.21c, d)

$$\Phi(x,t) = 0 \quad \Rightarrow \quad \sin(k\,x) = 0 \quad \Rightarrow \quad n \in |Z:$$

$$k\,x_n = \pi\,n \quad \Rightarrow \quad x_n = \frac{n\,\pi}{k} = \frac{n\,\lambda}{2} \tag{3.21a–f}$$

spaced by a multiple of half a wavelength (3.21e, f). Whereas the harmonic (3.1a) and wave (3.11a) equations have all derivatives of the same second order, the diffusion equation has derivatives of different orders, the first-order for time and the second-order for position, leading to very different properties (Subsections 3.1.4–3.1.5).

3.1.4 One-Dimensional Cartesian Diffusion Equation (Fourier 1818)

The heat or **diffusion equation** (1.327b) in one-dimensional Cartesian coordinates is (Note1.4) the linear second-order p.d.e. with constant coefficients (3.22a) that involves a first-order derivative with regard to time, a second-order derivative with regard to position and the **diffusivity** χ as the coefficient:

$$\chi \frac{\partial^2 \Phi}{\partial x^2} = \frac{\partial \Phi}{\partial t}; \quad \frac{1}{\chi T} \frac{dT}{dt} = \frac{1}{X} \frac{d^2 X}{dx^2} = - a^2, \qquad (3.22\text{a--c})$$

since the diffusion equation (3.22a) is a p.d.e. of the first (second) order in time (position), the method of separation of variables (3.11b) leads (3.22b, c) to a linear o.d.e. with constant coefficients of the first (3.22b) $\equiv$ (3.23a) [second (3.22c) $\equiv$ (3.23c)] order whose solution(s) is (3.23b) [are (3.23d), that are linearly independent)]:

$$\frac{dT}{dt} = - a^2 \chi T: \quad T(t) = e^{-a^2 \chi t}; \quad \frac{d^2 X}{dx^2} = - a^2 X: \quad X_{\pm} = e^{\pm iax}. \qquad (3.23\text{a--d})$$

The product (3.24a) of (3.23b, d) specifies (3.11b) two solutions (3.24b) of the one-dimensional Cartesian diffusion equation:

$$\Phi_{\pm}(x,t) = X_{\pm}(x) \, T(t) = \exp(\pm i \, a \, x - a^2 \, \chi \, t), \qquad (3.24\text{a, b})$$

where a is an arbitrary constant.

The solutions (3.24b) remain valid (3.25a–c) multiplying by an arbitrary function of a and integrating over a:

$$\bar{\Phi}(x,t) \equiv \int \Phi_{\pm}(x,t) \, C_{\pm}(a) \, da = \int C_{\pm}(a) \exp(\pm i \, a \, x - a^2 \, \chi \, t) \, da. \qquad (3.25\text{a--c})$$

The integrals (3.25a–c) for the one-dimensional diffusion equation (3.22a) involve exponentials with different powers of a, and do not lead to similarity solutions like (3.5a, b) [(3.14a, b)] for the Cartesian two-dimensional harmonic (3.1a) [one-dimensional wave (3.11a)] equation, suggesting that diffusion excludes permanent waveform solutions because dissipation alone leads ultimately to the decay to zero for large distances or long extents of time. It can be confirmed that (3.25c) leads to non-similarity solutions, for example, taking the particular case of unit coefficient (3.26a) and integration over the whole real line (3.26b) leading to the Gaussian integral (3.26c) $\equiv$ (III.1.13):

$$C_{\pm}(a) = 1: \qquad \bar{\Phi}(x,t) = \int_{-\infty}^{+\infty} \exp(\pm i \, a \, x - a^2 \, \chi \, t) \, db$$

$$= \sqrt{\frac{\pi}{\chi \, t}} \, \exp\left(- \frac{x^2}{4 \, \chi \, t}\right). \qquad (3.26\text{a--d})$$

The solution (3.26d) of the Cartesian one-dimensional diffusion equation (3.22a): (i) is real unlike the solution (3.7a–c) of the two-dimensional Cartesian harmonic equation; (ii) is not a similarity solution unlike the solutions (3.14a, b) of the Cartesian one-dimensional wave equation (3.11a); and (iii) leads to non-permanent shapes with decay with distance and time as shown next (Subsection 3.1.5).

3.1.5   SPACE-TIME DECAY OF NON-PERMANENT SOLUTIONS

The particular solution (3.26d) of the Cartesian one-dimensional diffusion equation (3.22a) is considered first (second) as a function of position (time) at a fixed time (position). At (Figure 3.2a) a fixed time (3.27a) as a function of position: (i) is a Gaussian (3.27b) with variance increasing linearly with time (3.27c), so that it spreads with time (Figure 3.2a) as the maximum reduces (3.27d):

$$t = const: \qquad \bar{\Phi}(x,t) = \frac{\sqrt{2\,\pi}}{\sigma}\,\exp\!\left(-\frac{x^2}{2\,\sigma^2}\right), \qquad \sigma^2 = 2\,\chi\,t,$$

$$\Phi_{max}(x,t) = \Phi(0,t) = \sqrt{\frac{\pi}{\chi\,t}}\,; \tag{3.27a--d}$$

(ii) the integral over all space (3.28a) is a constant (3.28b–d), hence independent of time, implying a conservation law:

$$A(t) \equiv \int_{-\infty}^{+\infty} \bar{\Phi}(x,t)\,dx = \sqrt{\frac{\pi}{\chi\,t}} \int_{-\infty}^{+\infty} \exp\!\left(-\frac{x^2}{4\,\chi\,t}\right) dx = \sqrt{\frac{\pi}{\chi\,t}}\,\sqrt{\pi\,4\,\chi\,t} = 2\,\pi.$$
$$\tag{3.28a--d}$$

As (see Figure 3.2b) *a function of time at fixed position (3.29a), the solution: (i) is always zero at zero (3.29b) or infinite (3.29c) time; (ii) goes* (see Figure 3.2b) *at time (3.19d) through a maximum (3.29e) that reduces with distance:*

$$x = const: \qquad \bar{\Phi}(x,0) = 0 = \bar{\Phi}(x,\infty), \qquad t_* = \frac{x^2}{2\,\chi}, \qquad \bar{\Phi}_* = \Phi(x,t_*) = \frac{1}{x}\sqrt{\frac{2\,\pi}{e}}.$$
$$\tag{3.29a--d}$$

The preceding statements have used the following results: (a) the maximum of (3.26d) as a function of time at a fixed position is specified by the zero derivative with regard to time (3.30a, b):

$$0 = \frac{\partial}{\partial t}\left[\sqrt{\frac{\pi}{\chi\,t}}\,\exp\!\left(-\frac{x^2}{4\,\chi\,t}\right)\right] = \sqrt{\frac{\pi}{\chi\,t}}\,\exp\!\left(-\frac{x^2}{4\,\chi\,t}\right)\frac{x^2 - 2\,\chi\,t}{4\,\chi\,t^2}; \tag{3.30a, b}$$

(b) the root of (3.30b) for finite time is (3.29c) and substitution in (3.26d) gives (3.29d); and (c) the Gaussian integral (III.1.10b) = (3.31)

$$\int_{-\infty}^{+\infty} \exp(-a\,x^2)\,dx = \sqrt{\frac{\pi}{a}}, \tag{3.31}$$

is used to evaluate the integral in (3.28c) with (3.32a), leading to (3.32b-e):

$$a = \frac{1}{4 \chi t}: \qquad A(t) = \sqrt{\frac{\pi}{\chi t}} \int_{-\infty}^{+\infty} \exp(- a x^2) \, dx$$

$$= \sqrt{\frac{\pi}{\chi t}} \times \sqrt{\frac{\pi}{a}} = \frac{\pi}{\sqrt{a \chi t}} = 2 \pi, \qquad (3.32a\text{-}e)$$

in agreement with (3.32e) ≡ (3.28d).

The harmonic/wave/diffusion equations are simple examples of elliptic/hyperbolic/parabolic second-order p.d.e., respectively, and are compared next (Subsection 3.1.7) based on a common generalisation (Subsection 3.1.6).

3.1.6 COMPARISON OF THE HARMONIC, WAVE, AND DIFFUSION EQUATIONS

A generalisation of the harmonic (3.1a), wave (3.11a), and diffusion (3.22a) equations is the **telegraph** or **wave-diffusion equation** (3.33) that is a linear second-order p.d.e.:

$$A \frac{\partial^2 \Phi}{\partial x^2} + B \frac{\partial^2 \Phi}{\partial y^2} + C \frac{\partial \Phi}{\partial y} = 0, \qquad (3.33)$$

also known as the **telegraph equation,** since it applies to electromagnetic wave propagation along a telegraph wire with finite electrical conductivity, thus non-zero electrical resistance leading to dissipation and partial loss of signal. It is classified replacing the partial derivatives by the corresponding coordinates (3.34a):

$$\left\{ \frac{\partial}{\partial x} , \frac{\partial}{\partial y} \right\} \rightarrow \{x, y\}: \qquad A x^2 + B y^2 + C y = 1, \qquad (3.34a, b)$$

leading to a quadratic equation (3.34b) that suggests a classification in terms of **quadric curves,** namely, the ellipse, hyperbola, and parabola. The first case (3.35a–c) corresponds to the harmonic equation (3.35d) ≡ (3.1a) that is of **elliptic type** (3.35e):

$$C = 0, \quad A = \frac{1}{a^2}, \quad B = \frac{1}{b^2}: \quad \frac{\partial^2 \Phi}{\partial(ax)^2} + \frac{\partial^2 \Phi}{\partial(by)^2} = 0, \quad \rightarrow \quad \frac{x^2}{a^2} + \frac{y^2}{b^2} = 1,$$
$$(3.35a\text{-}e)$$

and has solutions (3.6a–d; 3.7a–c) involving complex variables. The second case (3.36a–c) corresponds to the wave equation (3.36d) ≡ (3.11a) with wave speed (3.36e) that is of the **hyperbolic type** (3.36f):

$$C = 0, \quad A = \frac{1}{a^2}, \quad B = -\frac{1}{b^2}:$$

$$\frac{\partial^2 \Phi}{\partial x^2} - c^2 \frac{\partial^2 \Phi}{\partial y^2} = 0, \quad c = \frac{a}{b}, \quad \frac{x^2}{a^2} - \frac{y^2}{b^2} = 1, \qquad (3.36a\text{-}f)$$

and has permanent solutions (3.14a, b) travelling uniformly in space-time (3.15a–d). The third case (3.37a–c) corresponds to the diffusion equation (3.37d) ≡ (3.22a) with diffusivity (3.37e) that is of the **parabolic type** (3.37f):

$$B = 0, \quad A > 0 > C: \qquad \frac{\partial \Phi}{\partial y} = \chi \frac{\partial^2 \Phi}{\partial x^2}, \qquad \chi = -\frac{A}{C} > 0, \qquad y = \chi \, x^2, \qquad (3.37\text{a–f})$$

and has solutions (3.26d) decaying in space-time (3.27a–d; 3.28a–d; 3.29a–d).

3.1.7 ELLIPTIC, HYPERBOLIC, AND PARABOLIC P.D.E.

It can be shown that the general second-order linear p.d.e. with variable coefficients:

$$0 = A(x,y) \frac{\partial^2 \Phi}{\partial x^2} + B(x,y) \frac{\partial^2 \Phi}{\partial y^2} + C(x,y) \frac{\partial^2 \Phi}{\partial x \, \partial y} + D(x,y) \frac{\partial \Phi}{\partial x}$$

$$+ E(x,y) \frac{\partial \Phi}{\partial y} + F(x,y) \, \Phi, \tag{3.38}$$

can be classified (Section 3.5) into elliptic/hyperbolic/parabolic types. The harmonic (3.35a–e)/wave (3.36a–f)/diffusion (3.37a–f) equations are simple examples of linear second-order p.d.e. with constant coefficients that are of the elliptic/hyperbolic/parabolic types, respectively, and can be solved by the method of separation of variables (Subsections 3.1.1/3.1.2/3.1.3–3.1.4). The method of separation of variables: (i) applies to some differential equations with variable coefficients, for example the generalised equation of mathematical physics in Cartesian, polar, cylindrical, spherical, hyperspherical, and hypercylindrical coordinates (notes IV.8.1–IV.8.15 and Subsection 3.1.8); and (ii) may or may not apply to linear p.d.e. with more than two terms, even for constant coefficients, as shown by the next two examples.

Attempting a solution of the wave-diffusion equation (3.33) by separation of variables (3.1b) leads to (3.39):

$$-\frac{A}{X} \frac{d^2 X}{dx^2} = \frac{B}{Y} \frac{d^2 Y}{dy^2} + \frac{C}{Y} \frac{dY}{dy} = a^2, \tag{3.39}$$

which is separable, even though it has derivatives of different orders, namely, first and second. In contrast, attempting the same solution by separation of variables (3.1b) of the linear second-order p.d.e. with constant coefficients and all derivatives of the same order (3.40):

$$A \frac{\partial^2 \Phi}{\partial x^2} + B \frac{\partial^2 \Phi}{\partial y^2} + C \frac{\partial^2 \Phi}{\partial x \, \partial y} = 0, \tag{3.40}$$

leads to (3.41):

$$\frac{A}{X} \frac{d^2 X}{dx^2} + \frac{B}{Y} \frac{d^2 Y}{dy^2} + \frac{C}{X \, Y} \frac{dX}{dx} \frac{dY}{dy} = 0, \tag{3.41}$$

which is generally not separable, so the method fails. In contrast, a linear p.d.e. with variable coefficients and derivatives of different orders may have solution by the separation of variables as shown in the next example (Subsection 3.1.8).

3.1.8 SEPARABLE AND NON-SEPARABLE P.D.E.

An example is considered the linear p.d.e. with variable coefficients and derivatives of the second and first-order (3.42a) that correspond to a diffusion equation (3.22a) with variable coefficients:

$$0 = A(x)\frac{\partial^2\Phi}{\partial x^2} + C(t)\frac{\partial\Phi}{\partial t}; \qquad \frac{A(x)}{X(x)}\frac{d^2X}{dx^2} = -\frac{C(t)}{T(t)}\frac{dT}{dt} = a^2. \qquad (3.42a\text{--}c)$$

Seeking a solution by separation of variables (3.11b) ≡ (3.43a) leads to (3.42b, c), specifying a first- (3.43b) and second-order (3.43c) o.d.e.:

$$\Phi(x,t) = X(x)\,T(t): \qquad C(t)\frac{dT}{dt} + a^2\,T = 0 = A(x)\frac{d^2X}{dx^2} - a^2\,X. \qquad (3.43a\text{--}c)$$

Thus, *the one-dimensional diffusion equation with variable coefficients (3.42a) has a solution by the separation of variables (3.43a) where the temporal (spatial) factor satisfies the o.d.e. (3.43b) [(3.43c)].*
Consider the linear p.d.e. (3.44c) of diffusion type (3.42a) with variable coefficients (3.44a, b):

$$A(x) = x^2, \qquad C(t) = \frac{b}{t}: \qquad\qquad x^2\frac{\partial^2\Phi}{\partial x^2} + \frac{b}{t}\frac{\partial\Phi}{\partial t} = 0. \qquad (3.44a\text{--}c)$$

Seeking a solution (3.43a) by the separation of variables the: (i) first-order factor satisfies the o.d.e. (3.43b; 3.44b) ≡ (3.45a, b) with solution (3.45c, d):

$$\frac{b}{t}\frac{dT}{dt} = -\,a^2\,T \quad\Rightarrow\quad \frac{dT}{T} = -\,\frac{a^2}{b}\,t\,dt \qquad (3.45a,\ b)$$

$$\Rightarrow\quad \log T = -\frac{a^2\,t^2}{2\,b} \quad\Rightarrow\quad T(t) = \exp\!\left(-\frac{a^2\,t^2}{2\,b}\right); \qquad (3.45c,\ d)$$

(ii) the second-order factor satisfies the o.d.e. (3.43c; 3.44a) ≡ (3.46a) that has power solutions (3.46b) with exponent c satisfying (3.46c–f), leading to the solutions (3.46g):

$$x^2\frac{d^2X}{dx^2} - a^2\,X = 0: \qquad X(x) = x^c, \qquad 0 = x^c\left[c\,(c-1) - a^2\right], \qquad (3.46a\text{--}c)$$

$$0 = c^2 - c - a^2 = (c - c_+)(c - c_-), \qquad 2\,c_{\pm} = 1 \pm \sqrt{1 + 4\,a^2}\,, \qquad (3.46d\text{--}f)$$

$$X_{\pm}(x) = \sqrt{x}\ x^{\,\pm\sqrt{1/4 + a^2}}. \qquad (3.46g)$$

A linear combination of (3.46g) with coefficients $E_{\pm}(a)$ and arbitrary functions of a, multiplied (3.43a) by (3.45d), and integrated over a, specifies the general integral (3.47):

$$\Phi(x,t) = \int \sqrt{x}\,\exp\!\left(-\frac{a^2\,t^2}{2\,b}\right)\left[E_+(a)\,x^{\sqrt{1/4-a^2}} + E_-(a)\,x^{-\sqrt{1/4-a^2}}\right]da, \qquad (3.47)$$

of the linear diffusion equation (3.44c) with variable coefficients. A linear p.d.e. of any order with constant coefficients and derivatives all of the same order can always be solved by the method of similarity functions (Sections 1.6–1.9); if the derivatives are of different orders, the method of similarity functions can still be applied, but the final result may not be a similarity solution (Section 3.2).

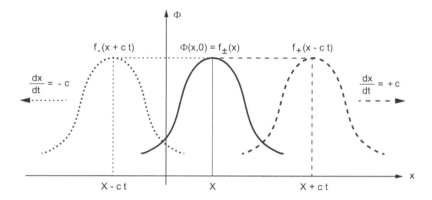

FIGURE 3.1 A similarity solution in space-time, for example for the one-dimensional Cartesian wave equation, is a permanent waveform that travels without change of shape with constant phase speed.

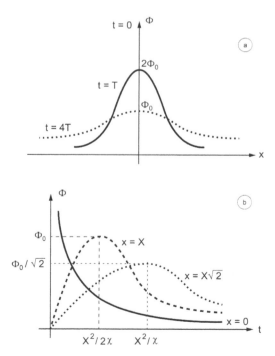

FIGURE 3.2 The solution of the diffusion equation for a point source at the origin can be seen as: (a) a Gaussian decay as a function of distance, with a lower peak and broader shape as time elapses; (b) a sharp signal at the origin as a function of time, with the peak decaying as distance increases for the same time.

3.2 METHOD OF SIMILARITY FUNCTIONS

The method of similarity functions applies to linear partial differential equations with constant coefficients in the case (a) of all derivatives of the same order (Sections 1.3–1.9) leading to an algebraic equation whose roots specify the constant in the similarity solution, for example for the Cartesian two-dimensional harmonic (one-dimensional wave) equation [Subsection 3.2.1 (3.2.2)]. The method of similarity functions can also be applied to linear partial differential equations with constant coefficients in the case (b) of derivatives not all of the same order, leading to a linear ordinary differential equation with constant coefficients whose solutions lead to non-similarity solutions (Subsection 3.2.4), for example the Cartesian one-dimensional diffusion equation (Subsection 3.2.3). Other examples of using the method of similarity functions to obtain non-similarity solutions of linear partial differential equations with constant coefficients and derivatives of different orders include: (i) the telegraph equation that combines the wave and diffusion equations (Subsection 3.2.5) and has derivatives of the second order with regard to position and the first and second orders with regard to time; (ii/iii) the bar (beam) equation that applies to the transverse vibrations of an elastic rod in the absence (presence) of axial tension [Subsection 3.2.6 (3.2.7)] that involve derivatives of the second order with regard to time and fourth (and second) order with regard to position.

3.2.1 SIMILARITY SOLUTIONS FOR THE HARMONIC EQUATION

A **similarity function** in two variables (3.48b) is a function of a single variable (3.48c) that is a linear combination of the two variables (3.48a):

$$\xi = x + a\ y: \qquad \Phi(x,y) = f(x + a\ y) = f(\xi). \qquad (3.48a\text{–}c)$$

If the function is differentiable (3.49a), then it specifies all partial derivatives (3.49b, c):

$$f \in \mathcal{D}^{\,n+m}(IR): \qquad \frac{\partial^{n+m}\Phi}{\partial x^n\,\partial y^m} = \frac{d^{n+m}f}{d\xi^{n+m}}\left(\frac{\partial\xi}{\partial x}\right)^n\left(\frac{\partial\xi}{\partial y}\right)^m = a^m\ f^{\,n+m}(\xi). \qquad (3.49a\text{–}c)$$

For example, seeking for the Cartesian two-dimensional harmonic equation (3.1a) ≡ (3.50a) solutions as similarity functions (3.48c; 3.49b) leads to (3.50b); a non-trivial solution (3.50c) leads to an algebraic equation (3.50d) with imaginary roots (3.50e, f):

$$0 = \frac{\partial^2\Phi}{\partial x^2} + \frac{\partial^2\Phi}{\partial y^2} = \frac{d^2f}{d\xi^2}\left(1 + a^2\right): \quad \frac{d^2f}{d\xi^2} \neq 0 \ \Rightarrow\ a^2 = -1, \quad a_\pm = \pm\,i. \qquad (3.50a\text{–}f)$$

The roots (3.50e, f) specify (3.46b) to the particular integrals (3.51a, b) and, hence, lead to the general integral (3.51c, d) of the two-dimensional Cartesian harmonic equation:

$$\Phi_\pm(x,y) = f_\pm(x + a_\pm\ y) = f_\pm(x \pm i\ y):$$
$$\Phi(x,y) = \Phi_+(x,y) + \Phi_-(x,y) = f_+(x + i\ y) + f_-(x - i\ y), \quad (3.51\text{a--d})$$

involving similarity solutions in agreement with (3.51d) $\equiv$ (3.7b) $\equiv$ (1.232a–c).

Besides (3.8a, b; 3.9a–c; 3.10a–d), another set of examples of solutions of the two-dimensional Cartesian harmonic equation (3.1a) $\equiv$ (3.50a) is specified by the sinusoidal functions of complex variable cosine (3.52a–c) [sine (3.53a–c)]:

$$\cos(z, z^*) = \cos(x \pm i\ y) = \cos x \cos(\pm\ i\ y) - \sin x \sin(\pm\ i\ y)$$
$$= \cos x \cosh y \mp i \sin x \sinh y, \quad (3.52\text{a--c})$$

$$\sin(z, z^*) = \sin(x \pm i\ y) = \sin x \cos(\pm\ i\ y) + \cos x \sin(\pm\ i\ y)$$
$$= \sin x \cosh y \pm i \cos x \sinh y, \quad (3.53\text{a--c})$$

where were used (3.54a–c; 3.55a–c):

$$\cos(\pm\ i\ y) \equiv \frac{e^{\pm i^2 y} + e^{\mp i^2 y}}{2} = \frac{e^{\mp y} + e^{\pm y}}{2} \equiv \cosh y, \quad (3.54\text{a--c})$$

$$\sin(\pm\ i\ y) \equiv \frac{e^{\pm i^2 y} + e^{\mp i^2 y}}{2i} = -i\,\frac{e^{\mp y} - e^{\pm y}}{2} \equiv \pm\,i \sinh y. \quad (3.55\text{a--c})$$

The real and imaginary parts of (3.52c) $\equiv$ (3.56a, b) and (3.53c) $\equiv$ (3.56c, d) specify four harmonic functions that are solutions of the two-dimensional Cartesian Laplace equation (3.1a) $\equiv$ (3.50a):

$$\Phi(x,y) = \cos x \cosh y\ ,\ \ \sin x \sinh y\ ,\ \ \sin x \cosh y\ ,\ \cos x \sinh y, \quad (3.56\text{a--d})$$

as can be checked substituting (3.56a–d) in (3.1a) $\equiv$ (3.50a). The method of separation of variables (similarity functions) applies [Section 3.1 (3.2)] both to the harmonic [Subsection 3.1.1 (3.2.1)] and wave [Subsection 3.1.2 (3.2.2)] equations.

3.2.2 SIMILARITY SOLUTIONS FOR THE WAVE EQUATION

Another example of application of the method of similarity solutions is the one-dimensional Cartesian wave equation (3.11a) $\equiv$ (3.57a), whose solution as similarity function (3.57b, c)

$$\frac{\partial^2 \Phi}{\partial t^2} = c^2\,\frac{\partial^2 \Phi}{\partial x^2}: \qquad\qquad \Phi(x,t) = g(x + b\ t) = g(\zeta), \quad (3.57\text{a--c})$$

involves the variable (3.58a) and the derivatives (3.58b–d):

$$\zeta \equiv x + b\,t: \qquad \frac{\partial^{n+m}\Phi}{\partial x^n\,\partial t^m} = \frac{d^{n+m}g}{d\zeta^{n+m}}\left(\frac{\partial\zeta}{\partial x}\right)^n\left(\frac{\partial\zeta}{\partial t}\right)^m = b^m\,g^{n+m}(\zeta). \qquad (3.58a\text{–}d)$$

Substitution of (3.57b; 3.58d) in (3.57a) leads to (3.59a, b) that has a non-trivial solution (3.59c) if the constant b satisfies the algebraic equation (3.37d) that has roots (3.59e,f):

$$0 = \frac{\partial^2\Phi}{\partial t^2} - c^2\frac{\partial^2\Phi}{\partial x^2} = \frac{d^2g}{d\zeta^2}(b^2 - c^2): \quad \frac{d^2g}{d\zeta^2} \neq 0 \;\Rightarrow\; b^2 = c^2 \;\Rightarrow\; b_\pm = \mp\,c.$$
$$(3.59a\text{–}f)$$

Substitution of (3.59e, f) in (3.57b) leads to two permanent waveforms propagating with velocity c in the positive and negative x-direction (Figure 3.1) as particular integrals (3.60a, b) and to their sum (3.60c, d) as the general integral of the Cartesian one-dimensional wave equation (3.14a) $\equiv$ (3.57a):

$$\Phi_\pm(x,t) = g_\pm(x + b_\pm\,t) = g_\pm(x \mp c\,t):$$
$$\Phi(x,t) = \Phi_+(x,t) + \Phi(x,t) = g_+(x - c\,t) + g_-(x + c\,t), \qquad (3.60a\text{–}d)$$

involving similarity solutions in agreement with (3.60d) $\equiv$ (3.16d).

Using the circular cosine (3.61a) instead of the circular sine (3.17) for a permanent wave propagating with velocity c in the positive $+$ or negative $-$ x-direction with amplitudes $A_\pm$

$$\Psi_\pm(x,t) = A_\pm\cos\left[k(x \mp c\,t)\right] = A_\pm\sin\left[k(x \mp c\,t) + \frac{\pi}{2}\right]$$
$$= A_\pm\sin\left[k\,(x \mp c\,t) + \frac{k\,\lambda}{4}\right] = A_\pm\sin\left[k\left(x + \frac{\lambda}{4} \mp c\,t\right)\right] = \Phi_\pm\left(x + \frac{\lambda}{4},\,t\right),$$
$$(3.61a\text{–}e)$$

is equivalent (3.61b–d) to a spatial displacement (3.61e) by a quarter of a wavelength (3.19a). The **period** (3.62a) is the time taken to travel one wavelength at the wave speed, and is related to the **frequency** (3.62b) in the same way as the wavelength is related to the wavenumber (3.19a), implying (3.62c, d):

$$\tau \equiv \frac{\lambda}{c}: \qquad \omega \equiv \frac{2\pi}{\tau} = \frac{2\pi c}{\lambda} = k\,c:$$
$$\{\Phi_\pm(x,t),\ \Psi_\pm(x,t)\} = A_\pm\cos,\sin\left[\omega\left(\frac{x}{c} \mp t\right)\right], \qquad (3.62a\text{–}f)$$

and the sinusoidal propagating waves specified by the circular sine (3.17) [cosine (3.61a)] in terms of wavenumber can alternatively use the frequency (3.62e) [(3.62f)]. Thus, *the sinusoidal propagating waves specified by the circular sine (cosine) in: (i) terms of the wavenumber (3.17) [(3.61a)], are equivalent to a spatial translation by one-quarter of the wavelength (3.61e); (ii) terms of the frequency (3.62e) [(3.62f)], are equivalent to a temporal delay by a quarter period (3.63a–e):*

$$\Psi_\pm(x,t) = A_\pm \cos\left[\omega\left(\frac{x}{c}-t\right)\right] = A_\pm \sin\left[\omega\left(\frac{x}{c}\mp t\right)+\frac{\pi}{2}\right]$$

$$= A_\pm \sin\left[\omega\left(\frac{x}{c}\mp t\right)+\frac{\omega\,\tau}{4}\right] = A_\pm \sin\left[\omega\left(\frac{x}{c}\mp t+\frac{\tau}{4}\right)\right] = \Phi_\pm\left(x,\ t\mp\frac{\tau}{4}\right).$$

$$(3.63a–e)$$

If the derivatives in the linear partial differential equation with constant coefficients are not all of the same order, instead of an algebraic equation for the constant in the similarity solution such as (3.51d, 3.60d), a linear ordinary differential equation is obtained for the similarity function (subsection 3.2.3).

3.2.3   Non-Similarity Solution for the Diffusion Equation

In the case of the one-dimensional Cartesian diffusion equation (3.22a) ≡ (3.64a), that is, a linear partial differential equation with constant coefficients with derivatives of a different order with regard to time and position, respectively one and two, the similarity function (3.57b; 3.58d) leads to a linear ordinary differential equation of the second order with constant coefficients (3.64b, c):

$$0 = \chi\,\frac{\partial^2\Phi}{\partial x^2} - \frac{\partial\Phi}{\partial t} = \chi\,\frac{d^2g}{d\zeta^2} - b\,\frac{dg}{d\zeta} = \chi\,g'' - b\,g'.\qquad (3.64a–c)$$

The solution of (3.64c) is an exponential (3.65a) whose coefficient μ satisfies (3.65b) and, hence, is a root (3.65e, f) of the characteristic polynomial (3.65c, d):

$$g(\zeta) = e^{\mu\,\zeta}:\qquad 0 = e^{\mu\,\zeta}\left(\chi\,\mu^2 - b\,\mu\right),\qquad 0 = \mu\left(\chi\,\mu - b\right) \equiv P_2(\mu),$$

$$\mu_{1,2} = 0,\ \frac{b}{\chi}.\qquad (3.65a–f)$$

The zero root (3.65e) ≡ (3.60a) leads to a constant solution (3.66a) and the nonzero root (3.65f) ≡ (3.66c) leads to the exponential solution (3.66d, e):

$$\mu_1 = 0:\qquad g_1(\zeta) = const;\qquad \mu_2 = \frac{b}{\chi}:\qquad g_2(\zeta) = e^{\mu_2\,\zeta} = e^{b\,\zeta/\chi}.\qquad (3.66a–e)$$

Substituting (3.66b) [(3.66e)] with (3.58a) in (3.57b, c), multiplying by an arbitrary function of b, and integrating over b leads to (3.67a) [(3.67b)]:

$$\Phi_1(x,t) = const,\qquad \Phi_2(x,t) = \int B(b)\exp\left[(b/\chi)\,(x+b\,t)\right]db.\qquad (3.67a, b)$$

where: (i) the constant solution (3.67a) trivially satisfies (3.64a); and (ii) the solution (3.67b) with the changes of notation (3.68a, b) leads to (3.68c, d)

$b = \pm i \chi a,$ $\quad B_\pm(b) \equiv \pm i \chi B(\pm i \chi a) \equiv C_\pm(a,\chi):$

$$\Phi_2(x,t) = \int C_\pm(a,\chi)\exp(\pm i\, a\, x - a^2\, \chi\, t)\, da, \qquad (3.68a\text{–}d)$$

that coincides with $(3.25c) \equiv (3.68c, d)$. The sum of $(3.67a)$ and $(3.68d)$ specifies *the general integral (3.69) of the one-dimensional Cartesian diffusion equation (3.22a)* $\equiv (3.64a)$:

$$\Phi(x,t) = C_1 + \int C_2(a)\exp(\pm i\, a\, x - a^2\, \chi\, t)\, da, \qquad (3.69)$$

that is not a similarity solution, and includes the particular case (3.70a) corresponding to (3.26a–d), with $C_1 = 0$ *and* $C_2 = 1$:

$$\Phi(x,t) = C_1 + C_2 \sqrt{\frac{\pi}{\chi t}}\exp\left(-\frac{x^2}{4\chi t}\right): \qquad \lim_{t\to\infty}\Phi(x,t) = C_1 = \lim_{x\to\pm\infty}\Phi(x,t); \qquad (3.70a\text{–}c)$$

the value $C_1 = 0$ in $(3.70b)$ leads to $(3.29c)$.

Choosing in $(3.25c) \equiv (3.68d)$ the function $(3.26a)$ $[(3.71a)]$ and integration over the whole real line leads to the solution $(3.26c, d)$ $[(3.71b)]$ of the one-dimensional diffusion equation $(3.22a) \equiv (3.64a)$:

$$C_2(a) = a^{2n}: \qquad \Phi_n(x,t) = \int_{-\infty}^{+\infty} a^{2n}\exp(\pm i\, a\, x - \chi\, a^2\, t)\, da. \qquad (3.71a, b)$$

The integral $(3.71b) \equiv (3.71c)$ is uniformly convergent with regard to the parameter $b = -\chi t$ and, hence $(I.13.40)$, the derivatives $\partial/\partial b$ can be taken outside the integral $(3.71d)$:

$$\Phi_n(x,t) = \int_{-\infty}^{+\infty} \frac{\partial^n}{\partial(-\chi t)^n}\exp(\pm i\, a\, x - \chi\, a^2\, t)\, db$$

$$= \frac{(-)^n}{\chi^n}\frac{\partial^n}{\partial t^n}\int_{-\infty}^{+\infty}\exp(\pm i\, a\, x - \chi\, a^2\, t)\, db, \qquad (3.71c, d)$$

showing that differentiation $\partial/\partial(-\chi t)$ with regard to $-\chi t$ is equivalent to multiplication by a^2. The integral in $(3.71d)$ is evaluated by $(3.26d)$ leading to $(3.72b)$ for $(3.72a)$ all non-negative integers n:

$$n \in I\, N: \qquad \Phi_n(x,t) = \frac{(-)^n}{\chi^n}\sqrt{\frac{\pi}{\chi}}\frac{\partial^n}{\partial t^n}\left[\frac{1}{\sqrt{t}}\exp\left(-\frac{x^2}{4\chi t}\right)\right]. \qquad (3.72a, b)$$

Thus, *the one-dimensional Cartesian diffusion equation (3.22a) ≡ (3.64a) has solutions (3.72b) for all non-negative integer values (3.72a) of n, of which the simplest are: (i) (3.26b) for n = 0 ; (ii) (3.73) for n = 1:*

$$\Phi_1(x,t) = \frac{1}{2}\sqrt{\frac{\pi}{(\chi\,t)^3}}\left(1 - \frac{x^2}{2\,\chi\,t}\right)\exp\left(-\frac{x^2}{4\,\chi\,t}\right). \tag{3.73}$$

The method of similarity functions is applied next to the general linear partial differential equation with constant coefficients (Subsection 3.2.4) regardless of whether the derivatives are all (are not all) of the same order and the final particular and/or general integrals are (are not) similarity solutions [Subsections 3.2.1 – 3.2.2 (3.2.3)].

3.2.4 SIMILARITY METHOD FOR SIMILARITY AND NON-SIMILARITY SOLUTIONS

The general linear partial differential equation with constant coefficients (3.74a) may involve derivatives up to different orders $N\,(M)$ for the independent variables $x\,(y)$:

$$0 = \sum_{n=0}^{N}\sum_{m=0}^{M} A_{n,m}\,\frac{\partial^{n+m}\Phi}{\partial x^n\,\partial y^m} = \sum_{n=0}^{N}\sum_{m=0}^{M} A_{n,m}\,a^m\,\frac{d^{n+m} f}{d\xi^{n+m}}, \tag{3.74a, b}$$

and for a similarity function (3.48b; 3.49c) leads to a linear ordinary differential equation with constant coefficients (3.74b). The latter has exponential solutions (3.75a), where the constants μ, a must be such that (3.75b):

$$f(\xi) = e^{\mu\,\xi}: \qquad 0 = e^{\mu\,\xi}\sum_{n=0}^{N}\sum_{m=0}^{M} A_{n,m}\,\mu^{n+m}\,a^m = e^{\mu\,\xi}\,P(a,\mu), \tag{3.75a–c}$$

the characteristic polynomial in two variables (3.75c) ≡ (3.76a) vanishes (3.76b):

$$0 = \sum_{n=0}^{N} \Sigma_{m=0}^{M}\,A_{n,m}\,\mu^{n+m}\,a^m \equiv P(a,\mu). \tag{3.76a, b}$$

Factorising (3.77a, b), the characteristic polynomial (3.76b) for its roots assumed to be single relative to μ leads (3.75b, c) to (3.77c) as particular integrals of the linear ordinary differential equation with constant coefficients (3.74b):

$$S \le N + M: \qquad P(a,\mu) = A_{N,M}\prod_{s=1}^{S}\left[\mu - \mu_s(a)\right]: \qquad f_s(\xi) = \exp\{\,\mu_s(a)\,\xi\,\}. \tag{3.77a–c}$$

Substituting (3.77c) in (3.48a–c), multiplying by an arbitrary function of a, and integrating over a leads to *the particular integrals (3.78a) of the unforced linear partial differential equation with constant coefficients (3.74a) of order N (M) in the independent variable x (y):*

$$\Phi_s(x,y) = \int C_s(a) \exp\left[\mu_s(a)(x+a\ y)\right] da: \quad \Phi(x,y) = \sum_{s=1}^{S} \Phi_s(x,t); \quad (3.78a, b)$$

the particular integrals (3.78a) are linearly independent for distinct roots μ_s of the characteristic polynomial, and for roots of multiplicity r factors, $1, x, x^2, \ldots, x^{r-1}$ are inserted in (3.78a). The general integral is (3.78b), a sum of S linearly independent particular integrals. If all derivatives are of the same order (3.79a, b), the roots (3.77a) of the characteristic polynomial (3.76a, b) are (3.79c-e) independent of μ and the particular integrals (3.78a) become (3.79f, g) similarity solutions:

$$n+m = N: \qquad\qquad 0 = \sum_{n=0}^{N} A_{n,N-n} \frac{\partial^N \Phi}{\partial x^n\, \partial y^{N-n}}, \qquad\qquad (3.79a, b)$$

$$0 = P(a,\mu) = \mu^N \sum_{n=0}^{N} A_{n,N-n}\, a^{N-n} = \mu^N\, A_{0,N} \prod_{n=1}^{N} (a-a_n), \quad (3.79c\text{–}e)$$

$$\Phi_n(x,y) = \int C_n(a_n) \exp\left[\mu(x+a_n\ y)\right] da = f_n(x+a_n\ y). \qquad (3.79f, g)$$

The method of similarity functions is applied next to the non-similarity solutions for linear partial differential equations with constant coefficients with derivatives of different orders, up to two (four) for the telegraph (bar and beam) equations [Subsections 3.2.5 (3.2.6–3.2.7)].

3.2.5 Telegraph or Wave-Diffusion Equation (Maxwell 1873, Heaviside 1892)

The telegraph equation in Cartesian one-dimensional form (3.80) combines the wave (3.59a) ≡ (3.11a) and diffusion (3.64a) ≡ (3.22a) equations and is a linear partial differential equation whose constant coefficients are the wave speed c and diffusivity χ, and involves derivatives of the second (and first) order with regard to position (time):

$$\frac{1}{c^2} \frac{\partial^2 \Phi}{\partial t^2} - \frac{\partial^2 \Phi}{\partial x^2} + \frac{1}{\chi} \frac{\partial \Phi}{\partial t} = 0, \qquad\qquad (3.80)$$

so that it reduces to the one-dimensional Cartesian diffusion (wave) equation (3.59a) ≡ (3.81b) [(3.64a) ≡ (3.82b)] for infinite wave speed (3.81a) [diffusivity (3.82a)]:

$$0 = \lim_{c \to \infty}\left(\chi \frac{\partial^2 \Phi}{\partial x^2} - \frac{\partial \Phi}{\partial t} - \frac{\chi}{c^2} \frac{\partial^2 \Phi}{\partial t^2} \right) = \chi \frac{\partial^2 \Phi}{\partial x^2} - \frac{\partial \Phi}{\partial t}, \qquad (3.81a, b)$$

$$0 = \lim_{\chi \to \infty}\left(\frac{1}{c^2} \frac{\partial^2 \Phi}{\partial t^2} - \frac{\partial^2 \Phi}{\partial x^2} + \frac{1}{\chi} \frac{\partial \Phi}{\partial t} \right) = \frac{1}{c^2} \frac{\partial^2 \Phi}{\partial t^2} - \frac{\partial^2 \Phi}{\partial x^2}. \qquad (3.82a, b)$$

The similarity solution (3.57b, c; 3.58d) for the telegraph equation (3.80) leads to the linear ordinary differential equation (3.83a):

$$0 = \left(b^2 - c^2\right)\frac{d^2g}{d\zeta^2} + \frac{c^2\,b}{\chi}\frac{dg}{d\zeta}, \qquad \mu\left[\frac{c^2\,b}{\chi} + \left(b^2 - c^2\right)\mu\right] = 0, \quad \text{(3.83a, b)}$$

whose solution is an exponential (3.65a) with coefficient μ satisfying (3.83b).

The zero root (3.66a) leads to a constant solution (3.66b; 3.67b); the non-zero root (3.84a) leads to the similarity function (3.84b):

$$\mu_2 = \frac{c^2\,b/\chi}{c^2 - b^2}, \qquad g_2\left(\zeta\right) = \exp\left(\mu_2\,\zeta\right) = \exp\left(\frac{\zeta}{\chi}\frac{c^2\,b}{c^2 - b^2}\right), \qquad \text{(3.84a, b)}$$

and non-similarity solution (3.85):

$$\Phi_2(x,t) = \int C_2(b)\exp\left(\frac{x + b\,t}{\chi}\frac{c^2\,b}{c^2 - b^2}\right)db. \qquad \text{(3.85)}$$

Thus, *the telegraph equation (3.80) is a linear partial differential equation with constant wave speed c and diffusivity χ, involving second- (and first-) order derivatives with regard to position (time) whose general integral (3.86) is the sum of (3.67a) and (3.85):*

$$\Phi(x,t) = C_1 + \int C_2(b)\exp\left(\frac{x + b\,t}{\chi}\frac{c^2\,b}{c^2 - b^2}\right)db. \qquad \text{(3.86)}$$

For $c \to \infty$ in (3.86), it is regained the non-similarity solution (3.67b) of the diffusion equation (3.81a, b); for $\chi \to \infty$, the relation (3.83b) leads to (3.59e) and the similarity solutions (3.60a–d) of the wave equation. Next (Subsection 3.2.6), a fourth-order linear partial differential equation with constant coefficients is considered.

3.2.6 TRANSVERSE VIBRATIONS OF AN ELASTIC BAR (BERNOULLI 1744, EULER 1744)

The transverse vibrations of an elastic bar are specified by (note IV.6.7) the one-dimensional Cartesian stiff **bar equation** (3.87b) $\equiv$ (IV.6.782f):

$$q^2 = \frac{E\,I}{\rho}: \qquad\qquad \frac{\partial^2\Phi}{\partial t^2} + q^2\frac{\partial^4\Phi}{\partial x^4} = 0, \qquad\qquad \text{(3.87a, b)}$$

that is a linear partial differential equation of the second (fourth) order in time (position) with a constant **stiffness coefficient** (3.87a) specified (note IV.6.7) by the mass density per unit length ρ, Young's modulus of the material E, and moment of inertia of the cross-section I. The similarity function (3.57b, c; 3.58d) leads to a linear ordinary differential equation with constant coefficients (3.88a) that has exponential solutions (3.65a) with the exponent satisfying (3.88b):

$$0 = b^2 \frac{d^2 g}{d\zeta^2} + q^2 \frac{d^4 g}{d\zeta^4}, \qquad \mu^2 \left(b^2 + q^2 \, \mu^2 \right) = 0. \qquad \text{(3.88a, b)}$$

The double zero root (3.66a) ≡ (3.89a) leads to a constant (3.89b) ≡ (3.66b) and linear (3.89c) solution:

$$\mu_1 = 0 = \mu_2: \qquad \Phi_1(x,t) = C_1, \qquad \Phi_2(x,t) = \int C_2(b)\,(x + b\,t)\,db; \qquad \text{(3.89a–c)}$$

the imaginary symmetric roots (3.90a, b) lead to the solutions (3.90c, d):

$$\mu_\pm = \pm\, i\, \frac{b}{q}: \qquad \Phi_\pm(x,t) = \int C_\pm(b)\exp\left(\pm\, i\, b\, \frac{x + b\,t}{q} \right)\, db, \qquad \text{(3.90a–d)}$$

where (3.90c, d) may be replaced by the real solutions:

$$\Phi_{3,4}(x,t) = \int C_{3,4}(b)\,\cos,\sin\left(b\, \frac{x + b\,t}{q} \right)\, db. \qquad \text{(3.90e, f)}$$

Thus, *the general integral of the stiff bar equation (3.87b) that is a linear partial differential equation with constant coefficients of the second (fourth) order in time (position) is the sum (3.91) of the particular integrals (3.89b, c; 3.90e, f):*

$$\Phi(x,t) = C_1 + \Phi_2(x,t) + \Phi_3(x,t) + \Phi_4(x,t). \qquad \text{(3.91)}$$

The wave (stiff bar) equations [Subsection 3.2.2 (3.2.6)] are combined in the stiff beam equation (Subsection 3.2.7).

3.2.7 STIFF BAR AND BEAM EQUATIONS

The transverse vibrations of an elastic string (notes IV.6.3 and IV.7.8–IV.7.9) are specified by the wave equation (IV.7.393c) ≡ (3.92b) ≡ (3.59a), where the wave speed (3.92a) is the square root of the longitudinal tension T divided by the mass density per unit length ρ:

$$c^2 = \frac{T}{\rho}: \qquad \rho\, \frac{\partial^2 \Phi}{\partial t^2} = T\, \frac{\partial^2 \Phi}{\partial x^2}. \qquad \text{(3.92a, b)}$$

In the case of an elastic **beam**, that is an elastic bar (3.87a, b) subject to an axial tension, the inertia force on the l.h.s. of (3.87a) ≡ (3.93) balances both the transverse tension on the r.h.s. of (3.92b) and the bending stiffness that is the second term of (3.87b), and both appear in the r.h.s. of (3.93):

$$\rho\, \frac{\partial^2 \Phi}{\partial t^2} = T\, \frac{\partial^2 \Phi}{\partial x^2} - E\, I\, \frac{\partial^4 \Phi}{\partial x^4}. \qquad \text{(3.93)}$$

This leads to the stiff **beam equation** (3.94c) that is a linear partial differential equation whose constant coefficients are the elastic wave speed (3.92a) ≡ (3.94a) and stiffness coefficient (3.87a) ≡ (3.93b) involving second- (and fourth-) order derivatives with regard to time (position):

$$c^2 = \frac{T}{\rho}, \qquad q^2 = \frac{E I}{\rho}: \qquad \frac{\partial^2 \Phi}{\partial t^2} - c^2 \frac{\partial^2 \Phi}{\partial x^2} + q^2 \frac{\partial^4 \Phi}{\partial x^4} = 0. \qquad (3.94a\text{–}c)$$

A similarity function (3.57b, c; 3.58d) leads to a fourth-order linear ordinary differential equation with constant coefficients (3.95a) that has exponential solutions (3.65a) with a parameter satisfying (3.95b):

$$\left(b^2 - c^2\right) \frac{d^2 g}{d\zeta^2} + q^2 \frac{d^4 g}{d\zeta^4} = 0: \qquad \mu^2 \left(b^2 - c^2 + q^2 \mu^2\right) = 0. \qquad (3.95a, b)$$

The double-root zero (3.89a) leads to the solutions (3.89b, c) and the symmetric roots (3.96a, b) lead to the solutions (3.96c, d):

$$\mu_{\pm} = \pm i \frac{\sqrt{b^2 - c^2}}{q}: \qquad \Phi_{\pm}(x,t) = \int C_{\pm}(b) \exp\left(\pm i \frac{x + b t}{q} \sqrt{b^2 - c^2}\right) db,$$
$$(3.96a\text{–}d)$$

that may be replaced by the real solutions:

$$\Phi_{3,4}(x,t) = \int C_{3,4}(b) \cos, \sin\left(\frac{x + bt}{q} \sqrt{b^2 - c^2}\right) db. \qquad (3.96e, f)$$

Thus, *the general integral of the stiff beam equation (3.94c) that is a linear partial differential equation with constant coefficients and derivatives of the second (and fourth) order with regard to time (position) is the sum (3.97) of the particular integrals (3.89b, c; 3.96e, f):*

$$\Phi(x,t) = C_1 + \Phi_2(x,t) + \Phi_3(x,t) + \Phi_4(x,t), \qquad (3.97)$$

involving the stiffness parameter (3.94b) and elastic wave speed (3.94a). In the case of a bar, the absence of longitudinal tension $T = 0$ leads to zero $c = 0$ elastic wave speed (3.94a), the stiff beam (3.94c) becomes the stiff bar equation (3.87b), and the particular integrals (3.96c-f) simplify (3.90c-f). Besides [Section 3.1 (3.2)] the method I (II) of separation of variables (similarity functions), a method III of symbolic operators of derivatives (Section 3.3) can also be used to solve linear partial differential equations, most readily in the case of constant coefficients.

3.3 METHOD OF SYMBOLIC DIFFERENTIATION OPERATORS

The method of symbolic differentiation operators applies to linear partial differential equations, most readily in the case of constant coefficients, and derivatives of the same order, for example the harmonic (wave) equation [Subsection 3.3.1 (3.3.2)]. The wave equation has second-order derivatives with regard to position (time) and, thus, has a unique solution satisfying two boundary (initial) conditions [Subsections 3.3.3 (3.3.4)]. The method of symbolic differentiation operators also applies to linear partial differential equations with constant coefficients and derivatives of different orders, for example the diffusion equation, leading to unique solutions satisfying one (two) initial (boundary) condition(s) [Subsection 3.3.5 (3.3.6)] that are equivalent (Subsection 3.3.8) in spite of involving one (two) arbitrary functions. The one-dimensional Cartesian diffusion equation has both singular (analytic) solutions [Subsections 3.3.5–3.3.7 (3.3.8)]. The method of symbolic differentiation operators applies readily to the linear first-order unforced partial differential equation with constant coefficients including a non-differentiated term (Subsection 3.3.9) and relates to unidirectional and bidirectional waves with damping (Subsection 3.3.10). The method also extends to higher orders (Subsection 3.3.11); in particular, the conditions of factorisation of a second-order equation (Subsection 3.3.12) into the square (product) of first-order equations [Subsection(s) 3.3.14 (3.3.12–3.3.13)]. The four methods I–IV of solution of linear partial differential equations with constant coefficients (Sections 1.6–1.9 and 3.1–3.3) are compared (Subsection 3.4.15) before introducing method V, which will be developed in more detail (Sections 3.4–3.9).

3.3.1 SYMBOLIC DIFFERENTIAL OPERATORS FOR THE HARMONIC EQUATION

In the case of the two-dimensional Cartesian harmonic or Laplace equation (3.1a) ≡ (3.50a) ≡ (3.98a), the factorisation (3.98b) ≡ (1.231b)

$$0 = \frac{\partial^2 \Phi}{\partial x^2} + \frac{\partial^2 \Phi}{\partial y^2} = \left(\frac{\partial}{\partial y} - i \frac{\partial}{\partial x} \right)\left(\frac{\partial}{\partial y} + i \frac{\partial}{\partial x} \right) \Phi(x,y), \qquad \text{(3.98a, b)}$$

leads to two first-order partial differential equations (3.99a) that may be seen as first-order ordinary differential equations (3.99a) in the second independent variable y, with the partial derivative with regard to the other first independent variable x included in the parameter (3.99c):

$$0 = \frac{\partial \Phi_\pm}{\partial y} \mp i \frac{\partial \Phi_\pm}{\partial x} = \frac{d\Phi_\pm}{dy} \mp \lambda \, \Phi_\pm, \qquad \mu \equiv i \frac{d}{dx}. \qquad \text{(3.99a–c)}$$

The solution of (3.99b) is (3.100b) where $f_\pm$ is the value for all x and $y = 0$; then, use of the series (II.3.1) ≡ (3.100a) for the exponential in (3.100b) leads to a Taylor series (3.100c) ≡ (I..23.32b) for the functions (3.100d):

$$e^{\pm\mu y} = \sum_{n=0}^{\infty} \frac{\left(\pm\mu\, y\right)^n}{n!};$$

$$\Phi_\pm(x,y) = e^{\pm\mu\, y} f_\pm(x) = \sum_{n=0}^{\infty} \frac{\left(\pm i\, y\right)^n}{n!} \frac{d^n}{dx^n}\left[f_\pm(x)\right] = f_\pm\left(x \pm i\, y\right) \qquad \text{(3.100a–d)}$$

assumed to be analytic (3.101a) in the general integral (3.101b):

$$f_\pm(z) \in \mathcal{A}\,(IC): \qquad \Phi(x,y) = f_+\left(x+i\, y\right) + f_-\left(x-i\, y\right). \qquad \text{(3.101a, b)}$$

Thus, *the general integral of the two-dimensional Cartesian harmonic equation (3.98a) ≡ (3.50a) ≡ (3.1a) ≡ (1.230d–f) is the sum (3.101b) of two analytic functions (3.101a) of a complex variable and its conjugate, in agreement with (1.232a–c) ≡ (3.7a–c) ≡ (3.51a–d) ≡ (3.101a, b),* which all arrive at the same result using four different methods.

The general form of the harmonic equation (3.98a) in N – dimensions involves the Laplace operator (3.102a, b):

$$0 = \nabla^2\Phi = \sum_{n=1}^{N} \frac{\partial^2\Phi}{\partial x_n^2}, \qquad \text{(3.102a, b)}$$

and in two-dimensional (3.102e) polar coordinates leads to (I.11.28b, c) ≡ (III.6.45a, b) ≡ (3.102d, e):

$$N = 2: \qquad -\frac{1}{r^2}\frac{\partial^2\Phi}{\partial\varphi^2} = \frac{1}{r}\frac{\partial}{\partial r}\left(r\frac{\partial\Phi}{\partial r}\right) = \frac{\partial^2\Phi}{\partial r^2} + \frac{1}{r}\frac{\partial\Phi}{\partial r}. \qquad \text{(3.102c–e)}$$

The complex variable (its conjugate) has the Cartesian representation (3.103a) [(3.103c)] equivalent to the polar representation (3.103b) [(3.103d)]:

$$z = x + i\, y = r\, e^{i\varphi}, \qquad z^* = x - i\, y = r\, e^{-i\varphi} \qquad \text{(3.103a–d)}$$

and, thus, *(3.104b) is the general integral of the two-dimensional polar harmonic equation (3.102c, d) and equals the sum of two analytic functions (3.102a) of a complex variable (3.103b) and its conjugate (3.103d):*

$$f_\pm \in \mathcal{A}\,(\mathrm{I}\,C): \qquad \Phi(r,\varphi) = f_+\left(r\, e^{i\,\varphi}\right) + f_-\left(r\, e^{-i\,\varphi}\right). \qquad \text{(3.104a, b)}$$

Some examples of harmonic functions are given in Cartesian (polar) coordinates [Subsections 3.1.1 and 3.2.1 (3.3.1)].

The functions (3.105a–d; 3.106a–c; 3.107a–f):

$$n \in I \, N: \qquad z^n, (z^*)^n \equiv (r \, e^{\pm i\varphi})^n = r^n \, e^{\pm in\varphi} = r^n \left[\cos(n\,\varphi) \pm i \sin(n\,\varphi)\right], \quad (3.105a\text{--}d)$$

$$\log z \,, \;\; \log z^* = \log(r \, e^{\pm i\varphi}) = \log r \pm i \; \varphi, \qquad (3.106a\text{--}c)$$

$$a = \alpha + i\,\beta: \qquad z^a, (z^*)^a \equiv \exp\left\{\log\left[z^a \,, (z^*)^a\right]\right\} = \exp\left[a\,(\log z \,, \; \log z^*)\right]$$

$$= \exp\left[(\alpha + i\,\beta)\,(\log r \pm i\,\varphi)\right]$$

$$= \exp\left[(\alpha \log r \mp \beta\,\varphi) + i\,(\beta \log r \pm \alpha\,\varphi)\right]$$

$$= r^\alpha \, \exp\left(\mp \beta\,\varphi\right) \exp\left[i\,(\beta \log r \pm \alpha\,\varphi)\right], \qquad (3.107a\text{--}f)$$

have *real and imaginary parts (3.108a–f) that are solutions of the polar harmonic equation (3.102c, d):*

$$\Phi(r,\varphi) = r^n \cos(n\,\varphi)\,, \;\; r^n \sin(n\,\varphi)\,, \;\; \log r \,, \;\; \varphi, \qquad (3.108a\text{--}d)$$

$$r^\alpha \, \exp(\mp\,\beta\,\varphi) \cos(\beta \log r \pm \alpha\,\varphi), \qquad (3.108e)$$

$$r^\alpha \, \exp(\mp\,\beta\,\varphi) \sin(\beta \log r \pm \alpha\,\varphi), \qquad (3.108f)$$

as can be confirmed substituting (3.107a–f) in (3.102d). Method III of symbolic operators of derivatives applies both to the harmonic (wave) equation [Subsection 3.3.1 (3.3.2)].

3.3.2 SYMBOLIC DIFFERENTIAL OPERATORS FOR THE WAVE EQUATION

The one-dimensional Cartesian wave operator *(3.109a) ≡ (3.59a) ≡ (3.11a) has the factorisation (3.109b):*

$$0 = \frac{\partial^2 \Phi}{\partial t^2} - c^2 \, \frac{\partial^2 \Phi}{\partial x^2} = \left(\frac{\partial}{\partial t} - c\frac{\partial}{\partial x}\right)\left(\frac{\partial}{\partial t} + c\frac{\partial}{\partial x}\right)\Phi(x,t); \qquad (3.109a, b)$$

each factor (in 3.109b) can be written as a linear first-order ordinary differential equation (3.110a) with time t as the independent variable, where the coefficient (3.110b) involves the derivative with regard to the other independent variable position x that appears in the solution (3.110c):

$$\frac{d\Phi_\pm}{dt} \mp \mu \, \Phi_\pm = 0, \qquad \mu \equiv c\,\frac{d}{dx}: \qquad \Phi(x,t) = e^{\mp t\mu} g(x), \qquad (3.110a\text{--}c)$$

where $g(x)$ is the value at all positions at $t = 0$ zero time. Using the power series (II.3.1) ≡ (3.111a) for the exponential, the **symbolic differentiation operator** in (3.110c) is interpreted as (3.111b):

$$e^{\mp t\mu} = \sum_{n=0}^{\infty} \frac{(t\,\mu)^n}{n!}: \qquad \exp\left(\mp c\,t\,\frac{d}{dx}\right) = 1 + \sum_{n=1}^{\infty} \frac{(\mp c\,t)^n}{n!}\,\frac{d^n}{dx^n}, \qquad (3.111a, b)$$

and the solution becomes (3.112c), that is, the Taylor series (3.112b) ≡ (I.23.32b) for the analytic function (3.112a):

$$\Phi(x,t) \in \mathcal{A}\,(\mathbb{R}): \qquad \Phi_{\pm}(x,t) = 1 + \sum_{n=1}^{\infty} \frac{(\mp c\,t)^n}{n!}\,\frac{d^n[\Phi(x,0)]}{dx^n} = g_{\pm}(x \mp c\,t).$$

$$(3.112a\text{–}c)$$

This proves that *the general integral of the one-dimensional Cartesian wave equation (3.109a) ≡ (3.59a) ≡ (3.11a) is the sum (3.113b) of two analytic functions (3.113a) of the variables $x \mp c\,t$* :

$$g_{\pm} \in \mathcal{A}^2\,(\mathbb{R}): \qquad \Phi(x,t) = g_+(x - c\,t) + g_-(x + c\,t). \qquad (3.113a, b)$$

The results (3.113a, b) [(3.60a–d) ≡ (3.16a–d)] agree except for requiring analytic (twice differentiable) functions.

The propagating sinusoidal sine (3.19a–e) ≡ (3.114a, b) ≡ (III.6.829) [cosine (3.61a–e) ≡ (3.115a, b)] with (3.62a–d) wavenumber k and frequency ω:

$$\Phi_{\pm}(x,t) = A_{\pm}\sin(k\,x \mp \omega\,t) = A_{\pm}\left[\sin(k\,x)\cos(\omega\,t) \mp \cos(k\,x)\sin(\omega\,t)\right],$$

$$(3.114a, b)$$

$$\Psi_{\pm}(x,t) = A_{\pm}\cos(k\,x \mp \omega\,t) = A_{\pm}\left[\cos(k\,x)\cos(\omega\,t) \pm \sin(k\,x)\sin(\omega\,t)\right],$$

$$(3.115a, b)$$

with the same phase [(3.114a, b) or (3.115a, b)] and the same amplitude (3.116a, b), we lead to: (i) superposition to standing modes like (3.20a–e) ≡ (3.116d), namely (3.116c, d) [(3.116e, f)]

$$A_+ = A_- \equiv A: \qquad \Phi_+(x,t) + \Phi_-(x,t) = 2\,A\sin(k\,x)\cos(\omega\,t), \qquad (3.116a\text{–}c)$$

$$\Phi_-(x,t) - \Phi_+(x,t) = 2\,A\cos(k\,x)\sin(\omega\,t), \qquad (3.116d)$$

$$\Psi_+(x,t) + \Psi_-(x,t) = 2\,A\cos(k\,x)\cos(\omega\,t), \qquad (3.116e)$$

$$\Psi_+(x,t) - \Psi_-(x,t) = 2\,A\sin(k\,x)\sin(\omega\,t)\,; \qquad (3.116f)$$

(ii) in all cases (3.116c–f), the amplitude is double of (3.114a, b; 3.115a, b); and (iii) the standing modes have fixed nodes, where the wavefield vanishes at the positions

(3.21a–f; 3.117a–e) [times (3.118a–d; 3.119a–d)] spaced by half of a wavelength (wave period):

$$n \in | Z: \qquad \cos(k\,x) = 0 \quad \Rightarrow \quad k\,x_n = \left(n + \frac{1}{2}\right)\pi \qquad\qquad (3.117a\text{–}c)$$

$$\Rightarrow \quad x_n = \frac{(2\,n+1)\,\pi}{2\,k} = \left(n + \frac{1}{2}\right)\frac{\lambda}{2}, \qquad (3.117d,\, e)$$

$$\sin(\omega\,t) = 0 \quad \Rightarrow \quad \omega\,t_n = n\,\pi \quad \Rightarrow \quad t_n = \frac{n\,\pi}{\omega} = \frac{n\,\tau}{2}, \quad (3.118a\text{–}d)$$

$$\cos(\omega\,t) = 0 \quad \Rightarrow \quad \omega\,t_n = \left(n + \frac{1}{2}\right)\pi \quad \Rightarrow \quad t_n = \frac{(2\,n+1)\,\pi}{2\,\omega} = \left(n + \frac{1}{2}\right)\frac{\tau}{2}.$$
$$(3.119a\text{–}d)$$

Since the wave equation (3.109a) has highest order of derivation equal to two, the two arbitrary functions in the solution (3.113a, b) are determined by two independent and compatible initial (boundary) conditions [Subsections 3.3.3 (3.3.4)].

3.3.3 Two Initial Conditions for Waves

If Φ is interpreted as the **transverse displacement** (3.120a) of waves in an elastic string (3.11a) ≡ (3.59a) ≡ (3.109a), the derivative, with regard to position (3.120b) [time (3.120c)], is the **slope (transverse velocity)**:

$$\Phi(x,t) \equiv \zeta(x,t), \qquad \frac{\partial \Phi(x,t)}{\partial x} = s(x,t), \qquad \frac{\partial \Phi(x,t)}{\partial t} = v(x,t). \quad (3.120a\text{–}c)$$

Specifying the displacement (3.120a) [velocity (3.120c)] for all positions at zero time $t = 0$ specifies (3.113b) as the **initial conditions** (3.121a) [(3.121b)], where the prime denotes the derivative with regard to the argument:

$$\zeta(x,0) = g_+(x) + g_-(x), \qquad v(x,0) = c\left[g'_-(x) - g'_+(x)\right], \qquad (3.121a,\, b)$$

$$2\,g'_\pm(x) = \zeta'(x,0) \mp \frac{v(x,\,0)}{c}, \qquad (3.121c)$$

that may be solved (3.121c) for $g_\pm$, leading to (3.122a, b):

$$g_\pm(x \mp c\,t) = \frac{1}{2}\int^{x \mp c\,t}\left[\zeta'(\xi,0) \mp \frac{v(\xi,0)}{c}\right]d\xi = \frac{\zeta(x \mp c\,t,\,0)}{2} \mp \frac{1}{2c}\int^{x \mp c\,t} v(\xi,0)\,d\xi.$$
$$(3.122a,\, b)$$

Thus, *the general integral of the one-dimensional Cartesian wave equation (3.123a) with initial conditions specifying the displacement (3.123b) and velocity (3.123c) at all positions at time $t = 0$:*

$$\frac{\partial^2 \Phi}{\partial t^2} - c^2\,\frac{\partial^2 \Phi}{\partial x^2} = 0, \qquad \Phi(x,0) = \zeta(x,0), \qquad \frac{\partial \Phi}{\partial t}(x,0) = v(x,0) \quad (3.123a\text{–}c)$$

is obtained substituting (3.122b) in (3.113b). For example, the even displacement (3.124a, b) without velocity (3.124c) at initial time $t = 0$ leads for all other times to waves (3.124e, f) with wavenumber k and frequency ω (3.124d), which are symmetric functions of position (3.124g) and time (3.124h):

$$\Phi(x,0) = A\cos(k\ x) = \Phi(-\ x\ ,\ 0), \qquad \frac{\partial \Phi}{\partial t}(x,0) = 0, \qquad \omega \equiv k\ c:$$

$$\Phi(x,t) = \frac{A}{2}\left[\cos(k\ x - \omega\ t) + \cos(k\ x + \omega\ t)\right]$$

$$= A\cos(k\ x)\cos(\omega\ t) = \Phi(-\ x\ ,\ t) = \Phi(x\ ,- t). \qquad (3.124a-h)$$

The converse case of zero displacement (3.125a) and even non-zero velocity (3.125b, c) leads to waves (3.125d–g) that are symmetric (skew-symmetric) functions of position (3.125h) [time (3.125i)]:

$$\Phi(x,0) = 0, \qquad \frac{\partial \Phi}{\partial t}(x\ ,\ 0) = B\cos(k\ x) = \frac{\partial \Phi}{\partial t}(-\ x\ ,\ 0):$$

$$\Phi(x,t) = \frac{B}{2\ c}\left(\int^{x\ +c\ t} - \int^{x\ -\ c\ t}\right)\cos(k\ \xi)\ d\xi$$

$$= \frac{B}{2\ c\ k}\left\{\sin[k\ (x + c\ t)] - \sin[k\ (x - ct)]\right\}$$

$$= \frac{B}{2\ \omega}\left[\sin(k\ x + \omega\ t) - \sin(k\ x - \omega\ t)\right]$$

$$= \frac{B}{\omega}\cos(k\ x)\sin(\omega\ t) = \Phi(-\ x\ ,\ t) = -\Phi(x\ ,- t).$$
$$(3.125a-i)$$

Since the wave equation is of the same second order for position (time), similar methods may be used to apply initial (boundary) conditions [Subsections 3.3.3 (3.3.4)].

3.3.4   TWO BOUNDARY CONDITIONS FOR WAVES

Specifying the displacement (3.120a) [slope (3.120c)] for all time at the origin leads (3.113h) to the **boundary conditions** (3.126a, b)

$$\Phi(0,t) = g_+(-\ c\ t) + g_-(c\ t), \qquad s(0,t) = g'_-(c\ t) + g'_+(-\ c\ t), \qquad (3.126a, b)$$

which can be rewritten (3.127b, c) using the variable (3.127a):

$$\eta = c\ t: \qquad \zeta\left(0\ ,\ \frac{\eta}{c}\right) = g_+(-\ \eta) + g_-(\eta), \qquad c\ s\left(0\ ,\ \frac{\eta}{c}\right) = g'_-(\eta) + g'_+(-\ \eta).$$
$$(3.127a-c)$$

Replacing (3.127c) with (3.128a):

$$g_-(\eta) - g_+(-\ \eta) = c\int^{\eta/c} s(0\ ,\xi)\ d\xi, \qquad (3.128a)$$

allows elimination with (3.127b), leading to (3.128b):

$$g_{\pm}(\eta) = \frac{1}{2}\varsigma\left(0, \mp \frac{\eta}{c}\right) \mp \frac{c}{2}\int^{\mp \eta/c} s(0,\xi)\,d\xi, \qquad (3.128b)$$

which is equivalent to (3.128b) ≡ (3.128c):

$$g_{\pm}(x \mp c\,t) = \frac{1}{2}\varsigma\left(0, \ t \mp \frac{x}{c}\right) \mp \frac{1}{2}\int^{t \mp x/c} s(0,\xi)\,d\xi. \qquad (3.128c)$$

Thus, *the general solution of the one-dimensional Cartesian wave equation (3.129a) with two boundary conditions specifying the displacement (3.129b) and slope (3.129c) at the origin for all time:*

$$\frac{\partial^2 \Phi}{\partial t^2} - c^2 \frac{\partial^2 \Phi}{\partial x^2} = 0, \qquad \Phi(0,t) = \varsigma(0,t), \qquad \frac{\partial \Phi}{\partial x}(0,t) = s(0,t), \quad (3.129a\text{–}c)$$

is obtained substituting (3.128c) in (3.113b). For example, the displacement (3.130a, b) odd function of time without slope (3.130c) at the origin for all time leads to waves (3.130d–f) at all positions that are odd (even) functions of time (3.130g) [position (3.130h)]:

$$\Phi(0,t) = E\sin(\omega\,t) = -\,\Phi(0,-t), \qquad \frac{\partial \Phi}{\partial x}(0,t) = 0:$$

$$\Phi(x,t) = \frac{E}{2}\left\{\sin\left[\omega\left(t - \frac{x}{c}\right)\right] + \sin\left[\omega\left(t + \frac{x}{c}\right)\right]\right\}$$

$$= \frac{E}{2}\left[\sin(\omega\,t - k\,x) + \sin(\omega\,t + k\,x)\right]$$

$$= E\sin(\omega\,t)\cos(k\,x) = -\,\Phi(x,-t) = \Phi(-x,t). \qquad (3.130a\text{–}h)$$

The converse case of zero displacement (3.131a) and odd slope (3.131b, c), for all time at the origin, leads to waves (3.131d–g) for all other positions that are odd functions of position (3.131h) and time (3.131i):

$$\Phi(0,t) = 0, \qquad \frac{\partial \Phi}{\partial x}(0,t) = F\sin(\omega\,t) = -\frac{\partial \Phi}{\partial x}(0,-t):$$

$$\Phi(x,t) = \frac{F\,c}{2}\left\{\int^{t+x/c} - \int^{t-x/c}\right\}\sin(\omega\,\xi)\,d\xi$$

$$= \frac{F\,c}{2\,\omega}\left\{\cos\left[\omega\left(t - \frac{x}{c}\right)\right] - \cos\left[\omega\left(t + \frac{x}{c}\right)\right]\right\}$$

$$= \frac{F}{2\,k}\left[\cos(\omega\,t - k\,x) - \cos(\omega\,t + k\,x)\right]$$

$$= \frac{F}{k}\sin(\omega\,t)\sin(k\,x) = -\,\Phi(-x,\,t) = -\,\Phi(x,-t). \qquad (3.131a\text{–}i)$$

The Cartesian one-dimensional diffusion (wave) equations is of: (i) the second order in position and, thus, has two boundary conditions [Subsection 3.3.6 (3.3.4)]; and (ii) of the first (second) order in time and, thus, has one (two) initial condition(s) [Subsection 3.3.5 (3.3.3)].

3.3.5   ONE INITIAL CONDITION FOR THE DIFFUSION EQUATION

Taking in the Cartesian one-dimensional diffusion equation (3.132a) $\equiv$ (3.64a) $\equiv$ (3.22a) time as the independent variable (3.132b) and position as a parameter (3.132c):

$$\frac{\partial \Phi}{\partial t} = \chi \frac{\partial^2 \Phi}{\partial x^2}: \qquad \frac{d\Phi}{dt} = \mu \, \Phi, \qquad \mu \equiv \chi \frac{d^2}{dx^2}, \qquad (3.132a\text{--}c)$$

leads to a linear ordinary first-order differential equation (3.131b) whose general integral is (3.133a):

$$\Phi(x,t) = e^{t\mu} \, \Phi(x,0) = \exp\left(\chi \, t \, \frac{d^2}{dx^2} \right) \Phi(x,0) = \sum_{n=0}^{\infty} \frac{(\chi \, t)^n}{n!} \frac{d^{2n}}{dx^{2n}} \big[\Phi(x,0)\big],$$
$$(3.133a\text{--}c)$$

where: (i) it is used the exponential series (3.100a) in (3.133b); and (ii) the leading term in (3.133c) is the initial value at all positions. Thus, *the general solution of the Cartesian one-dimensional diffusion equation (3.132a) $\equiv$ (3.134a) with one initial condition (3.134b) specifying the value at all positions at zero time is (3.133c) $\equiv$ (3.134c):*

$$\frac{\partial \Phi}{\partial t} = \chi \frac{\partial^2 \Phi}{\partial x^2}, \qquad \lim_{t \to 0} \Phi(x,t) = \Phi(x,0):$$

$$\Phi(x,t) = \sum_{n=0}^{\infty} \frac{(\chi \, t)^n}{n!} \frac{d^{2n}}{dx^{2n}} \big[\Phi(x,0)\big]. \qquad (3.134a\text{--}c)$$

For example, *the initial condition (3.135a) leads to the solution (3.135b) for all time with an exponential decay*

$$\Phi(x,0) = A\cos(k \, x): \qquad \Phi(x,t) = A\cos(k \, x)\exp(-\chi \, k^2 \, t) \qquad (3.135a, b)$$

as follows, substituting (3.135a) in (3.134c), leading to (3.136a–c) $\equiv$ (3.135b):

$$\Phi(x,t) = A \sum_{n=0}^{\infty} \frac{(\chi \, t)^n}{n!} \frac{d^{2n}}{dx^{2n}} \big[\cos(k \, x)\big] = A\cos(k \, x) \sum_{n=0}^{\infty} \frac{(-\chi \, t \, k^2)^n}{n!}$$

$$= A\cos(k \, n)\exp(-\chi \, k^2 \, t). \qquad (3.136a\text{--}c)$$

The same result (3.135b) can be obtained as follows: (i) substitution of (3.137a) in the Cartesian one-dimensional diffusion equation (3.134a) leads to (3.137b); (ii) the linear first-order differential equation with constant coefficients (3.137b) has an exponential solution (3.137c):

$$\Phi(x,t) = T(t)\cos(k\ x), \qquad \frac{dT}{dt} = -\chi\ k^2\ T, \qquad T(t) = A\exp(-\chi\ k^2\ t);$$
$$(3.137a\text{–}c)$$

and (iii) substitution of (3.137c) in (3.137a) leads to (3.135b). The one-dimensional Cartesian diffusion equation (3.134a) is of the first (second) order in time (position) and, thus, has a unique solution [Subsection 3.3.5 (3.3.6)] with one initial (two boundary) condition(s).

3.3.6   TWO BOUNDARY CONDITIONS FOR THE DIFFUSION EQUATION

Taking position as the variable and time as a parameter in the Cartesian one-dimensional diffusion equation (3.132a) ≡ (3.138a) leads to a linear ordinary second-order differential equation (3.138b) with coefficient (3.138c):

$$\frac{\partial^2 \Phi}{\partial x^2} = \frac{1}{\chi}\frac{\partial \Phi}{\partial t}; \qquad \frac{d^2\Phi}{dx^2} = \mu\ \Phi, \qquad \mu \equiv \frac{1}{\chi}\frac{d}{dt}. \qquad (3.138a\text{–}c)$$

Two linearly independent solutions of (3.138b) are (3.139a), leading (3.111a) to (3.139b):

$$\Phi_{\pm}(x,t) = \exp\left(\pm x\ \sqrt{\mu}\right) F(t) = \sum_{n=0}^{\infty} \frac{\left(\pm x/\sqrt{\chi}\right)^n}{n!}\frac{d^{n/2}F}{dt^{n/2}}. \qquad (3.139a,\ b)$$

Another pair of linearly independent solutions is specified by half the sum (3.140a, b) and difference (3.141a, b) of (3.139a, b):

$$\Phi_1(x,t) = \frac{1}{2}\left[\Phi_+(x,t) + \Phi_-(x,t)\right] = \sum_{n=0}^{\infty} \frac{\left(x^2/\chi\right)^n}{(2\ n)!}\frac{d^n F}{dt^n}, \qquad (3.140a,\ b)$$

$$\Phi_2(x,t) = \frac{1}{2}\left[\Phi_+(x,t) - \Phi_-(x,t)\right] = \sum_{n=0}^{\infty} \frac{\left(x/\sqrt{\chi}\right)^{2n+1}}{(2\ n+1)!}\frac{d^n}{dt^n}\left(\frac{d^{1/2}F}{dt^{1/2}}\right). \qquad (3.141a,\ b)$$

The solution (3.141b) ≡ (3.141d) can be written in terms of ordinary derivatives, including the derivative of half order in the function (3.141c):

$$G(t) \equiv \frac{1}{\sqrt{\chi}}\frac{d^{1/2}F}{dt^{1/2}}; \qquad \Phi_2(x,t) = x\sum_{n=0}^{\infty} \frac{\left(x^2/\chi\right)^{2n}}{(2\ n+1)!}\frac{d^n G}{dt^n}. \qquad (3.141c,\ d)$$

The solutions of (3.140b; 3.141b) of the one-dimensional Cartesian diffusion equation (3.138a) involve the following: (i) derivatives of positive integer order in (3.140b), starting with the value of the function at the origin for all time (3.142a, b); (ii) the derivatives of order integer plus one-half are derivatives of the positive integer order of the derivative of order one-half (3.141c) that specifies the derivative with regard to position at the origin for all time (3.142c, d):

$$\Phi(0,t) = \lim_{x \to 0} \Phi(x,t) = F(t), \qquad \frac{\partial \Phi(0,t)}{\partial x} \equiv \lim_{x \to 0} \frac{\partial \Phi(x,t)}{\partial x} = G(t). \qquad (3.142a\text{–}d)$$

Thus, it is not necessary to use the explicit definition of derivative of order one-half or positive integer plus one-half in the two boundary conditions (3.142a, c).

The choice of the linearly independent particular integrals (3.140b) [(3.141b)] involving only ordinary derivatives of positive integer order of the functions (3.142b) [(3.142d)] avoids the derivatives of order integer plus one half in (3.141b) and does not change the general integral:

$$\Phi(x,t) = \Phi_+(x,t) + \Phi_-(x,t) = \Phi_1(x,t) + \Phi_2(x,t), \qquad (3.143a, b)$$

which remains the sum (3.143a) [(3.143b)] of (3.139a, b) [(3.140a; 3.141a)]. It has been shown that *the general solution of the Cartesian one-dimensional diffusion equation (3.22a) ≡ (3.64a) ≡ (3.132a) ≡ (3.138a) ≡ (3.144a) with two boundary conditions specifying the value (3.144b) and the spatial gradient (3.144c) for all time at the origin:*

$$\frac{\partial \Phi}{\partial t} = \chi \frac{\partial^2 \Phi}{\partial x^2}: \qquad \Phi(0,t) = F(t), \qquad \frac{\partial \Phi}{\partial x}(0,t) = G(t), \qquad (3.144a\text{–}c)$$

is (3.143b; 3.140b; 3.141b; 3.142d) ≡ (3.145):

$$\Phi(x,t) = \sum_{n=0}^{\infty} \frac{(x^2/\chi)^n}{(2n)!} \frac{d^n F}{dt^n} + x \sum_{n=0}^{\infty} \frac{(x^2/\chi)^n}{(2n+1)!} \frac{d^n G}{dt^n}. \qquad (3.145)$$

An example follows.

The boundary value (3.146a) and gradient (3.146b) for all time lead to the solution (3.146c) of the one-dimensional diffusion equation (3.144a) at all positions:

$$\Phi(0,t) = A\, e^{-at}, \qquad \frac{\partial \Phi}{\partial x}(0,t) = B\, e^{-bt}: \qquad (3.146a, b)$$

$$\Phi(x,t) = A\, e^{-at} \cos\left(x\sqrt{\frac{a}{\chi}} \right) + \sqrt{\frac{\chi}{b}}\, B\, e^{-bt} \sin\left(x\sqrt{\frac{b}{\chi}} \right), \qquad (3.146c)$$

as follows from substitution of (3.146a, b) in (3.145), leading to (3.147a–c) ≡ (3.146c):

$$\Phi(x,t) = A \sum_{n=0}^{\infty} \frac{(x/\sqrt{\chi})^{2n}}{(2n)!} \frac{d^n(e^{-at})}{dt^n} + B\sqrt{\chi} \sum_{n=0}^{\infty} \frac{(x/\sqrt{\chi})^{2n+1}}{(2n+1)!} \frac{d^n(e^{-bt})}{dt^n}$$

$$= A\, e^{-at} \sum_{n=0}^{\infty} (-)^n \frac{(x\sqrt{a/\chi})^{2n}}{(2n)!} + B\sqrt{\frac{\chi}{b}}\, e^{-bt} \sum_{n=0}^{\infty} (-)^n \frac{(x\sqrt{b/\chi})^{2n+1}}{(2n+1)!}$$

$$= A\, e^{-at} \cos\left(x\sqrt{\frac{a}{\chi}} \right) + B\sqrt{\frac{\chi}{b}}\, e^{-bt} \sin\left(x\sqrt{\frac{b}{\chi}} \right). \qquad (3.147a\text{–}c)$$

The same result (3.147a–c) is obtained next in an alternative way.

The solution of the Cartesian one-dimensional diffusion equation (3.144a) with boundary conditions (3.146a, b) is sought in the form (3.148a) leading, on substitution in (3.144a), to (3.148b):

$$\Phi(x,t) = X(x)\,e^{-at} + Y(x)\,e^{-bt}:$$

$$0 = e^{-at}\left(\chi\,\frac{d^2X}{dx^2} + a\,X\right) + e^{-bt}\left(\chi\,\frac{d^2Y}{dx^2} + b\,Y\right). \qquad (3.148\text{a, b})$$

From (3.148b), follow the linear ordinary second-order differential equations with constant coefficients (3.149a) [(3.150a)] with solutions (3.149b) [(3.150b)], where $\left(C_{11},\ C_{12},\ C_{21},\ C_{22}\right)$ are arbitrary constants:

$$\frac{d^2X}{dx^2} + \frac{a}{\chi}\,X = 0: \qquad X(x) = C_{11}\cos\left(x\,\sqrt{\frac{a}{\chi}}\right) + C_{12}\sin\left(x\,\sqrt{\frac{a}{\chi}}\right), \qquad (3.149\text{a, b})$$

$$\frac{d^2Y}{dx^2} + \frac{b}{\chi}\,Y = 0: \qquad Y(x) = C_{21}\cos\left(x\,\sqrt{\frac{b}{\chi}}\right) + C_{22}\sin\left(x\,\sqrt{\frac{b}{\chi}}\right). \qquad (3.150\text{a, b})$$

Substituting (3.149b, 3.150b) in (3.148a), the boundary conditions (3.146a) [(3.146b)] lead to (3.151a, b) [(3.152a, b)] and, hence, to (3.151c, d) [(3.152c, d)]

$$A\,e^{-at} = \Phi(0,t) = C_{11}\,e^{-at} + C_{21}\,e^{-bt}: \qquad C_{11} = A, \qquad C_{21} = 0 \qquad (3.151\text{a–d})$$

$$B\,e^{-bt} = \frac{\partial\Phi}{\partial x}(0,t) = C_{12}\,e^{-at}\sqrt{\frac{a}{\chi}} + C_{22}\,e^{-bt}\sqrt{\frac{b}{\chi}}: \qquad (3.152\text{a, b})$$

$$C_{12} = 0, \qquad C_{22} = B\,\sqrt{\frac{\chi}{b}}. \qquad (3.152\text{c, d})$$

Substitution of (3.151c, d) [(3.152c, d)] in (3.149b) [(3.150b)] leads to (3.153a) [(3.153b)]:

$$X(x) = A\cos\left(x\,\sqrt{\frac{a}{\chi}}\right), \qquad Y(x) = B\,\sqrt{\frac{\chi}{b}}\,\sin\left(y\,\sqrt{\frac{b}{\chi}}\right), \qquad (3.153\text{a, b})$$

that substituted in (3.148a) leads to (3.146c). The solutions of the one-dimensional Cartesian diffusion equation (3.134a) ≡ (3.144a) with one initial (3.134b) [two boundary (3.144b, c)] condition(s) involve one (3.134c) [two (3.145)] function(s), and are shown next to be equivalent (Subsection 3.3.7).

3.3.7 EQUIVALENCE OF ONE INITIAL AND TWO BOUNDARY CONDITIONS

The general integral (3.134c) [(3.143b; 3.140b; 3.141b)] of the Cartesian one-dimensional diffusion equation (3.134a) is an analytic function of both variables (3.154a) with convergent double Maclaurin series (3.154b) with coefficients (3.154c):

$$\Phi(x,t) \in A\left(|R^2\right): \qquad \Phi(x,t) = \sum_{n,m=0}^{\infty} B_{n,m} \frac{x^n}{n!} \frac{t^m}{m!}, \qquad B_{n,m} \equiv \lim_{x,t \to 0} \frac{\partial^{n+m} \Phi}{\partial x^n \, \partial t^m}.$$
$$(3.154a\text{-}c)$$

From (3.154c), follow the initial condition (3.155a) specified by the Maclaurin series (3.155c) for the analytic function of one variable (3.155b) with coefficient (3.155d, e):

$$t = 0: \qquad \Phi(x,0) \in A\,(|\,R), \qquad\qquad \Phi(x,0) = \sum_{n=0}^{\infty} B_n \frac{x^n}{n!},$$

$$B_n \equiv B_{n,0} = \lim_{x \to 0} \frac{d^n}{dx^n}\left[\Phi(x,0)\right]. \qquad (3.155a\text{-}e)$$

Substituting (3.155d) in (3.134c) specifies the general integral (3.156b, c):

$$k = m - 2\,n:$$

$$\Phi(x,t) = \sum_{n=0}^{\infty} \frac{(\chi\,t)^n}{n!} \frac{d^{2n}}{dx^{2n}} \left\{ \sum_{m=0}^{\infty} B_m \frac{x^m}{m!} \right\} = \sum_{n=0}^{\infty} \sum_{m=2n}^{\infty} \frac{(\chi\,t)^n}{n!} B_m \frac{x^{m-2n}}{(m-2\,n)!}$$

$$= \sum_{n,k=0}^{\infty} \frac{(\chi\,t)^n}{n!} B_{2n+k} \frac{x^k}{k!} = \Phi_1(x,t) + \Phi_2(x,t), \qquad (3.156a\text{-}e)$$

where: (i) the change of variable of summation (3.156a) is made in (3.156d); and (ii) in (3.156e) the terms with even Φ_1 (odd Φ_2) powers of x are separated.

Thus, the even (odd) powers of x in (3.156d) lead to (3.157a–f) [(3.158a–f)]:

$$s = n + k:$$

$$\Phi_1(x,t) = \sum_{k,n=0}^{\infty} \frac{x^{2k}}{(2\,k)!} B_{2n+2k} \frac{(\chi\,t)^n}{n!} = \sum_{k=0}^{\infty} \frac{(x^2/\chi)^k}{(2\,k)!} \frac{d^k}{dt^k} \left\{ \sum_{n=0}^{\infty} B_{2n+2k} \frac{(\chi\,t)^{n+k}}{(n+k)!} \right\}$$

$$= \sum_{k=0}^{\infty} \frac{(x^2/\chi)^k}{(2\,k)!} \frac{d^k F}{dt^k} = \Phi_1(-x,t), \qquad F(t) = \sum_{s=0}^{\infty} B_{2s} \frac{(\chi\,t)^s}{s!},$$

$$(3.157a\text{-}f)$$

$s = n + k$:

$$\Phi_2(x,t) = \sum_{k,n=0}^{\infty} \frac{x^{2k+1}}{(2k+1)!} B_{2n+2k+1} \frac{(\chi t)^n}{n!}$$

$$= \sqrt{\chi} \sum_{k=0}^{\infty} \frac{\left(x/\sqrt{\chi}\right)^{2k+1}}{(2k+1)!} \frac{d^k}{dt^k} \left\{ \sum_{n=0}^{\infty} B_{2n+2k+1} \frac{(\chi t)^{n+k}}{(n+k)!} \right\}$$

$$= \sqrt{\chi} \sum_{k=0}^{\infty} \frac{\left(x/\sqrt{\chi}\right)^{2k+1}}{(2k+1)!} \frac{d^k G}{dt^k} = -\Phi_2(-x,t), \quad G(t) = \sqrt{\chi} \sum_{s=0}^{\infty} B_{2s} \frac{(\chi t)^s}{s!},$$

$$(3.158a\text{–}f)$$

which coincide with (3.157a, d, e) $\equiv$ (3.140a, b) [(3.158a, d, e) $\equiv$ (3.141a, b)]. In (3.157f) [(3.158f)], the sum from $s = k$ to $s = \infty$ can be replaced by $s = 0$ to $s = \infty$ since the terms from $s = 0$ to $s = k-1$ vanish under differentiation of order k in (3.157e) [(3.158e)]; thus, the functions (3.157f) [(3.158f)] are independent of k as required by the coincidence (3.157d, f) $\equiv$ (3.140a, b) [(3.158d, f) $\equiv$ (3.141a, b)]. It has been shown that *the general integral of the one-dimensional Cartesian diffusion equation (3.144a), as an analytic function (3.154a-c) of both independent variables position and time, can be represented equivalently by one function (3.156a–d; 3.155a–e) $\equiv$ (3.134c) [the sum (3.156e) of two functions (3.157a–e; 3.158a–e) $\equiv$ (3.140b; 3.141b)] to satisfy one initial (3.134b) [two boundary (3.144b, c)] conditions*. Not all solutions of the one-dimensional Cartesian diffusion equation (3.134a) are analytic functions of both variables (3.154a-c) as shown next (Subsection 3.3.8) by the particular case (3.26d).

3.3.8   ANALYTIC AND SINGULAR SOLUTIONS OF THE DIFFUSION EQUATION

The particular solution (3.26d) $\equiv$ (3.159a) of the one-dimensional Cartesian diffusion equation (3.22a) is: (i) an analytic function of position (3.159b), whose Maclaurin series (3.159c) converges because the remainder vanishes at high order (3.159d):

$$\Phi(x,t) = \sqrt{\frac{\pi}{\chi t}} \exp\left(-\frac{x^2}{4\chi t}\right) = \sqrt{\frac{\pi}{\chi t}} \sum_{n=0}^{\infty} \frac{(-)^n}{n!} \left(\frac{x^2}{4\chi t}\right)^n$$

$$= \sqrt{\frac{\pi}{\chi t}} \sum_{n=0}^{N} \frac{(-)^n}{n!} \left(\frac{x^2}{4\chi t}\right)^n + R_N(x,t), \qquad \lim_{N \to \infty} R_N(x,t) = 0;$$
$$(3.159a\text{–}d)$$

and (ii) a singular function (3.159a) of time that is infinitely differentiable at time zero where the derivatives of all orders vanish (3.160a, b), so that the Maclaurin series (3.160c) reduces to the remainder (3.160d) that never vanishes because it coincides with the function (3.160e), so that the Maclaurin series does not converge:

$$\lim_{t \to 0} \frac{\partial^n [\Phi(x,t)]}{\partial t^n} = \lim_{t \to 0} \exp\left(-\frac{x^2}{4 \chi t}\right) 0(t^{-n-1/2}) = 0, \qquad \text{(3.160a, b)}$$

$$\sum_{n=0}^{N} \frac{t^n}{n!} \lim_{t \to 0} \frac{\partial^n [\Phi(x,t)]}{\partial t^n} + R_N(x,t) = R_N(x,t) = \Phi(x,t). \qquad \text{(3.160c, d)}$$

*The particular solution (3.26d) of the one-dimensional Cartesian diffusion equation (3.22a) is an example of a function of time that is: (i) **smooth**, that is infinitely differentiable; and (ii) not **analytic**, because its Maclaurin series does not converge, since the remainder equals the function.* Another example of a smooth non-analytic function was given earlier (Subsection III.3.3.4). The solutions of the one-dimensional Cartesian harmonic/wave/diffusion equations (Subsections 3.3.1/3.3.2–3.3.3/3.3.4–3.3.8) used the method of symbolic differential operators that: (i) applies in its simplest form to a linear first-order partial differential equation (Subsection 3.3.9); and (ii) can be generalised to higher orders (Subsection 3.3.10).

3.3.9 First-Order Linear Partial Differential Equations with Constant Coefficients

The solution of linear partial differential equations with constant coefficients was given (Section 1.6–1.9) in the case when all derivatives were of the same order. The latter restriction is dropped (Sections 3.1–3.6), most simply, by considering a first-order equation:

$$A \frac{\partial \Phi}{\partial x} + B \frac{\partial \Phi}{\partial y} + C \, \Phi = 0, \qquad \text{(3.161)}$$

without a forcing term. Using the method of symbolic operators of derivatives (Subsection 3.3.1), the first-order partial differential equation may be reduced to a first-order ordinary differential equation by taking x as the variable (3.162a):

$$A \frac{d\Phi}{dx} + v \, \Phi = 0, \qquad v = C + B \frac{d}{dy}, \qquad \text{(3.162a, b)}$$

and treating y as a parameter in (3.162b). The solution of (3.162a) is (3.162c):

$$\Phi(x, y) = \left\{ \exp\left(-\frac{v \, x}{A}\right) \right\} f(y), \qquad \text{(3.162c)}$$

denoting the arbitrary constant of integration by f, and allowing dependence on the "other" independent variable y. Substituting the parameter v (3.162b) in (3.162c) leads to (3.162d):

$$\Phi(x, y) = \exp\left(-\frac{C}{A} x\right) \exp\left\{-\frac{B}{A} x \frac{d}{dy}\right\} f(y), \qquad \text{(3.162d)}$$

where: (i) the first exponential is an ordinary function; and (ii) the second exponential is a symbolic operator, which may be interpreted (3.162e, f):

$$\mu = -\frac{B}{A}: \qquad \left\{ \exp\left(\mu \ x \ \frac{d}{dy} \right) \right\} f(y) = \sum_{n=0}^{\infty} \frac{(\mu \ x)^n}{n!} \frac{d^n f}{dy^n} = f(y + \mu \ x), \qquad (3.162e, f)$$

using: (ii-a) the series expansion of the exponential (3.111a); and (ii-b) the Taylor series (I.23.32b) for the arbitrary differentiable function (3.163a). Substituting (3.162e, f) into (3.162d), it follows that:

$$f(z) \in \mathcal{D} (| \ C): \qquad \Phi(x,y) = \exp\left(-\frac{C}{A} \ x \right) f\left(y - \frac{B}{A} \ x \right), \qquad (3.163a, b)$$

the general integral of the linear first-order partial differential equation (3.161), with constant coefficients and no forcing term is (3.163b), involving an arbitrary differentiable function (3.163a); the function (3.162e) need not be analytic, since substitution of (3.163b) satisfies (3.161) needing only first-order derivatives as shown below (Subsection 3.3.10).

3.3.10   UNIDIRECTIONAL AND BIDIRECTIONAL DAMPED WAVES

Two particular cases of (3.163b) are *the linear first-order partial differential equations with constant coefficients (3.164b) [(3.164e)]:*

$$f \in \mathcal{D} (| \ R^2): \qquad \frac{\partial \Phi}{\partial t} - c \ \frac{\partial \Phi}{\partial x} + \alpha \ \Phi = 0: \qquad \Phi(x,t) = e^{-\alpha t} \ f\left(t - \frac{x}{c} \right), \qquad (3.164a\text{--}c)$$

$$g \in \mathcal{D} (| \ R^2): \qquad \frac{\partial \Phi}{\partial x} - \frac{1}{c} \ \frac{\partial \Phi}{\partial t} + \beta \ \Phi = 0: \qquad \Phi(x,t) = e^{-\beta x} \ g(x - c \ t), \qquad (3.164d\text{--}f)$$

*whose solution (3.164c) [(3.164f)] is a **unidirectional wave** propagating in the positive x-direction with **phase speed** c and **temporal (spatial) damping** α (β), where f (g) are differentiable functions (3.164a) [(3.164d)]. In the case of a one-dimensional bidirectional wave propagating in the positive and negative x-directions with the same wave speed c and temporal damping α in (3.165c):*

$$f, g \in \mathcal{D}^2 (| \ R^2): \qquad \Phi(x,t) = e^{-\alpha t} \left[f\left(t - \frac{x}{c} \right) + g\left(t + \frac{x}{c} \right) \right], \qquad (3.165a\text{--}c)$$

where f, g are twice differentiable functions (3.165a, b), the wave equation is (3.165d–f):

$$0 = \left\{ \left(\frac{\partial}{\partial t} - c \ \frac{\partial}{\partial x} + \alpha \right) \left(\frac{\partial}{\partial t} + c \ \frac{\partial}{\partial x} + \alpha \right) \right\} \Phi(x,t)$$

$$= \left\{ \left(\frac{\partial}{\partial t} + \alpha \right)^2 - c^2 \ \frac{\partial^2}{\partial x^2} \right\} \Phi(x,t)$$

$$= \frac{\partial^2 \Phi}{\partial t^2} - c^2 \ \frac{\partial^2 \Phi}{\partial x^2} + 2 \ \alpha \ \frac{\partial \Phi}{\partial t} + \alpha^2 \ \Phi. \qquad (3.165d\text{--}f)$$

The generalisation to one-dimensional bidirectional waves propagating in the positive and negative x-direction with different wave speeds $c_\pm$ and dampings $\alpha_\pm$ in:

$$\Phi(x,t) = e^{-\alpha_+ t} \, f\!\left(t - \frac{x}{c_+}\right) + e^{-\alpha_- t} \, g\!\left(t + \frac{x}{c_-}\right), \tag{3.166a}$$

where f, g are twice differentiable functions (3.165a, b), is the wave equation (3.166b, c)

$$0 = \left\{\left(\frac{\partial}{\partial t} - c_+ \frac{\partial}{\partial x} + \alpha_+\right)\left(\frac{\partial}{\partial t} + c_- \frac{\partial}{\partial x} + \alpha_-\right)\right\} \Phi(x,t)$$

$$= \frac{\partial^2 \Phi}{\partial t^2} - c_+ \, c_- \, \frac{\partial^2 \Phi}{\partial x^2} - (c_+ - c_-) \, \frac{\partial^2 \Phi}{\partial x \, \partial t}$$

$$+ (\alpha_+ + \alpha_-) \, \frac{\partial \Phi}{\partial t} - (c_+ \, \alpha_- - c_- \, \alpha_+) \, \frac{\partial \Phi}{\partial x} + \alpha_+ \, \alpha_- \, \Phi. \tag{3.166b, c}$$

The result (3.166b, c) relates to the factorisation of a linear second-order partial differential equation into the product of two first-order equations (Subsections 3.3.12–3.3.14). This result (3.163a, b) can be obtained alternatively via a change of variable (Subsection 3.3.11) that leads to the solution of a class of linear partial differential equations with constant coefficients and order N.

3.3.11   EXTENSION TO HIGHER-ORDER PARTIAL DIFFERENTIAL EQUATIONS VIA CHANGE OF DEPENDENT VARIABLE

In the case (3.167a), only first-order derivatives appear in (3.161) $\equiv$ (3.167b) and the solution (3.167c):

$$C = 0: \qquad A \, \frac{\partial \Psi}{\partial x} + B \, \frac{\partial \Psi}{\partial y} = 0, \qquad \Psi(x,y) = f\!\left(y - \frac{B}{A}\, x\right), \tag{3.167a–c}$$

coincides with the similarity solution (1.206a–c) of (1.204a, b) obtained by the method of similarity functions (Subsection 1.6.2). The implication is that *the change of dependent variable (3.168a) transforms the linear first-order unforced partial differential equation with constant coefficients (3.161) $\equiv$ (3.168b) into one involving only first-order derivatives (3.168c):*

$$\Psi(x,y) = e^{Cx/A} \, \Phi(x,y): \qquad A \, \frac{\partial \Phi}{\partial x} + B \, \frac{\partial \Phi}{\partial y} + C \, \Phi = 0$$

$$\Leftrightarrow \qquad A \, \frac{\partial \Psi}{\partial x} + B \, \frac{\partial \Psi}{\partial y} = 0. \tag{3.168a–c}$$

The proof of (3.168a–c) follows, noting that (3.168a) implies (3.169a, b)

$$\left\{ A\,\frac{\partial \Psi}{\partial x}\ ,\ B\,\frac{\partial \Psi}{\partial y} \right\} = e^{Cx/A} \left\{ C\,\Phi + A\,\frac{\partial \Phi}{\partial x}\ ,\ B\,\frac{\partial \Phi}{\partial y} \right\}, \qquad (3.169\text{a, b})$$

that substituted in (3.168c) implies (3.168b).

The same change of variable (3.168a) can be used to transform the partial differential equation (3.170) of order N

$$\left\{ \left(A\,\frac{\partial}{\partial x} + B\,\frac{\partial}{\partial y} + C \right)^{N} \right\} \Phi = 0, \qquad (3.170)$$

into one involving only N- th order derivatives (3.171a) $\equiv$ (1.213):

$$\left\{ \left(A\,\frac{\partial}{\partial x} + B\,\frac{\partial}{\partial y} \right)^{N} \right\} \Psi = 0: \qquad \Psi(x,y) = \sum_{n=0}^{N-1} y^{n}\, f_{n}\!\left(y - \frac{B\,x}{A} \right), \qquad (3.171\text{a, b})$$

whose solution (Subsection 1.6.5) is (3.171b) $\equiv$ (1.216a, b). Substituting (3.171b) into (3.168a), it follows that:

$$f_{1}\,,...,\,f_{N-1}(z) \in \mathcal{D}^{N}(|\,R): \qquad \Phi(x,y) = \exp\!\left(-\frac{C}{A}\,x \right) \sum_{n=0}^{N-1} y^{n}\, f_{n}\!\left(y - \frac{B}{A}\,x \right),$$
$$(3.172\text{a, b})$$

the general integral of the linear partial differential equation (3.170) of order N with constant coefficients is (3.171b) involving N arbitrary N-times differentiable (3.172a) similarity functions. For example, the partial differential equation (3.173e–g):

$A = B = C = 1, \qquad N = 3$:

$$0 = \left(\frac{\partial}{\partial x} + \frac{\partial}{\partial y} + 1 \right)^{3} \Phi$$

$$= \left\{ \left(\frac{\partial}{\partial x} + \frac{\partial}{\partial y} \right)^{3} + 3\left(\frac{\partial}{\partial x} + \frac{\partial}{\partial y} \right)^{2} + 3\left(\frac{\partial}{\partial x} + \frac{\partial}{\partial y} \right) + 1 \right\} \Phi$$

$$= \frac{\partial^{3}\Phi}{\partial x^{3}} + \frac{\partial^{3}\Phi}{\partial y^{3}} + 3\,\frac{\partial^{2}\Phi}{\partial x^{2}\,\partial y} + 3\,\frac{\partial^{3}\Phi}{\partial x\,\partial y^{2}} + 3\,\frac{\partial^{2}\Phi}{\partial x^{2}} + 3\,\frac{\partial^{2}\Phi}{\partial y^{2}}$$

$$+ 6\,\frac{\partial^{2}\Phi}{\partial x\,\partial y} + 3\,\frac{\partial\Phi}{\partial x} + 3\,\frac{\partial\Phi}{\partial y} + \Phi, \qquad (3.173\text{a–g})$$

that corresponds to (3.170; 3.172a, b) with (3.173a–d), has the solution (3.173d)

$$f,\ g,\ h \in \mathcal{D}^{3}(|\,R): \qquad \Phi(x,y) = e^{-x}\left[f(y-x) + y\, g(y-x) + y^{2}\, h(y-x) \right],$$
$$(3.173\text{h–l})$$

involving three arbitrary three times differentiable functions (3.173h–j). The solution (3.172a, b) of (3.170) is extended next to a class of second-order linear unforced partial differential equations with constant coefficients (Subsection 3.3.12) using a non-square (square) first-order factorisation [Subsection 3.3.13 (3.3.14)].

3.3.12 SECOND-ORDER PARTIAL DIFFERENTIAL EQUATION

Consider the general linear unforced second-order partial differential equation with constant coefficients (3.174a):

$$0 = A \frac{\partial^2 \Phi}{\partial x^2} + B \frac{\partial^2 \Phi}{\partial y^2} + C \frac{\partial^2 \Phi}{\partial x \, \partial y} + D \frac{\partial \Phi}{\partial x} + E \frac{\partial \Phi}{\partial y} + F \, \Phi$$

$$= \left\{ \left(A_1 \frac{\partial}{\partial x} + B_1 \frac{\partial}{\partial y} + C_1 \right) \left(A_2 \frac{\partial}{\partial x} + B_2 \frac{\partial}{\partial y} + C_2 \right) \right\} \Phi(x, y). \qquad \text{(3.174a, b)}$$

If it can be factorised as the product (3.174b) of two first-order equations (3.161), then the general integral (3.175c) would be the sum of two terms like (3.163a, b):

$f, \ g \in \mathcal{D}^2 (| R)$:

$$\Phi(x, y) = \exp\left(-\frac{C_1}{A_1} x \right) f\left(y - \frac{B_1}{A_1} x \right) + \exp\left(-\frac{C_2}{A_2} x \right) g\left(y - \frac{B_2}{A_1} x \right),$$
$$\text{(3.175a–c)}$$

where (3.175a, b) are arbitrary twice differentiable functions. To prove that (3.175a–c) is the general integral of (3.174a, b), it is sufficient to determine (A_1, B_1, C_1) and (A_2, B_2, C_2) from (A, B, C, D, E, F) by solving the relations (3.176a–f)

$$\{A, B, C, D, E, F\} = \{ A_1 A_2, \ B_1 B_2, \ A_1 B_2 + A_2 B_1, \ A_1 C_2 + A_2 C_1,$$
$$B_1 C_2 + B_2 C_1, \ C_1 C_2 \}, \qquad \text{(3.176a–f)}$$

that follow from the equality of (3.174a) $\equiv$ (3.174b).

The equations (3.176a, b, f) can be solved for (A_1, B_1, C_1) in (3.177a–c) and substituted in (3.176c, d, e), leading to (3.177d–f):

$$\{A_1, B_1, C_1\} = \left\{ \frac{A}{A_2}, \ \frac{B}{B_2}, \ \frac{F}{C_2} \right\}, \qquad \text{(3.177a–c)}$$

$$\{C, D, E\} = \left\{ A \frac{B_2}{A_2} + B \frac{A_2}{B_2}, \ A \frac{C_2}{A_2} + F \frac{A_2}{C_2}, \ B \frac{C_2}{B_2} + F \frac{B_2}{C_2} \right\}. \qquad \text{(3.177d–f)}$$

The equation (3.177d) ≡ (3.178a) is a quadric with roots (3.178b):

$$B\left(\frac{A_2}{B_2}\right)^2 - C\frac{A_2}{B_2} + A = 0: \qquad 2\,B\,\frac{A_2}{B_2} = C \pm \sqrt{C^2 - 4\,A\,B}, \qquad (3.178\text{a, b})$$

$$2\,F\,\frac{A_2}{C_2} = D \pm \sqrt{D^2 - 4\,A\,F}, \qquad 2\,F\,\frac{B_2}{C_2} = E \pm \sqrt{E^2 - 4\,B\,F}, \qquad (3.178\text{c, d})$$

and, likewise, (3.178c) [(3.178d)] follow from (3.177e) [(3.177f)]. The ratio of (3.178c, d) must be consistent with (3.178b) leading to (3.179a, b):

$$\frac{D \pm \sqrt{D^2 - 4\,A\,F}}{E \pm \sqrt{E^2 - 4\,B\,F}} = \frac{A_2}{B_2} = \frac{C \pm \sqrt{C^2 - 4\,A\,B}}{2\,B}, \qquad (3.179\text{a, b})$$

which is a necessary condition for the decomposition of (3.174a) into two factors (3.174b) to exist.

The condition (3.179a, b) implies that the three equations (3.178b–d) are redundant so that one coefficient can be chosen arbitrarily (3.180a) and: (i) three others (3.180b–d) follow from (3.178b, c; 3.177a); and (ii) the remaining two (3.180e, f) follow, substituting (3.180c, d) in (3.177b, c):

$$A_2 = 1, \qquad B_2 = \frac{2\,B}{C \pm \sqrt{C^2 - 4\,A\,B}}, \qquad C_2 = \frac{2\,F}{D \pm \sqrt{D^2 - 4\,A\,F}},$$
$$(3.180\text{a-c})$$

$$A_1 = A, \qquad B_1 = \frac{C \pm \sqrt{C^2 - 4\,A\,B}}{2}, \qquad C_1 = \frac{D \pm \sqrt{D^2 - 4\,A\,F}}{2}.$$
$$(3.180\text{d-f})$$

It has been shown that *the general linear unforced second-order partial differential equation with constant coefficients (3.174a) satisfying (3.179a, b)* ≡ *(3.181):*

$$2\,B\left[D \pm \sqrt{D^2 - 4\,A\,F}\,\right] = \left[C \pm \sqrt{C^2 - 4\,A\,B}\,\right]\left[E \pm \sqrt{E^2 - 4\,B\,F}\,\right], \qquad (3.181)$$

can be decomposed into the product (3.174b) of two first-order equations and, thus, has general integral (3.175a–c) with coefficients given by (3.180a–f). There is an exception to this result (3.181) when the two terms on the r.h.s. of (3.175c) reduce to one arbitrary function; this exception will be considered (Subsection 3.3.14) after giving next an example of a non-exceptional case (Subsection 3.3.13).

3.3.13   SECOND-ORDER DECOMPOSITION INTO FIRST-ORDER FACTORS

Consider *the linear unforced second-order partial differential equation with constant coefficients (3.182a) that can be decomposed into the product of two first-order equations (3.182b, c):*

$$0 = 2\,\frac{\partial^2 \Phi}{\partial x^2} + 2\,\frac{\partial^2 \Phi}{\partial y^2} + 5\,\frac{\partial^2 \Phi}{\partial x\,\partial y} + 4\,\frac{\partial \Phi}{\partial x} + 5\,\frac{\partial \Phi}{\partial y} + 2\,\Phi$$

$$= \left\{ \left(2\,\frac{\partial}{\partial x} + 4\,\frac{\partial}{\partial y} + 2 \right) \left(\frac{\partial}{\partial x} + \frac{1}{2}\,\frac{\partial}{\partial y} + 1 \right) \right\}\,\Phi(x,y)$$

$$= \left\{ \left(2\,\frac{\partial}{\partial x} + \frac{\partial}{\partial y} + 2 \right) \left(\frac{\partial}{\partial x} + 2\,\frac{\partial}{\partial y} + 1 \right) \right\}\,\Phi(x,y), \qquad (3.182a\text{–}c)$$

and, thus, has a general integral (3.183c) involving two arbitrary twice differentiable functions (3.183a, b):

$$f,\, g \in \mathcal{D}^2\,(|\,R): \qquad \Phi(x,y) = e^{-x}\left[f(y - 2\,x) + g\left(y - \frac{x}{2} \right) \right]. \qquad (3.183a\text{–}c)$$

The result can be proved as follows: (i) the equation (3.182b) is of the type (3.174a) with five coefficients (3.184a–e) that satisfy the necessary condition (3.181) for the factorisation (3.174b) if the sixth coefficient satisfies (3.184f):

$$\{A\,,\,B\,,\,D\,,\,E\,,\,F\} = \{2\,,\,2\,,\,4\,,\,5\,,\,2\}: \qquad 16 = \left(C \pm \sqrt{C^2 - 16} \right)(5 \pm 3); \quad (3.184a\text{–}f)$$

(ii) the condition (3.184f) is met (3.185a) by either (3.185b, d) [or (3.185c, e)], leading both to (3.185f):

$$C^2 - 16 = \left(C - \frac{16}{5 \pm 3} \right)^2 = \left\{ (C - 2)^2\,,\,(C - 8)^2 \right\} = \left\{ C^2 - 4\,C + 4\,,\,C^2 - 16\,C + 64 \right\}$$

$$C = 5; \qquad (3.185a\text{–}f)$$

(iii) the coefficients (3.184a–e; 3.185f) substituted in (3.174a) lead to the particular case (3.182a) that meets the factorisation condition (3.181); (iv) substitution of (3.184a–e; 3.185f) in (3.180d–f) with all upper (lower) signs specifies (3.186a) two alternatives (3.186b) [(3.186c)] for the factorisation:

$$\{A_1\,,\,A_2\,,\,B_1\,,\,B_2\,,\,C_1\,,\,C_2\} = \left\{ 2\,,\,1\,,\,\frac{5 \pm 3}{2}\,,\,\frac{4}{5 \pm 3}\,,\,2\,,\,1 \right\}$$

$$= \left\{ 2\,,\,1\,,\,4\,,\,\frac{1}{2}\,,\,2\,,\,1 \right\} \ or \ \{2\,,\,1\,,\,1\,,\,2\,,\,2\,,\,1\}; \qquad (3.186a\text{–}c)$$

(v) substitution of (3.186b) [(3.186c)] in (3.174b) leads to the factorisation (3.182b) [(3.182c)] and it can be checked directly that both agree with (3.182a) ≡ (3.182b)

[(3.182a) ≡ (3.182c)]; and (vi) substitution of (3.186b) [(3.186c)] in (3.175a–c) leads in both cases to (3.183a–c) merely interchanging the two arbitrary functions. The exceptional case in which (3.175a–c) does not hold is considered next (Subsection 3.3.14).

3.3.14 SECOND ORDER AS THE SQUARE OF THE FIRST ORDER

If the factors in (3.174b) are proportional (3.187a), then (3.175c) reduces to one arbitrary function, and cannot be the general integral that is given instead by (3.187b):

$$\frac{A_2}{A_1} = \frac{B_2}{B_1} = \frac{C_2}{C_1} \equiv \mu: \quad \Phi(x,y) = \exp\left(-\frac{C_1}{A_1}x\right)\left\{f\left(y - \frac{B_1}{A_1}x\right) + y\, g\left(y - \frac{B_1}{A_1}x\right)\right\}.$$

$$(3.187a, b)$$

In this case, the differential equation (3.174a, b) must be a square of the form (3.188a, b):

$$0 = \left\{\left(\sqrt{A}\,\frac{\partial}{\partial x} + \sqrt{B}\,\frac{\partial}{\partial y} + \sqrt{F}\right)^2\right\}\Phi(x,y)$$

$$= A\,\frac{\partial^2\Phi}{\partial x^2} + B\,\frac{\partial^2\Phi}{\partial y^2} + 2\,\sqrt{A\,B}\,\frac{\partial^2\Phi}{\partial x\,\partial y} + 2\,\sqrt{A\,F}\,\frac{\partial\Phi}{\partial x} + 2\,\sqrt{B\,F}\,\frac{\partial\Phi}{\partial y} + F\,\Phi,$$

$$(3.188a, b)$$

implying that only three coefficients (3.189a–c) are independent and the remaining three are determined by (3.189d–f):

$$\{A\,,\,B\,,\,F\}: \qquad \{C\,,\,D\,,\,E\} = \{2\,\sqrt{A\,B}\,,\,2\,\sqrt{A\,F}\,,\,2\,\sqrt{B\,F}\,\}. \qquad (3.189a\text{–}f)$$

and the general integral is (3.190c):

$$f\,,\,g \in D^2(|\,R): \quad \Phi(x,y) = \exp\left(-x\,\sqrt{\frac{F}{A}}\right)\left[f\left(y - x\,\sqrt{\frac{B}{A}}\right) + y\, g\left(y - x\,\sqrt{\frac{B}{A}}\right)\right],$$

$$(3.190a\text{–}c)$$

involving two arbitrary twice differentiable functions (3.190a, b).

It has been shown that *the linear unforced second-order partial differential equation with constant coefficients (3.174a) satisfying (3.189d–f) takes the form (3.188b) with factorisation into the square of first order (3.188a) and, hence, the general integral (3.190a–c).* The coefficients (3.189d–f) satisfy trivially the factorisation condition (3.181) that simplifies to (3.191a–c):

$$0 = 2\,B\,D - C\,E = 4\,B\,\sqrt{A\,F} - 4\,\sqrt{A\,B}\,\sqrt{B\,F} = 0. \qquad (3.191a\text{–}c)$$

As an example, *the linear unforced partial differential equation with constant coefficients, of second-order square type (3.192a, b):*

$$0 = \left\{ \left(\frac{\partial}{\partial x} - \frac{\partial}{\partial y} + 2 \right)^2 \right\} \Phi(x,y) = \frac{\partial^2 \Phi}{\partial x^2} + \frac{\partial^2 \Phi}{\partial y^2} - 2\frac{\partial^2 \Phi}{\partial x\,\partial y} + 4\frac{\partial \Phi}{\partial x} - 4\frac{\partial \Phi}{\partial y} + 4\,\Phi,$$

$$(3.192\text{a--c})$$

has general integral (3.193a–c):

$$f,\, g \in \mathcal{D}^2\,(\mathbb{R}): \qquad \Phi(x,y) = e^{-2x}\left[f(y+x) + y\, g(y+x) \right]. \qquad (3.193\text{a--c})$$

The preceding four methods I/II/III/IV of solution of linear partial differential equations with constant coefficients (Sections 1.6–1.9/3.1/3.2/3.3, respectively) are compared next (Subsection 3.3.14), before introducing a fifth method, V, that will be presented in more detail (Section 3.4) and extended to a wider variety of cases (Section 3.5–3.9).

3.3.15 FIVE METHODS OF SOLUTION OF PARTIAL DIFFERENTIAL EQUATIONS

The solution of a linear unforced partial differential equation with constant coefficients may be approached (Table 3.1) by five methods: (I) factorisation of the differential operator (Sections 1.6–1.9), which is always possible if all derivatives are of the same order, but may not be possible if there are derivatives of different orders; (II) separation of variables (Section 3.1) that does not apply to all linear partial differential equations with constant coefficients, but applies to some with variable coefficients; (III) similarity functions (Section 3.2) that reduce the partial to an ordinary differential equation that is easier to solve; (IV) symbolic differential operators (Section 3.3) that involve series expansions, that become quite cumbersome except in the simplest cases, and apply only to analytic functions; and (V) exponential solutions (Section 3.4) is the simplest because it reduces to the algebraic problem of finding the roots of a polynomial of several variables and, in addition, also applies to a variety of cases with a forcing term.

Method V of exponential solutions involves a characteristic polynomial and is both simpler and mostly more general than methods I–IV, making it preferable because: (i) unlike method I of factorisation, it applies to any linear partial differential equation with constant coefficients, regardless of whether there are derivatives of different orders or not; (ii) unlike method II of separation of variables, it is not limited to particular forms of linear partial differential equations, although it is restricted to constant coefficients; (iii) it has the same domain of application as method III of similarity functions because it requires a single step of finding the roots of a polynomial of two variables instead of two steps of (1) solving an ordinary differential equation with constant coefficients by (2) finding the roots of a characteristic polynomial of one variable; (iv) it is simpler than method IV of

TABLE 3.1
Five Methods of Solution of Linear Partial Differential Equations

Method	I	II	III	IV	V
Designation	Factorisation	Separation of variables	Similarity functions	Symbolic differential operators	Exponentials
Section	1.6–1.9, 3.3.9–3.3.14	3.1	3.2	3.3	3.4
$\Phi(x,y)$	$\displaystyle\prod_{n=1}^{N} f_n\!\left(\frac{\partial}{\partial x}, \frac{\partial}{\partial y}\right)$	$X(x)\,Y(y)$	$f(x+a\,y)$	$f\!\left(\mu\,\dfrac{d}{dy}\right)$	$\exp\left(a\,x+b\,y\right)$
Coefficients of p.d.e. constant	Not necessary	Not necessary	Yes	Yes	Yes
Solution involves	Lower order p.d.e.	2 o.d.e.	1 o.d.e.	Series expansion	Characteristic polynomial
Application	Simple	Medium	Medium	Difficult	Simple

Note: Five methods of solution of linear partial differential equations are compared on three criteria: (i) allowing derivatives of different order; (ii) allowing for variable coefficients; (iii) level of complexity.
Abbreviations: p.d.e. (o.d.e.) = partial (ordinary) differential equation.

symbolic operators of derivatives since it requires only algebraic operations with polynomials instead of manipulation of a series, which may raise convergence issues. The method V of exponential solutions or the characteristic polynomial also extends from unforced to forced linear partial differential equations with constant coefficients, and on account of all these advantages, is presented in more detail next (Section 3.4).

3.4 METHOD OF CHARACTERISTIC POLYNOMIAL OR EXPONENTIALS OF SEVERAL VARIABLES

The solution of a linear partial differential equation with constant coefficients is considered without (with) a forcing term [Subsections 3.4.1–3.4.15 (3.4.16–3.4.44)]. In the case of a linear unforced partial differential equation with constant coefficients, method V of exponential solutions or the characteristic polynomial of all independent variables (Subsection 3.4.1) provides the simplest alternative to (Table 3.1) the preceding four methods of: (I) factorisation of the differential operators (Sections 1.6–1.9); (II) separation of variables (Section 3.1); (III) similarity functions (Section 3.2); and (IV) symbolic operators of derivatives (Section 3.3). Method V of exponential solutions of several variables (Section 3.4) leads to a characteristic polynomial in several variables (Subsections 3.4.1–3.4.3) whose simple (multiple) roots specify the particular and general integrals [Subsection 3.4.4 (3.4.5)] for the unforced linear partial differential equation with constant coefficients and derivatives of any order. The method is applied as an example to: (i) the (Table 3.2) harmonic/wave/diffusion equations (Subsections 3.4.6/3.4.7/3.4.8–3.4.12); (ii) to (Table 3.3) the telegraph/bar/beam equations (Subsection 3.4.13/3.4.14/3.4.15); and (iii) a generalisation to a fourth-order equation of mathematical physics (notes 3.1–3.5) that includes the preceding (i) and (ii) as particular cases.

Method V of the characteristic polynomial of several variables also applies both to unforced (forced) linear [Subsections 3.4.1–3.4.15 (3.4.16–3.4.44)] partial differential equations with constant coefficients, most simply, when the forcing function is an exponential of all independent variables (Subsection 3.4.16) leading to non-resonant (resonant) solutions [Subsection(s) 3.4.17 (3.4.18–3.4.21)], for example, (Table 3.3) for the harmonic, wave, diffusion, telegraph, bar, and beam equations, that may have single (double) resonances [Subsections 3.4.18–3.4.19 (3.4.20–3.4.21)]. The forcing by exponentials (Subsections 3.4.16–3.4.21) extends to circular and hyperbolic cosines and sines and their products (Subsection 3.4.22), for example, are applied to the harmonic/wave/diffusion equation (Subsection 3.4.23 / 3.4.24 / 3.4.25). The doubly resonant solutions (Subsection 3.4.26) of the harmonic (wave) equation can be compared with solutions obtained by the method of complex conjugate (phase) variables [Subsection(s) 3.4.30 (3.4.27–3.4.29)]. An alternative method of the solution of forced linear partial differential equations with constant coefficients is to use the inverse characteristic polynomial of partial derivatives (Subsections 3.4.31–3.4.32) that applies when the forcing function is smooth (Subsections 3.4.33–3.4.34) or an exponential is multiplied by a smooth function (Subsection 3.4.35). This applies to forcing by the product of a power and an exponential (Subsection 3.4.35) and multiplied by a hyperbolic (circular) cosine or sine or their products (Subsection 3.4.36), for example, in the case of the harmonic/wave/diffusion equation (Subsection 3.4.37/3.4.38/3.4.39).

The method of the characteristic polynomial applies to forced linear differential equations with constant coefficients (Subsections 3.4.21–3.4.39) and if the derivatives are all of the same order, it can be combined (Subsections 3.4.40–3.4.44) with the method of factorisation (Sections 1.6–1.9), for single (multiple) roots [Subsections 3.4.40–3.4.42 (3.4.43–3.4.44)], for example, for the harmonic/wave (other second-order) equations [Subsection 3.4.42 (3.4.44)]. As for any linear ordinary (Chapter IV.1) or partial (Chapter V.3) differential equation, the complete integral of the forced equation (Subsection 3.4.45) is the sum of the general (particular) integral of the unforced (forced) equation [Subsections 3.4.1–3.4.15 (3.4.16–3.4.44)]. The arbitrary functions in the general or complete integral are determined by compatible and independent boundary and/or initial conditions. There are other methods to obtain forced solutions of linear partial differential equations, for example, the influence or Green function for impulsive forcing also specifies forcing by an integrable function, for example, for the diffusion equation (Subsections 3.4.8–3.4.12). The method of characteristic polynomials deserves detailed consideration because it extends (Subsection 3.4.46) to single equations (simultaneous systems) of linear partial differential equations with ordinary [Section 3.4 (3.6)] or homogeneous [Section 3.5 (3.7)] derivatives, and also to multiple finite difference equations with constant coefficients [Section 3.8 (3.9)].

3.4.1   UNFORCED LINEAR PARTIAL DIFFERENTIAL EQUATIONS WITH CONSTANT COEFFICIENTS

The linear unforced partial differential equation with constant coefficients (1.11a, b) ≡ (3.194a) and order $N(M)$ in the independent variable x (y) has solutions specified by exponentials of the independent variables (3.194b):

$$0 = \sum_{n=0}^{N} \sum_{m=0}^{M} A_{n,m} \frac{\partial^{n+m} \Phi}{\partial x^n \, \partial y^m}: \qquad\qquad \Phi(x,y) = e^{ax+by}, \qquad\qquad (3.194a, b)$$

as can be shown by four different methods (II to V). When it succeeds, method II (Section 3.1) of separation of variables (3.1b) ≡ (3.195a) leads to linear unforced ordinary differential equations with constant coefficients for each of the functions that have exponential solutions (3.195b, c), leading to (3.195d) ≡ (3.194b):

$$\Phi(x,y) = X(x)\,Y(y), \qquad X(x) = e^{ax}, \qquad Y(y) = e^{by}\backslash:$$

$$\Phi(x,y) = e^{ax+by}. \qquad\qquad (3.195a\text{–}d)$$

Method III (Section 3.2) of similarity functions (3.48a-c) ≡ (3.196a, b) leads to a linear unforced ordinary differential equation with constant coefficients whose solutions are exponentials (3.196c–d) corresponding to (3.196e) ≡ (3.194b) for (3.196f):

$$\xi = x + \mu\, y: \qquad \Phi(x,y) = f(\xi) = e^{a\xi} = e^{a(x+\mu y)} = e^{ax+by}, \qquad\qquad b \equiv \mu\, a.$$
$$(3.196a\text{–}f)$$

Method IV (Section 3.3) of symbolic operators of derivatives leads to solutions of the type (3.162b, c) ≡ (3.197a) that lead to solutions of the type (3.197a, b) ≡ (3.100a-d) involving exponentials (3.197c), leading to (3.197d, e) ≡ (3.194b) with (3.197f):

$$\Phi(x,y) = e^{\mu x} \exp\left(v \; x \; \frac{d}{dy} \right) f(y) = e^{\mu x} \; f(y + v \; x),$$

$$f(\xi) = e^{b\xi}: \qquad \Phi(x,y) = e^{\mu x + b(y + vx)} = e^{ax + by}, \qquad a = \mu + b \; v. \qquad (3.197\text{a–f})$$

These four methods (I to IV) suggest a simpler, more direct method V using a characteristic polynomial of several variables (Subsection 3.4.2).

3.4.2   CHARACTERISTIC POLYNOMIAL OF SEVERAL VARIABLES

The simplest method V is to note that a linear unforced ordinary (partial) differential equation with constant coefficients has: (i) solutions that are exponentials of the independent variables (3.194b) ≡ (3.195d) ≡ (3.196e) ≡ (3.197e) where it may be inserted a multiplying constant C in (3.198a):

$$\Phi(x,y) = C \; e^{ax+by}: \qquad \frac{\partial^{n+m} \Phi}{\partial x^n \, \partial y^m} = \frac{\partial^{n+m}}{\partial x^n \, \partial y^m} \left(C \; e^{ax+by} \right) = C \; a^n \; b^m \; e^{ax+by}$$

$$= a^n \; b^m \; \Phi(x,y); \qquad (3.198\text{a–d})$$

(ii) differentiation with regard to $x(y)$ is equivalent (3.198b–d) to multiplication by $a(b)$; and (iii) substitution of (3.198d) in (3.194a) ≡ (3.199a) leads to (3.199b):

$$0 = \left\{ \sum_{n=0}^{N} \sum_{m=0}^{M} A_{n,m} \frac{\partial^{n+m}}{\partial x^n \, \partial y^m} \right\} \Phi(x,y) = \Phi(x,y) \; P_{N,M}(a,b), \qquad (3.199\text{a, b})$$

involving the **characteristic polynomial** (3.199b) ≡ (3.200a):

$$P_{N,M}(a,b) = \sum_{n=0}^{N} \sum_{m=0}^{M} A_{n,m} \; a^n \; b^m = 0, \qquad (3.200\text{a, b})$$

that must vanish (3.200b) for the combination of constants (a, b) in (3.198a). The method V of the characteristic polynomial (3.200a, b) leads to the simplest approach (3.194b) to the solution of a linear unforced partial differential equation with constant coefficients (3.194a) and will be adopted as the baseline in the sequel (Sections 3.4–3.9), in combination with other methods, such as the symbolic operator of derivatives (Section 3.3). The starting point is the characteristic polynomial for a linear partial differential equation, starting with the unforced case (Subsection 3.4.3).

3.4.3 EXISTENCE OF A CHARACTERISTIC POLYNOMIAL

Consider the general linear partial differential equation of order N (M) with regard to the two independent variables x, y and one dependent variable Φ given by (1.10; 1.11a) ≡ (1.11b), without a forcing term (3.194a). If the coefficients are constant, the corresponding ordinary differential equation in d/dx (d/dy) would have (Section IV.1.4) exponential solutions e^{ax} (e^{by}), multiplied by a factor $f(y)$ $[g(x)]$, which could depend on the variable (3.201a, b):

$$f(y)\, e^{ax} = \Phi(x, y) = e^{by}\, g(x); \qquad (3.201a, b)$$

the common form of (3.201a, b) is an exponential in x, y in (3.202a–c):

$$f(y) = e^{by}, \qquad g(x) = e^{ax}: \qquad \Phi(x, y) = C\, e^{ax+by}, \qquad (3.202a–c)$$

where C is a constant in agreement with (3.202c) ≡ (3.198a). From (3.198b–d; 3.199a, b; 3.200a, b), it follows that the latter has the property of differentiation $\partial/\partial x$ $(\partial/\partial y)$ with regard to $x(y)$ being equivalent to multiplication by $a(b)$ in (3.198b–d). Thus, *the exponential in two variables (3.202c) is a solution of the linear unforced partial differential equation with constant coefficients (3.194a) iff,* that is if and only if, *the parameters a,b are roots of the* **characteristic polynomial** *(3.200a, b) in two variables.* The roots of the characteristic polynomial may be single (multiple), and are considered next [Subsection 3.4.4 (3.4.5)].

3.4.4 SINGLE ROOTS OF THE CHARACTERISTIC POLYNOMIAL

Suppose that the characteristic polynomial (3.200b) ≡ (3.203a) that is of degree $N(M)$ in $a(b)$ has $N(M)$ distinct roots $a_n(b)[b_m(a)]$, which may depend on $b(a)$ in (3.203b) [(3.203c)]:

$$0 = P_{N,M}(a) = a_0 \prod_{n=1}^{N} [a - a_n(b)] = b_0 \prod_{m=1}^{M} [b - b_m(a)], \qquad (3.203a–c)$$

where $a_0(b_0)$ is a constant. Each root in (3.203b) [(3.203c)] corresponds to a particular integral (3.204a) [(3.204b)] of (3.194a):

$$\Phi_n(x, y) = \int C_n(b)\, e^{by} \exp[x\, a_n(b)]\, db, \qquad (3.204a)$$

$$\Phi^m(x, y) = \int D_m(a)\, e^{ax} \exp[y\, b_m(a)]\, da, \qquad (3.204b)$$

where the exponential (3.202a) is multiplied by an arbitrary function of $b(a)$ and integrated in $db(da)$. If the roots $a_n(b_m)$ are distinct, the particular integrals (3.204a) [(3.204b)] are linearly independent; since (3.194a) is a linear unforced partial differential equation the sum of solutions (3.204a) [(3.204b)] is also a solution (3.205a) [(3.205b)]:

$$\Phi(x,y) = \sum_{n=1}^{N} \Phi_n(x,y) = \sum_{m=1}^{M} \Phi^m(x,y), \tag{3.205a, b}$$

and it is the general integral because it involves $N(M)$ arbitrary functions $C_n(b)[D_m(a)]$. It has been shown that *the general integral of the linear unforced partial differential equation with constant coefficients (3.194a) is given alternatively by (3.205a) [(3.205b)] as a sum of $N(M)$ linearly independent particular integrals (3.204a) [(3.204b)] each involving one integrable arbitrary function $C_n(D_m)$, and corresponding to a distinct root (3.203a) [(3.203b)] of the characteristic polynomial (3.200b).* The case of multiple roots of the characteristic polynomial is considered next (Subsection 3.4.5).

3.4.5   MULTIPLE ROOTS OF THE CHARACTERISTIC POLYNOMIAL

If the characteristic polynomial (3.200b) has $R(S)$ distinct roots of multiplicity $\alpha_r(\alpha_s)$ in $a(b)$ in (3.206a–c) [(3.206d–f)]:

$$\sum_{r=1}^{R} \alpha_r = N: \qquad 0 = P_{N,M}(a,b) = a_0 \prod_{r=1}^{R} [a - a_r(b)]^{\alpha_r}, \tag{3.206a–c}$$

$$\sum_{s=1}^{S} \beta_s = M: \qquad 0 = P_{N,M}(a,b) = b_0 \prod_{s=1}^{S} [b - b_s(a)]^{\beta_s}, \tag{3.206d–f}$$

then, to each distinct root $a_r(b_s)$ of multiplicity $\alpha_r(\beta_s)$ corresponds the particular integral (3.207) [(3.208)] involving $\alpha_r(\beta_s)$ arbitrary functions:

$$\Phi_r(x,y) = \int e^{by} \exp[x\, a_r(b)] \times \left\{ \sum_{p_r=1}^{\alpha_r} x^{p_r-1}\, C_{r,p_r}(b) \right\} db, \tag{3.207}$$

$$\Phi^s(x,y) = \int e^{ax} \exp[y\, b_s(a)] \times \left\{ \sum_{q_s=1}^{\beta_s} y^{q_s-1}\, D_{s,q_s}(a) \right\} da, \tag{3.208}$$

and the general integral (3.209a) [(3.209b)] is given by their sum:

$$\Phi(x,y) = \sum_{r=1}^{R} \Phi_r(x,y) = \sum_{s=1}^{S} \Phi^s(x,y). \tag{3.209a, b}$$

The proof of (3.207; 3.209a) [(3.208; 3.209b)] is made by noting that if $a_n(b_m)$ is a root of multiplicity $\alpha_r(\beta_s)$ in (3.206a–c) [(3.206d–f)], then, in (3.207) [(3.208)], $\alpha_r(\beta_s)$ terms appear (3.210a–c) [(3.211a–c)]:

$$p_r = 1,\ldots,\alpha_r: \qquad \left(\frac{\partial}{\partial a}\right)^{p_r-1} e^{ax+by} = e^{ax+by}\, x^{p_r-1}, \tag{3.210a–c}$$

$$q_s = 1,\ldots,\beta_s: \qquad \left(\frac{\partial}{\partial b}\right)^{q_s-1} e^{ax+by} = e^{ax+by}\, y^{q_s-1}, \tag{3.211a–c}$$

each multiplied by a distinct arbitrary function $C_{r,p_r}\left(D_{s,q_s}\right)$ of $b(a)$ in (3.207) [(3.208)]. This method V of solution of linear unforced partial differential equations of any order with constant coefficients is applied first to the harmonic/wave/diffusion equations (Subsections 3.4.6/ 3.4.7/3.4.8–3.4.10) to compare with the methods I/II/III/IV used before (Sections 1.6–1.9/3.1/3.2/3.3).

3.4.6 CHARACTERISTIC POLYNOMIAL FOR THE HARMONIC EQUATION

Substituting (3.198a) ≡ (3.212a) as a solution of the two-dimensional Cartesian harmonic equation (3.212b) ≡ (3.98a) ≡ (3.50a) ≡ (3.1a) ≡ (1.230e-f) leads to the algebraic relation in curved brackets in (3.212c):

$$\Phi(x,y) = C\, e^{ax+by}: \qquad 0 = \frac{\partial^2 \Phi}{\partial x^2} + \frac{\partial^2 \Phi}{\partial y^2} = e^{ax+by}\,(a^2+b^2) = \Phi(x,y)\, P_{2,2}(a,b), \tag{3.212a–d}$$

which specifies the characteristic polynomial (3.212d) ≡ (3.213b) with roots (3.213a, c):

$$0 = P_{2,2}(a,b) = a^2 + b^2: \qquad b_\pm(a) = \pm\, i\, a, \tag{3.213a–c}$$

leading (3.204b) to the particular integrals (3.214a–c):

$$\Phi_\pm(x,y) = \int \exp\left[a\,(x\pm i\,y)\right] C_\pm(a)\, da = f_\pm(x\pm i\,y) = f_+(z),\, f(z^*). \tag{3.214a–c}$$

Alternatively, the roots of (3.213a) are (3.215a) leading to the particular integrals (3.215b–d):

$$a_\pm(b) = \pm\, i\, b: \qquad \Phi^\pm(x,y) = \int \exp\left[b\,(y\pm i\,x)\right] C^\pm(b)\, db$$

$$= f^\pm(y\pm i\,x) = f^+(i\,z^*),\, f^-(-\,i\,z). \tag{3.215a–d}$$

In (3.214c) [(3.215d)] were used the complex variable (3.216a) and its conjugate (3.216b) [the related expressions (3.216c, d) and (3.216e, f)]:

$$z \equiv x + i\,y, \qquad z^* = x - i\,y:$$

$$y + i\,x = i(x - i\,y) = i\,z^*, \qquad y - i\,x = -\,i\,(x + i\,y) = -i\,z. \tag{3.216a–f}$$

Thus, *the general integral of the two-dimensional Cartesian harmonic equation (3.212b) is the sum (3.217a) [(3.217c)] of (3.214a–c) [(3.215a–d)]:*

$$\Phi(x,y) = f_+(x+i\ y) + f_-(x-i\ y) = f_+(z) + f_-(z^*)$$

$$= f^+(y+i\ x) + f^-(y-i\ x) = f^+(i\ z^*) + f^-(-\ i\ z), \qquad (3.217\text{a–d})$$

which may be expressed in terms of the complex variables (3.216a, b) [(3.216d, f)]. The result (3.217a) agrees with (3.217a, b) ≡ (3.101a, b) ≡ (3.51a–d) ≡ (3.7a–c) ≡ (1.232a–c) obtained respectively by methods V/IV/III/II/I, among which method V of characteristic polynomials is the simplest.

The real and imaginary parts of the complex functions (3.218a–c; 3.219a–c; 3.220a–f):

$$e^{\pm z}\ ,\ \ e^{\pm z^*} = e^{\pm(x\ \pm\ iy)} = e^{\pm x\ \pm\ iy} = e^{\pm x}\ (\cos y \pm i\sin y), \qquad (3.218\text{a–c})$$

$$e^{\pm iz}\ ,\ \ e^{\pm iz^*} = e^{\pm i(x\ \pm\ iy)} = e^{\pm y}\ e^{\pm ix} = e^{\pm y}\ (\cos x \pm i\sin x), \qquad (3.219\text{a–c})$$

$a = \alpha\ e^{i\beta}$:

$$a^z \equiv \exp\left[\log(a^z)\right] = \exp(z\log a) = \exp\left[(x+i\ y)\ (\log\alpha + i\ \beta)\right]$$

$$= \exp\left[x\log\alpha - \beta\ y + i\ (\beta\ x + y\log\alpha)\right]$$

$$= x^\alpha\ e^{-\beta y}\left[\cos(\beta\ x + y\log\alpha) + i\sin(\beta\ x + y\log\alpha)\right], \qquad (3.220\text{a–f})$$

specify the following *18 solutions (3.221a–n) of the two-dimensional Cartesian harmonic differential equation (3.212b):*

$$\Phi(x,y) = e^{\pm x}\cos, \sin y\ ;\ \ e^{\pm y}\cos, \sin x\ ; \cosh x(\cos,\sin)y\ ;\ \sinh x(\cos,\sin)y\ ;$$

$$\cos x(\cosh,\sinh)y\ ;\ \sin x(\cosh,\sinh)y\ ;\ x^\alpha\ e^{-\beta y}\cos, \sin(\beta\ x + y\log\alpha),$$
$$(3.221\text{a–n})$$

in addition to other earlier examples (3.108a–f; 3.56a–d; 3.10a–d). The solutions (3.221e–l) involving hyperbolic cosines or sines are linear combinations of the solutions (3.221a–d) involving exponentials, with the circular cosine or sine factors being the same. Method V of characteristic polynomials also applies to the one-dimensional Cartesian wave equation (Subsection 3.4.7).

3.4.7 CHARACTERISTIC POLYNOMIAL FOR THE WAVE EQUATION

Seeking an exponential solution in space–time (3.198a) ≡ (3.222a) for the one-dimensional wave equation (3.222b) ≡ (3.109a) ≡ (3.59a) ≡ (3.11a) leads to the algebraic relation (3.222c):

$$\Phi(x,t) = e^{at+bx}:\qquad 0 = \frac{\partial^2\Phi}{\partial t^2} - c^2\frac{\partial^2\Phi}{\partial x^2} = e^{at+bx}\ (a^2 - b^2\ c^2) = \Phi(x,t)\ P_{2,2}(a,b),$$
$$(3.222\text{a–d})$$

specifying the characteristic polynomial (3.222d) ≡ (3.223b) with roots (3.223a, c):

$$0 = P_{2,2}(a,b) = a^2 - b^2 c^2 \quad \Rightarrow \quad a_{\pm}(b) = \mp b c, \quad (3.223a\text{--}c)$$

yielding (3.204a) to the particular integrals (3.226a, b):

$$\Phi_{\pm}(x,t) = \int e^{b (x \mp ct)} C_{\pm}(b) \, db = f_{\pm}(x \mp c \, t); \quad (3.226a, b)$$

the alternative roots (3.227a) of the characteristic polynomial (3.223a) leads (3.204b) to the particular integrals (3.227b, c):

$$b_{\pm}(a) = \pm \frac{a}{c}: \qquad \Phi^{\mp}(x,t) = \int \exp\left[a\left(t \pm \frac{x}{c}\right)\right] C^{\mp}(a) \, da = f^{\pm}\left(t \mp \frac{x}{c}\right),$$
(3.227a--c)

Thus, *the general integral of the one-dimensional Cartesian wave equation (3.222b) is the sum (3.228a) [(3.228b)] of the particular integrals (3.226a, b) [3.227b, c)]*:

$$\Phi(x,t) = f_{+}(x - c \, t) + f_{-}(x + c \, t) = f^{+}\left(t - \frac{x}{c}\right) + f^{-}\left(t + \frac{x}{c}\right). \quad (3.228a, b)$$

The result (3.228a) agrees with (3.228a) ≡ (3.113a, b) ≡ (3.60a–d) ≡ (3.16a–d) obtained by methods V/IV/III/II, respectively, of which method V is the simplest.

The one-dimensional classical wave equation is generalised to any dimension replacing the second-order spatial derivatives by the Laplacian (3.102a), leading to (3.229a):

$$\nabla^2 \, \Psi = \frac{1}{c^2} \frac{\partial^2 \Psi}{\partial t^2}; \qquad \nabla^2 = \frac{1}{R^2} \frac{\partial}{\partial R} R^2 \frac{\partial}{\partial R} = \frac{\partial^2}{\partial R^2} + \frac{2}{R} \frac{\partial}{\partial R}; \quad (3.229a\text{--}c)$$

in the case of **spherical coordinates** with dependence only on the radial distance R from the origin, the Laplacian is given by (III.6.46a, b) ≡ (3.229b, c), leading by substitution in (3.206a) to the **spherical wave equation** (3.230a, b):

$$\frac{1}{c^2} \frac{\partial^2 \Psi}{\partial t^2} = \frac{1}{R^2} \frac{\partial}{R} \left(R^2 \frac{\partial \Psi}{\partial R}\right) = \frac{\partial^2 \Psi}{\partial R^2} + \frac{2}{R} \frac{\partial \Psi}{\partial R}. \quad (3.230a, b)$$

The spherical wave equation (3.230a, b) ≡ (3.231a, b) can be rewritten:

$$\frac{1}{c^2} \frac{\partial^2}{\partial t^2} (R \, \Psi) = R \frac{\partial^2 \Psi}{\partial R^2} + 2 \frac{\partial \Psi}{\partial R} = \frac{\partial^2}{\partial R^2} (R \, \Psi), \quad (3.231a, b)$$

showing that *multiplication by the radial distance (3.232b) leads from the spherical (3.230a, b) ≡ (3.231a, b) to the plane wave equation (3.232a) with solutions (3.228a) ≡ (3.232c)*:

$$\frac{1}{c^2} \frac{\partial^2 \Phi}{\partial t^2} = \frac{\partial^2 \Phi}{\partial R^2}: \qquad \Phi(R,t) = R \, \Psi(R,t) = f_{+}(R - c \, t) + f_{-}(R + c \, t).$$
(3.232a--c)

Thus, *the spherical wave equation (3.230a, b) ≡ (3.231a, b) has general integral (3.232c), for example, (3.233b, c) [(3.234a–d)] for sinusoidal propagating waves (standing modes) with (3.233a) frequency ω and wavenumber k:*

$$\omega = k\ c: \qquad\qquad \Psi(R,t) = \frac{A}{R}\ \cos,\sin(k\ R \mp \omega\ t), \qquad (3.233\text{a–c})$$

$$\Psi(R,t) = \frac{A}{R}\ \cos,\sin(k\ R)\cos,\sin(\omega\ t), \qquad (3.234\text{a–d})$$

*and amplitudes decaying with the inverse of the radial distance for **spherical waves**, instead of constant amplitude for plane propagating waves (3.19a–e; 3.61a–e) [standing modes (3.116c–f)].* Method V of characteristic polynomials applies to the harmonic (wave) equation [Subsection 3.4.6 (3.4.7)] and also to the diffusion equation (Subsection 3.4.8–3.4.12).

3.4.8 CHARACTERISTIC POLYNOMIAL FOR THE DIFFUSION EQUATION

Seeking an exponential solution in space-time (3.235a) ≡ (3.222a) for the one-dimensional Cartesian diffusion equation (3.235b) ≡ (3.132a) ≡ (3.64a) ≡ (3.22a) leads to the algebraic relation (3.235c):

$$\Phi(x,t) = e^{at+bx}: \qquad 0 = \chi\ \frac{\partial^2 \Phi}{\partial x^2} - \frac{\partial \Phi}{\partial t} = (\chi\ b^2 - a)\ e^{at+bx} = P_{1,2}(a,b)\ \Phi(x,t),$$

$$(3.235\text{a–d})$$

specifying the characteristic polynomial (3.235d) ≡ (3.236b) that has a single root (3.236a, c):

$$0 = P_{1,2}(a,b) = \chi\ b^2 - a :$$

$$a_0(b) = \chi\ b^2, \qquad \Phi_0(x,t) = \int \exp(b\ x + \chi\ b^2\ t)\ C(b)\ db, \qquad (3.236\text{a–d})$$

leading to the general integral (3.236d) involving one arbitrary function; this result (3.236d) agrees with (3.67b) [(3.25c)] with the changes $b \to \chi\ b$ ($\pm i\ a \to b$), showing the agreement of methods II, III, and V, of which method V is the simplest. An alternative set of two solutions is obtained using the roots (3.237a) of the characteristic polynomial (3.236a, b) leading by (3.204b) to (3.237b, c):

$$b_\pm(a) = \pm \sqrt{\frac{a}{\chi}}: \qquad \Phi_\pm(x,t) = \int \exp\left(a\ t \pm x\sqrt{\frac{a}{\chi}}\right) C_\pm(a)\ da. \qquad (3.237\text{a–c})$$

Thus, *the one-dimensional Cartesian diffusion equation (3.235a) has general integral (3.238a) [particular integral (3.238b)] involving one (3.236a–d) [sum of two (3.237a–c)] function specified by one (two) initial (boundary) condition(s):*

$$\Phi(x,t) = \Phi_0(x,t) = \Phi_+(x,t) + \Phi_-(x,t). \qquad (3.238\text{a, b})$$

The solution of an unforced linear differential equation specifies the forced solution through the influence or Green's function (Chapter III.7), which can be applied to ordinary (partial) differential equations as shown by examples [notes IV.1.5–IV.1.8 (Subsections 3.4.9–3.4.12)].

3.4.9 INFLUENCE FUNCTION DUE TO FORCING BY A UNIT IMPULSE

A particular solution of the one-dimensional Cartesian diffusion equation is (3.26a) ≡ (3.239a), which has the integral property (3.28d) ≡ (3.239b):

$$\Phi(x,t) = \sqrt{\frac{\pi}{\chi\,t}}\ \exp\left(-\frac{x^2}{4\,\chi\,t}\right); \qquad \int_{-\infty}^{+\infty} \Phi(x,t)\,dx = 2\pi. \qquad (3.239a, b)$$

The Dirac delta function or **unit impulse** (Chapters III.1 and III.3) has the properties (III.1.71a, b) ≡ (3.240a) and (III.1.76) ≡ (3.240c) where f is an integrable function (3.240b):

$$\int_{-\infty}^{+\infty} \delta(x)\,dx = 1; \qquad f \in E\,(\mathbb{R}): \qquad \int_{-\infty}^{+\infty} f(\eta)\,\delta(x-\eta)\,dx = f(x). \qquad (3.240a\text{–}c)$$

Comparing (3.240a) with (3.239b), it follows that *the solution of the one-dimensional Cartesian diffusion equation (3.235b) forced by an unit impulse or Dirac function at the origin (3.241a) is the **influence function** or Green function (3.241b)*

$$\left\{\chi\,\frac{\partial^2}{\partial x^2} - \frac{\partial}{\partial t}\right\}G(x,t) = \delta(x): \quad G(x,t) = \frac{1}{2\,\pi}\,\Phi(x,t) = \frac{1}{2\,\sqrt{\pi\,\chi\,t}}\,\exp\left(-\frac{x^2}{4\,\chi\,t}\right).$$
$$(3.241a, b)$$

The influence function may be generalised to (i) forcing by the unit impulse at any position and (ii) forcing by an arbitrary integrable function (Subsection 3.4.10).

3.4.10 CONVOLUTION INTEGRAL AND FORCING BY AN INTEGRABLE FUNCTION

The change of variable $x \to x - \eta$ shows that *in the case of forcing by a unit impulse at an arbitrary position (3.242a), the influence function for the one-dimensional Cartesian diffusion equation is (3.242b):*

$$\left\{\chi\,\frac{\partial^2}{\partial x^2} - \frac{\partial}{\partial t}\right\}G(x,t;\eta) = \delta(x-\eta): \qquad G(x,t;\eta) = \frac{1}{2\,\sqrt{\pi\,\chi\,t}}\,\exp\left[-\frac{(x-\eta)^2}{4\,\chi\,t}\right].$$
$$(3.242a, b)$$

In the case of forcing by an arbitrary integrable function (3.240b), the solution of the one-dimensional Cartesian diffusion equation (3.243a) is specified by (3.243d) the **convolution integral** *(3.243b, c) of the forcing and influence functions:*

$$f(x) = \chi \frac{\partial^2 \Psi}{\partial x^2} - \frac{\partial \Psi}{\partial t}: \qquad \Psi(x,t) = f(x) * G(x,t) = \int_{-\infty}^{+\infty} f(\eta)\, G(x,t;\eta)\, dx$$

$$= \frac{1}{2\sqrt{\pi \chi t}} \int_{-\infty}^{+\infty} f(\eta) \exp\left\{ -\frac{(x-\eta)^2}{4\chi t} \right\} dx.$$

$$(3.243a-d)$$

The proof of (3.243d) follows (Chapter III.7) from the substitution of (3.242a) in (3.240c) leading to (3.244a):

$$f(x) = \int_{-\infty}^{+\infty} f(\eta) \left\{ \chi \frac{\partial^2}{\partial x^2} - \frac{\partial}{\partial t} \right\} G(x,t;\eta)\, d\eta$$

$$= \left\{ \chi \frac{\partial^2}{\partial x^2} - \frac{\partial}{\partial t} \right\} \int_{-\infty}^{+\infty} f(\eta)\, G(x,t;\eta)\, d\eta = \left\{ \chi \frac{\partial^2}{\partial x^2} - \frac{\partial}{\partial t} \right\} f(x) * G(x,t),$$

$$(3.244a-c)$$

and then taking the diffusion operator outside the integral in $d\eta$ (3.244b, c) assuming that it is uniformly convergent (Section I.13.8) in (x,t). The convolution integral (3.243d) can be evaluated in terms of the error function (Subsection 3.4.11) in the case of forcing by a unit jump.

3.4.11   FORCING BY A UNIT JUMP AND ERROR FUNCTION

As an example, it is considered forcing the one-dimensional Cartesian diffusion equation (3.243a) ≡ (3.245a) by a **unit step** or Heaviside unit function (III.1.28a-c) ≡ (3.245b-d):

$$\chi \frac{\partial^2 \Psi}{\partial x^2} - \frac{\partial \Psi}{\partial t} = H(x) = \begin{cases} 0 & \text{if} & x < 0, \\ 1/2 & \text{if} & x = 1/2, \\ 1 & \text{if} & x > 0. \end{cases} \qquad (3.245a-d)$$

Substituting (3.245b) in (3.243b-d) leads to the forced solution (3.246a-c):

$$\Psi(x,t) = H(x) * G(x,t) = \int_{-\infty}^{+\infty} H(\eta)\, G(x,t;\eta)\, d\eta = \frac{1}{2\sqrt{\pi \chi t}} \int_{0}^{\infty} \exp\left[-\frac{(x-\eta)^2}{4\chi t} \right] d\eta.$$

$$(3.246a-c)$$

The change of variable (3.247a) leads from (3.246c) to the integral (3.247b):

$$\xi = \frac{x - \eta}{2\sqrt{\chi t}}: \qquad\qquad \Psi(x,t) = \frac{1}{\sqrt{\pi}} \int_{-\infty}^{x/2\sqrt{\chi t}} e^{-\xi^2} \, d\xi. \qquad\qquad (3.247a, b)$$

The **error function** is defined by (3.248a) and varies (3.248b) [(3.248d) ≡ (III.1.1a)] from zero (to unity) at minus (plus) infinity with the value one-half at the origin (3.248c) ≡ (III.1.1b):

$$\mathrm{erf}(\zeta) \equiv \frac{1}{\sqrt{\pi}} \int_{-\infty}^{\zeta} e^{-\xi^2} \, d\xi: \qquad \mathrm{erf}(-\infty) = 0, \qquad \mathrm{erf}(0) = \frac{1}{2}, \qquad \mathrm{erf}(+\infty) = 1.$$
$$(3.248a\text{–}d)$$

Using (3.248a) in (3.247b) shows that *the solution of the one-dimensional Cartesian diffusion equation forced by a spatial unit step (3.249a) ≡ (3.245a, b) is the integral (3.246c) ≡ (3.247b) expressible in terms of the error function (3.249b):*

$$\chi \frac{\partial^2 \Psi}{\partial x^2} - \frac{\partial \Psi}{\partial t} = H(x): \qquad\qquad \Psi(x,t) = \mathrm{erf}\left(\frac{x}{2\sqrt{\chi t}}\right). \qquad\qquad (3.249a, b)$$

This solution of the diffusion equation illustrates the signal decay process by dissipation (Subsection 3.4.12).

3.4.12 DECAY OF A SIGNAL BY DISSIPATION

The solution (3.249b) is plotted in the Figure 3.3 as a function of position for several times: (i) for $t = 0$ with $x < 0$ ($x > 0$) the value is zero (3.250a, b) ≡ (3.248b) [unity (3.250c, d) ≡ (3.248d)]:

$$\lim_{t \to 0} \Psi(x,t) = \begin{cases} \mathrm{erf}(-\infty) = 0 & \text{if} \quad x < 0, & (3.250a, b) \\ \mathrm{erf}(+\infty) = 1 & \text{if} \quad x > 0, & (3.250c, d) \end{cases}$$

corresponding to a unit jump about the mean value one-half (3.251a, b):

$$\Psi(0,t) = \mathrm{erf}(0) = \frac{1}{2}, \qquad \lim_{t \to 0} \Psi(x,t) = H(x), \qquad (3.251a\text{-}c)$$

so that it coincides (3.251c) with the unit jump (3.245b–d) ≡ (3.250b, d; 3.251b) at time $t = 0$; (ii) for all non-zero t , the value at the origin is one-half (3.251b, c) ≡ (3.248c); (iii) the slope at the origin is (3.252a, b):

$$\frac{\partial}{\partial t}[\Psi(x,t)] = \frac{1}{\sqrt{\pi}} \exp\left(-\frac{x^2}{4\chi t}\right) \frac{d}{dx}\left(\frac{x}{2\sqrt{\chi t}}\right) = \frac{1}{2\sqrt{\pi \chi t}} \exp\left(-\frac{x^2}{4\chi t}\right)$$
$$(3.252a, b)$$

and decreases as t increases; (iv) as $x \to 0$ the slope (3.253a)

$$\lim_{x \to 0} \frac{\partial}{\partial x}\big[\Psi(x,t)\big] = \frac{1}{2\sqrt{\pi\,\chi\,t}} = \begin{cases} \infty & \text{if} \quad t = 0, \\ 0 & \text{if} \quad t = \infty, \end{cases} \qquad (3.253\text{a--c})$$

varies from infinity (3.253b) in the case of a jump (3.250a–d; 3.251a–c) to zero (3.253c) in the uniform case (3.254a, b):

$$\lim_{t \to \infty} \Psi(x,t) = \operatorname{erf}(0) = \frac{1}{2}; \qquad (3.254\text{a, b})$$

and (v) the values zero (3.255a, b) [unity (3.255c, d)] at position minus (plus) infinity for all time

$$\lim_{x \to -\infty} \Psi(x,t) = \operatorname{erf}(-\infty) = 0, \qquad \lim_{x \to +\infty} \Psi(x,t) = \operatorname{erf}(+\infty) = 1 \quad (3.255\text{a--d})$$

lead to an arithmetic average of one-half (3.256):

$$\frac{1}{2}\big[\,\Psi(-\infty,t) + \Psi(+\infty,t)\,\big] = \frac{1}{2}, \qquad (3.256)$$

which coincides with the value at the origin for all time (3.256) $\equiv$ (3.251b). Thus, *the solution (3.246c) $\equiv$ (3.247b) $\equiv$ (3.249b; 3.248a) of the one-dimensional Cartesian diffusion equation (3.249a) $\equiv$ (3.245a) forced (3.245b–d) by a unit jump (Figure 3.3): (i) as a function of position x, varies between zero (3.255a, b) [unity (3.255c, d)] at minus (plus) infinity and, thus, has the mean value one-half (3.256), which coincides with the value at the origin (3.251b) for all time; and (ii) as a function of position, it varies from a unit jump (3.245b–d) $\equiv$ (3.250a–d; 3.251a, b) at time zero with infinite slope (3.253b), with decreasing slope at the origin (3.253b) [everywhere (3.252b)] as time increases, until the slope is zero (3.253c) at the constant mean value one-half (3.254a, b) $\equiv$ (3.256) at infinite time.* Method V of the characteristic polynomial is applied next to the telegraph equation (Subsection 3.4.13) that includes as particular cases the wave (diffusion) equation [Subsection(s) 3.4.7 (3.4.8–3.4.12)].

3.4.13 CHARACTERISTIC POLYNOMIAL FOR THE WAVE-DIFFUSION EQUATION

The wave-diffusion or telegraph equation (3.257b) $\equiv$ (3.80) with wave speed c and diffusivity χ has exponential solutions in space-time (3.257a) $\equiv$ (3.212a) $\equiv$ (3.222a):

$$\Phi(x,t) = e^{at+bx}: \qquad 0 = \frac{\partial^2 \Phi}{\partial x^2} - \frac{1}{c^2}\frac{\partial^2 \Phi}{\partial t^2} - \frac{1}{\chi}\frac{\partial \Phi}{\partial t}, \qquad (3.257\text{a, b})$$

leading to the algebraic relation (3.257d):

$$0 = e^{at+bx}\left(b^2 - \frac{a^2}{c^2} - \frac{a}{\chi}\right) = \Phi(x,t)\,P_{2,2}(a,b), \qquad (3.257\text{c, d})$$

specifying the characteristic polynomial (3.227c) ≡ (3.258b) with roots (3.258a, c):

$$0 = P_{2,2}(a,b) = b^2 - \frac{a^2}{c^2} - \frac{a}{\chi}: \qquad\qquad b_{\pm}(a) = \pm \sqrt{\frac{a}{\chi} + \frac{a^2}{c^2}}. \qquad (3.258a\text{--}c)$$

The roots (3.258a, c) of the characteristic polynomial (3.258b) lead (3.204b) to the particular integrals (3.259a, b):

$$\Phi_{\pm}(x,t) = \int e^{at} \exp\left\{\pm x \sqrt{\frac{a}{\chi} + \frac{a^2}{c^2}}\right\} C_{\pm}(a)\, da. \qquad (3.259a, b)$$

The alternative roots (3.260a) of the characteristic polynomial (3.258b) lead (3.204a) to the particular integrals (3.260b, c):

$$a_{\pm}(b) = -\frac{c^2}{2\,\chi} \pm \frac{c^2}{2} \sqrt{\frac{1}{\chi^2} + 4\frac{b^2}{c^2}}:$$

$$\Phi^{\pm}(x,t) = \int \exp\left[b\, x + t\, a_{\pm}(b)\right] C^{\pm}(b)\, db. \qquad (3.260a\text{--}c)$$

Thus, *the general integral of the one-dimensional Cartesian wave-diffusion equation (3.257b) is the sum (3.261a) [(3.261b)] of the particular integrals (3.259a, b) [(3.260a–c)]:*

$$\Phi(x,t) = \Phi_{+}(x,t) + \Phi_{-}(x,t) = \Phi^{+}(x,t) + \Phi^{-}(x,t). \qquad (3.261a, b)$$

The case of absence of dissipation (3.262a) corresponds (3.257b) to the wave equation (3.222b) and leads (3.259a–c) to (3.262b):

$$\chi = \infty: \qquad \Phi_{\pm}(x,t) = \int \exp\left[a\left(t \pm \frac{x}{c}\right)\right] C_{\pm}(a)\, da = h_{\pm}\left(t \pm \frac{x}{c}\right), \qquad (3.262a\text{--}c)$$

which is equivalent to (3.262c) ≡ (3.227c). The case of instantaneous propagation (3.263a) corresponds (3.257a) to the diffusion equation (3.235b) and leads (3.259a, b) to (3.263b, c):

$$c = \infty: \qquad \Phi_{\pm}(x,t) = \int e^{at} \exp\left(\pm x \sqrt{\frac{a}{\chi}}\right) C_{\pm}(a)\, da, \qquad (3.263a\text{--}c)$$

in agreement with (3.263b, c) ≡ (3.237b, c). Method V of the characteristic polynomial is exemplified by second- (fourth-) order linear unforced partial differential equations, namely, the harmonic/wave/diffusion/telegraph (bar/beam) equation [Subsections 3.4.6/3.4.7/3.4.8–3.4.12/3.4.13 (3.4.14/3.4.15)].

3.4.14   Characteristic Polynomial for the Elastic Bar Equation

The linear unsteady transverse displacement of a stiff elastic bar satisfies (3.264b) $\equiv$ (3.87b) $\equiv$ (IV.6.782f):

$$q^2 = \frac{E\,I}{\rho}: \qquad\qquad \frac{\partial^2 \Phi}{\partial t^2} + q^2\,\frac{\partial^4 \Phi}{\partial x^4} = 0, \qquad\qquad (3.264\text{a, b})$$

where the **stiffness parameter** (3.87a) $\equiv$ (3.264a) involves Young's modulus of material E, the moment of inertia of the cross-section I, and the mass density per unit length ρ. Assuming a constant (3.265a) stiffness parameter, the linear unforced partial differential equation with derivatives of order two (four) with regard to time (position) has exponential solutions (3.265b) $\equiv$ (3.222a), leading to the algebraic relation (3.265c):

$$q = const: \qquad \Phi(x,t) = e^{at+bx}, \qquad 0 = e^{at+bx}\left(a^2 + q^2\ b^4\right) = \Phi(x,t)\,P_{2,4}(a,b),$$
$$(3.265\text{a–d})$$

which specifies the characteristic polynomial (3.265d) $\equiv$ (3.266b):

$$0 = P_{2,4}(a,b) = a^2 + q^2\ b^4: \qquad a_\pm(b) = \pm\,i\,q\,b^2, \qquad b_{\pm\pm}(a) = (\pm\,1\pm i)\sqrt{\frac{a}{2q}},$$
$$(3.266\text{a–d})$$

where (Section I.5.3) were used (3.267a-d):

$$\left[b_\pm(a)\right]^2 = \pm\,i\,\frac{a}{q}, \qquad b_{\pm\pm}(a) = \sqrt{\pm\,i\,\frac{a}{q}} = (\pm\,1\pm i)\sqrt{\frac{a}{2q}}, \qquad (3.267\text{a–d})$$

with two (3.266c) [four (3.266d)] roots (3.260a) for a (b).

The two roots (3.266c) lead (3.204a) to two particular integrals (3.268a, b) $\equiv$ (3.90c, d):

$$\Phi^\pm(x,t) = \int \exp\left(b\,x \pm i\,q\,b^2\,t\right) C^\pm(b)\,db \qquad\qquad (3.268\text{a, b})$$

and involve two arbitrary functions specified by initial conditions. The four roots (3.266d) lead (3.204b) to four particular integrals (3.269a–d):

$$\Phi_{\pm\pm}(x,t) = \int \exp\left(a\,t + x\,(\pm\,1\pm i)\sqrt{\frac{a}{2q}}\right) C_{\pm\pm}(a)\,da, \qquad (3.269\text{a–d})$$

involving four arbitrary functions $C_{\pm\pm}(a)$ determined from initial conditions. Thus, *the uniform (3.264a; 3.265a) bar equation (3.264b) is a linear unforced partial differential equation with constant coefficients of order two (four) in time (position), whose general integral is the sum (3.270a) [(3.270b)] of two (3.267a, b) [four (3.269a–d)] particular integrals involving two (four) arbitrary functions determined by two initial (four boundary) conditions:*

$$\Phi(x,t) = \Phi^+(x,t) + \Phi^-(x,t) = \Phi_{++}(x,t) + \Phi_{+-}(x,t) + \Phi_{-+}(x,t) + \Phi_{--}(x,t).$$
$$(3.270\text{a, b})$$

A generalisation of the bar equation (Subsection 3.4.14) is the beam equation allowing for axial tension (Subsection 3.4.15).

3.4.15 CHARACTERISTIC POLYNOMIAL FOR THE ELASTIC BEAM EQUATION

In the presence of **longitudinal tension** T the elastic bar equation (3.264b) is generalised to the **elastic beam equation** (3.271b) ≡ (3.94c) ≡ (IV.6.782f) involving the elastic wave speed (3.271a) ≡ (3.94a) and stiffness parameter (3.264a) ≡ (3.94b):

$$c = \sqrt{\frac{T}{\rho}}: \qquad \frac{\partial^2 \Phi}{\partial t^2} - c^2 \frac{\partial^2 \Phi}{\partial x^2} + q^2 \frac{\partial^4 \Phi}{\partial x^4} = 0. \qquad (3.271a, b)$$

Seeking an exponential solution in space-time (3.272a) ≡ (3.222a) leads in (3.272b) to the algebraic relation (3.272c):

$$\Phi(x,t) = e^{at+bx}: \qquad 0 = \left[a^2 - b^2 \; c^2 + q^2 \; b^4 \right] e^{at+bx} = \Phi(x,t) \; P_{2,4}(a,b), \qquad (3.272a\text{--}d)$$

specifying the characteristic polynomial (3.272d) ≡ (3.273b) that is quadratic (biquadratic) in $a\,(b)$:

$$0 = P_{2,4}(a,b) = a^2 - b^2 \; c^2 + q^2 \; b^4: \qquad a_\pm(b) = \pm i \; b \; \sqrt{q^2 \; b^2 - c^2}. \qquad (3.273a\text{--}c)$$

Choosing two roots (3.273a, c) in a leads (3.204a) to two particular integrals (3.274a, b) involving arbitrary functions specified by two initial conditions:

$$\Phi^\pm(x,t) = \int \exp\left[b \left(x \pm i \; t \; \sqrt{q^2 \; b^2 - c^2} \right) \right] C_\pm(b) \; db. \qquad (3.274a, b)$$

Choosing the four roots (3.275a–d) of the characteristic polynomial (3.273a):

$$\left[b_\pm(a) \right]^2 = \frac{c^2}{2 \; q^2} \pm \frac{i}{2 \; q^2} \sqrt{4 \; a^2 \; q^2 - c^4} \qquad (3.275a\text{--}d)$$

leads to four particular integrals (3.276a–d) involving four arbitrary functions specified by four initial conditions:

$$\Phi_{\pm\pm}(x,t) = \int \exp\left[a \; t \pm x \; b_\pm(a) \right] C_{\pm\pm}(a) \; da. \qquad (3.276a\text{--}d)$$

Thus, *the general integral of the beam equation (3.271b) is the sum of two (3.270a) [four (3.270b)] particular integrals (3.274a, b) [(3.276a–d)] involving two (four) arbitrary functions determined by two (four) initial (boundary) conditions.* In the absence of stiffness $q = 0$, the beam equation (3.271b) reduces to the wave equation (3.222b) and the two particular integrals (3.274a, b) simplify to (3.226a, b) in the general integral (3.228a) ≡ (3.270a). In the absence of tension $T = 0$, the wave speed (3.271a) is zero $c = 0$, the beam equation (3.271b) reduces to the bar equation (3.264b), and the two (four) particular integrals (3.274a, b) [(3.275a–d; 3.276a–d)] simplify to (3.267a, b) [(3.269a-d)]. The method of the characteristic polynomial applies both to unforced (some cases of forced) linear partial differential equations with constant coefficients [Subsections 3.4.1–3.4.15 (3.4.16–3.4.44)].

3.4.16   RESONANT AND NON-RESONANT FORCING BY AN EXPONENTIAL

Proceeding with the method of the characteristic polynomial (3.200b), it is applied next to the linear partial differential equation with constant coefficients and with forcing term (1.11a, b) ≡ (3.277a, b):

$$B(x,y) = \sum_{n=1}^{N} \sum_{m=1}^{M} A_{n,m} \frac{\partial^{n+m} \Phi}{\partial x^n \partial y^m} = \left\{ P_{N,M} \left(\frac{\partial}{\partial x}, \frac{\partial}{\partial y} \right) \right\} \Phi(x,y). \qquad (3.277a, b)$$

The complete integral (Section IV.1.3) of the equation with a forcing term is the sum of: (i) the general integral of the equation without a forcing term, which was determined before (Subsections 3.4.1–3.4.15); and (ii) a particular integral of the equation with a forcing term, which is considered next (Subsections 3.4.16–3.4.44). The simplest forcing term in (3.277a) is an exponential (3.278a):

$$B(x,y) = B \; e^{ax+by}; \qquad\qquad \Phi(x,y) = C \; e^{ax+by}, \qquad (3.278a, b)$$

in this case, the solution may be expected to be another exponential (3.278b). Substitution of (3.278a, b) into (3.277a, b), and use of (3.198b–d) leads to (3.279a, b):

$$B \; e^{ax+by} = \left\{ P_{N,M} \left(\frac{\partial}{\partial x}, \frac{\partial}{\partial y} \right) \right\} C \; e^{ax+by} = P_{N,M}(a,b) \, \Phi(x,y), \qquad (3.279a, b)$$

implying that the solution is an exponential (3.280b), provided that the characteristic polynomial does not vanish (3.280a):

$$P_{N,M}(a,b) \neq 0: \qquad\qquad \Phi(x,y) = \frac{B \; e^{ax+by}}{P_{N,M}(a,b)}. \qquad (3.280a, b)$$

If the characteristic polynomial (3.179a, b) vanishes, then b (a) is a root of multiplicity r (s), in (3.281a, b);

$$P_{N,M}(\alpha,\beta) = (\alpha - a)^r \; (\beta - b)^s \; Q_{N-r,M-s}(\alpha,\beta), \qquad Q_{N-r,M-s}(\alpha,\beta) \neq 0, \qquad (3.281a, b)$$

then (3.282a) does not vanish and the solution (3.282b) is modified (Subsection 1.8.1) by applying L'Hôpital's rule (Section I.9.8) to (3.280b) that differentiates the numerator and denominator r, s times with regard to a, b, leading to (3.282a, b):

$$E(a,b) \equiv \frac{\partial^{r+s}}{\partial a^r \, \partial b^s} \left[P_{N,M}(a,b) \right] \neq 0: \qquad \Phi(x,y) = \frac{B}{E(a,b)} \; x^r \; y^s \; e^{ax+by}. \qquad (3.282a, b)$$

Thus, the resonant solution (3.282b) is obtained from the non-resonant solution (3.280b) by differentiating r times with regard to a and s times with regard to b the numerator (denominator), leading to (3.282c) [(3.280c; 3.281a, b) $\equiv$ (3.282d)]:

$$\frac{\partial^{r+s}}{\partial a^r \, \partial b^s} \left(e^{ax+by}\right) = x^r \, y^s \, e^{ax+by}, \qquad E(a,b) = r!s! Q_{N-r,M-s}(a,b). \quad (3.282c, d)$$

An alternative proof of (3.282b) is to differentiate (3.279a, b) with regard to a (b) until the r.h.s. does not vanish, requiring r (s) differentiations so that the r.h.s. (l.h.s.) leads to (3.282d–f):

$$Bx^r y^s \, e^{ax+by} = \frac{\partial^{r+s}}{\partial a^r \, \partial b^s} \left(B \, e^{ax+by}\right) = \Phi(x,y) \, \frac{\partial^{r+s}}{\partial a^r \, \partial b^s} \left[P_{N,M}(a,b) \right] = E(a,b)\Phi(x,y),$$
$$(3.282d–f)$$

where (3.282f) $\equiv$ (3.282b). Thus, *a particular integral of the linear partial differential equation with constant coefficients (3.277a, b) and exponential forcing term (3.278a) is (3.280b) [(3.282b)] if a,b are not roots (3.280a) [are roots of multiplicity r,s (3.281a, b; 3.282a)] of the characteristic polynomial.* Forcing by an exponential is considered next in the six cases of the harmonic, wave, diffusion, telegraphy, bar, and beam equations for all possibilities of non-resonance (Subsection 3.4.17) or single (double) resonance [Subsections 3.4.18–3.4.19 (3.4.20)].

3.4.17 NON-RESONANT EXPONENTIAL FORCING OF SIX PARTIAL DIFFERENTIAL EQUATIONS

Exponential forcing (3.283a) is considered (Table 3.3) for: (i) the two-dimensional Cartesian harmonic equation (3.283a) $\equiv$ (3.212b); (ii–iii) the one-dimensional Cartesian wave (diffusion) equation (3.283b) $\equiv$ (3.222b) [(3.283c) $\equiv$ (3.235b)]; (iv) the wave-diffusion equation (3.283d) $\equiv$ (3.257b); and (v–vi) the bar (beam) equation (3.283e) $\equiv$ (3.264b) [(3.283f) $\equiv$ (3.271b)]:

$$B \, e^{at+bx} = \left\{ \frac{\partial^2}{\partial t^2} + \frac{\partial^2}{\partial x^2} \,, \quad \frac{\partial^2}{\partial t^2} - c^2 \frac{\partial^2}{\partial x^2} \,, \quad \chi \frac{\partial^2}{\partial x^2} - \frac{\partial}{\partial t} \,, \quad \frac{\partial^2}{\partial x^2} - \frac{1}{c^2}\frac{\partial^2}{\partial t^2} - \frac{1}{\chi}\frac{\partial}{\partial t} \,, \right.$$
$$\left. \frac{\partial^2}{\partial t^2} + q^2 \frac{\partial^4}{\partial x^4} \,, \quad \frac{\partial^2}{\partial t^2} - c^2 \frac{\partial^2}{\partial x^2} + q^2 \frac{\partial^4}{\partial x^4} \right\} \Phi(x,t). \qquad (3.283a–f)$$

The non-resonant forced solution (3.280a, b): (i) is (3.284b) for the harmonic equation (3.283a) with non-zero (3.284a) characteristic polynomial (3.213b):

$$a \neq \pm i \, b: \qquad\qquad \Phi(x,t) = B \, \frac{e^{at+bx}}{a^2 + b^2}; \qquad\qquad (3.284a, b)$$

(ii) is (3.285b) for the wave equation (3.283b) with non-zero (3.285a) characteristic polynomial (3.223b):

$$a \neq \pm b \, c: \qquad\qquad \Phi(x,t) = B \, \frac{e^{at+bx}}{a^2 - b^2 \, c^2}; \qquad\qquad (3.285a, b)$$

(iii) is (3.286b) for the diffusion equation (3.283c) with non-zero (3.286a) characteristic polynomial (3.236b):

$$a \neq \chi \, b^2: \qquad \qquad \Phi(x,t) = B \, \frac{e^{at+bx}}{\chi \, b^2 - a}; \qquad \qquad (3.286a, b)$$

(iv) is (3.287b) for the wave-diffusion equation (3.283d) with non-zero (3.287a) characteristic polynomial (3.258b):

$$a \neq -\frac{c^2}{2\chi} \pm \frac{c^2}{2} \sqrt{\frac{1}{\chi^2} + 4 \, \frac{b^2}{c^2}}: \qquad \Phi(x,t) = B \, \frac{e^{at+bx}}{b^2 - a^2/c^2 - a/\chi}; \qquad (3.287a, b)$$

(v) is (3.288b) for the bar equation (3.283e) with non-zero (3.288a) characteristic polynomial (3.266b):

$$a \neq \pm i \, q \, b^2: \qquad \qquad \Phi(x,t) = B \, \frac{e^{at+bx}}{a^2 + q^2 \, b^4}; \qquad \qquad (3.288a, b)$$

and (vi) is (3.289b) for the beam equation (3.283f) with non-zero (3.289a) characteristic polynomial (3.273b):

$$a \neq \pm i \, b \, \sqrt{q^2 \, b^2 - c^2}: \qquad \Phi(x,t) = B \, \frac{e^{at+bx}}{a^2 - c^2 \, b^2 + q^2 \, b^4}. \qquad (3.289a, b)$$

Of the six cases, four (two) lead with real (imaginary) parameters to single resonance (Subsections 3.4.18–3.4.19). The single resonances are considered for time t (position x), taking the roots of the characteristic polynomial for a (b) in Subsection 3.4.18 (3.4.19).

3.4.18 SINGLE RESONANCE FOR EXPONENTIAL FORCING

The integrals with singly resonant exponential forcing (3.282a, b), with $(r,s) = (0,1)$ or $(r,s) = (1,0)$, correspond to single roots of the characteristic polynomial. Starting with single resonances with regard to time t, for roots of the characteristic polynomial with regard to a, the resonant solutions: (i) are (3.290d–f) for the forced harmonic equation (3.283a), with (3.290a) implying (3.290b, c), by use of (3.213a; 3.284b):

$$a_{\pm}(b) = \pm i \, b, \qquad E(a,b) = \frac{d}{da}(a^2 + b^2) = 2a:$$

$$\Phi_{\pm}(x,t) = \lim_{a \to \pm ib} Re\left\{ \frac{B \, t}{2 \, a} \exp(a \, t + b \, x) \right\}$$

$$= \frac{B \, t}{2 \, b} \, e^{bx} \, Re\{\mp i \, e^{\pm \, ibt}\} = \frac{B \, t}{2 \, b} \, e^{bx} \sin(b \, t); \quad (3.290a–f)$$

(ii) are (3.291e) for the forced wave equation (3.283b) with (3.291a), implying (3.291b–d) by use of (3.223a; 3.285b):

$$a_\pm(b) = \pm\, b\, c, \qquad E(a,b) = \frac{d}{da}(a^2 - b^2\, c^2) = 2a = \pm\, 2\, b\, c:$$

$$\Phi_\pm(x,t) = \pm\, \frac{B\, t}{2\, b\, c}\, \exp\big[\, b\,(x \pm c\, t)\big]; \qquad (3.291a\text{–}e)$$

(iii) is (3.292d) for the forced diffusion equation (3.283c) with (3.292a), implying (3.292b, c) by use of (3.236a; 3.286b):

$$a(b) = \chi\, b^2, \qquad E(a,b) = \frac{d}{da}(\chi\, b^2 - a) = -\, 1:$$

$$\Phi(x,t) = -\, B\, t\exp\big[\, b\,(x + \chi\, b\, t)\big]; \qquad (3.292a\text{–}d)$$

(iv) for the forced wave-diffusion equation (3.283d) are (3.293e) with (3.293a), implying (3.293b–d) by use of (3.258a; 3.287b):

$$a_\pm(b) = -\, \frac{c^2}{2\, \chi} \pm \frac{c^2}{2}\sqrt{\frac{1}{\chi^2} + 4\,\frac{b^2}{c^2}},$$

$$E(a,b) = \frac{d}{da}\left(b^2 - \frac{a^2}{c^2} - \frac{a}{\chi}\right) = -\, \frac{2\, a}{c^2} - \frac{1}{\chi} = \mp\sqrt{\frac{1}{\chi^2} + 4\,\frac{b^2}{c^2}}:$$

$$\Phi^\pm(x,t) = \mp\, B\, t\left(\frac{1}{\chi^2} + 4\,\frac{b^2}{c^2}\right)^{-1/2} \times \exp\left\{b\, x - \frac{c^2\, t}{2\, \chi}\left[1 \mp \sqrt{1 + \frac{4\, b^2\, \chi^2}{c^2}}\,\right]\right\};$$
$$(3.293a\text{–}e)$$

(v) are (3.294d–f) for the forced elastic bar equation (3.283e) with (3.294a), implying (3.294b–e) by use of (3.266a; 3.288b):

$$a_\pm(b) = \pm\, i\, q\, b^2, \qquad E(a,b) = \frac{d}{da}(a^2 + q^2\, b^4) = 2a = \pm\, i\, 2\, q\, b^2:$$

$$\Phi_\pm(x,t) = \mp\, \frac{i\, B\, t}{2\, q\, b^2}\, \exp\big(b\, x \pm i\, q\, b^2\, t\big) = \frac{B\, t}{2\, q\, b^2}\, e^{bx}\, \sin\big(q\, b^2\, t\big);$$
$$(3.294a\text{–}e)$$

and (vi) are (3.295e) for the forced beam equation (3.283f) with (3.295a), implying (3.295b–d) by use of (3.273a; 3.289b):

$$a_\pm(b) = \pm\, i\, b\,\sqrt{q^2\, b^2 - c^2},$$

$$E(a,b) = \frac{d}{da}(a^2 - c^2\, b^2 + q^2\, b^4) = 2\, a = \pm\, i\, 2\, b\,\sqrt{q^2\, b^2 - c^2}: \qquad (3.295a\text{–}d)$$

$$\Phi_\pm(x,t) = \mp\, \frac{i\, B\, t}{b\,\sqrt{q^2\, b^2 - c^2}}\, \exp\big(b\, x \pm i\, 2\, b\, t\,\sqrt{q^2\, b^2 - c^2}\,\big). \qquad (3.295e)$$

$$= \frac{B\, t}{b\,\sqrt{q^2\, b^2 - c^2}}\, e^{bx}\, \sin\big(2\, b\, t\,\sqrt{q^2\, b^2 - c^2}\,\big). \qquad (3.295f)$$

In (3.294e) [(3.295f)] are taken the real parts of (3.294d) [(3.295e)].

The six cases of single resonance can be considered alternatively with regard to time (Subsection 3.4.18) or position (Subsection 3.4.19).

3.4.19   SINGLE RESONANCE WITH REGARD TO TIME OR POSITION

Each of the preceding six cases of single resonance has one alternative form exchanging between time and position: (i) for the forced harmonic equation (3.283a), the time resonance (3.290a–f) has alternative spatial resonance (3.296d–f) for (3.296a), implying (3.296b, c) by use of (3.213a; 3.284b):

$$b_\pm(a) = \pm\, i\, a, \qquad E(a,b) = \frac{d}{db}(a^2 + b^2) = 2b:$$

$$\Phi^\pm(x,t) = \lim_{b \to \pm ia} Re\left\{ \frac{B\,x}{2\,b} \, e^{at+bx} \right\}$$

$$= \frac{B\,x}{2\,a} \, e^{at}\, Re(\mp\, i\, e^{\pm iax}) = \frac{B\,x}{2\,a} \, e^{at}\, \sin(a\,x); \quad (3.296a\text{–}f)$$

(ii) for the forced wave equation (3.283a), the temporal resonance (3.291a–e) has alternative spatial resonance (3.297e) for (3.297a), implying (3.297b–d) by use of (3.223a; 3.285b):

$$b_\pm(a) = \pm\, \frac{a}{c}, \qquad E(a,b) = \frac{d}{db}(a^2 - b^2\, c^2) = -\, 2\, b\, c^2 = \mp\, 2\, a\, c:$$

$$\Phi^\pm(x,t) = \mp\, \frac{B\,x}{2\,a\,c} \, \exp\left[a\left(t \pm \frac{x}{c} \right) \right]; \qquad (3.297a\text{–}e)$$

(iii) for the forced diffusion equation (3.283c), the temporal resonance (3.292a–d) has alternative spatial resonance (3.298e) for (3.298a), implying (3.298b–d) by use of (3.236a; 3.286b):

$$b_\pm(a) = \pm\, \sqrt{\frac{a}{\chi}}, \qquad E(a,b) = \frac{d}{db}(\chi\, b^2 - a) = 2\, \chi\, b = \pm\, 2\, \sqrt{\chi\, a}:$$

$$\Phi^\pm(x,t) = \pm\, \frac{B\,x}{2\, \sqrt{\chi\, a}} \, \exp\left(a\, t \pm x\, \sqrt{\frac{a}{\chi}} \right); \qquad (3.298a\text{–}e)$$

(iv) for the forced wave-diffusion equation (3.283d) the temporal resonance (3.293a–e) has, for alternative, spatial resonance (3.299e) with (3.299a), implying (3.299b–d) by use of (3.258a; 3.287b):

$$b_\pm(a) = \pm\, \sqrt{\frac{a}{\chi} + \frac{a^2}{c^2}}, \qquad E(a,b) = \frac{d}{db}\left(b^2 - \frac{a^2}{c^2} - \frac{a}{\chi} \right) = 2b = \pm\, 2\sqrt{\frac{a}{\chi} + \frac{a^2}{c^2}}:$$

$$\Phi(x,t) = \pm\, \frac{B\,x}{2}\left(\frac{a}{\chi} + \frac{a^2}{c^2} \right)^{-1/2} \exp\left[a\, t \pm x\, \sqrt{\frac{a}{\chi} + \frac{a^2}{c^2}} \right];$$

$$(3.299a\text{–}e)$$

(v) for the forced elastic bar equation (3.283e) the temporal resonance (3.294a–e) has alternative spatial resonance (3.300g–j) for (3.300a, b), implying (3.300c–f) by use of (3.266c, d) and (3.288b):

$$b_{\pm\pm}(a) = (\pm 1 \pm i)\sqrt{\frac{a}{2q}} = \pm e^{\pm i\pi/4}\sqrt{\frac{a}{q}}, \qquad (3.300a, b)$$

$$E(a,b) = \frac{d}{db}(a^2 + q^2\ b^4) = 4\ q^2\ b^3 = \pm 4\ e^{\pm i3\pi/4}\sqrt{a^3\ q} = 2\ (\pm 1 \pm i)\sqrt{2\ a^3 q}: \qquad (3.300c–f)$$

$$\Phi^{\pm\pm}(x,t) = Re\left\{\frac{B\ x}{4\ q^2\ b^3}\ \exp(a\ t \pm b\ x)\right\}, \qquad (3.300g)$$

$$= \pm\frac{B\ x}{4\ \sqrt{a^3\ q}}\ Re\left\{\exp\left[a\ t \mp i\ \frac{3\ \pi}{4} + (\pm 1 \pm i)\ x\sqrt{\frac{a}{2\ q}}\ \right]\right\} \qquad (3.300h)$$

$$= \pm\frac{B\ x}{4\ \sqrt{a^3\ q}}\ \exp\left(a\ t \pm x\sqrt{\frac{a}{2\ q}}\ \right)\cos\left(x\sqrt{\frac{a}{2\ q}}\ -\ \frac{\pi}{4}\ \right) \qquad (3.300i)$$

$$= \pm\frac{B\ x}{4\ \sqrt{2\ a^3\ q}}\ \exp\left(a\ t \pm x\sqrt{\frac{a}{2\ q}}\ \right)\left[\cos\left(x\sqrt{\frac{a}{2\ q}}\right) + \sin\left(x\sqrt{\frac{a}{2\ q}}\right)\right] \qquad (3.300j)$$

and (vi) for the forced beam equation (3.283f) the temporal resonance (3.295a–e) has alternative spatial resonance (3.301d) with (3.301a), implying (3.301b, c) by use of (3.273a; 3.275a; 3.289b):

$$\left[\pm b_{\pm}(a)\right]^2 = \frac{c^2}{2\ q^2} \pm \frac{i}{2\ q^2}\sqrt{4\ a^2\ q^2 - c^4}, \qquad (3.301a)$$

$$E(a,b) = \frac{d}{db}(a^2 - b^2\ c^2 + q^2\ b^4) = 4\ q^2\ b^3 - 2\ b\ c^2: \qquad (3.301b, c)$$

$$\Phi_{\pm\pm}(x,t) = Re\left\{\frac{B\ x}{4\ q^2\ b^3 - 2\ b\ c^2}\ \exp(a\ t \pm x\ b_{\pm})\right\}, \qquad (3.301d)$$

where (3.301a) must be substituted in (3.301d) for the solution to involve explicitly a. Only one case of double root can lead to a double resonance (Subsection 3.4.20).

3.4.20 Doubly Resonant Exponential Forcing

The harmonic (3.283a) ≡ (3.212b) / wave (3.283b) ≡ (3.222b) / diffusion (3.283c) ≡ (3.235b) / wave-diffusion (3.283d) ≡ (3.257b) / bar (3.283e) ≡ (3.264b) equations have characteristic polynomials respectively $(3.213b) \equiv (3.302a) / (3.223b) \equiv (3.302b) / (3.236b) \equiv (3.302c) / (3.258b) \equiv (3.302d) / (3.266b) \equiv (3.302e)$:

$$P(a,b) = \left\{ a^2 + b^2 \,, \ a^2 - b^2 c^2 \,, \ \chi \, b^2 - a \,, \ b^2 - \frac{a^2}{c^2} - \frac{a}{\chi} \,, \ a^2 + q^2 b^4 \,, \ a^2 - b^2 c^2 + q^2 b^4 \right\}$$

$$(3.302a\text{-}f)$$

that do not have double roots and, thus, no double resonance is possible. Of the six equations in (3.283a–f), only the beam equation $(3.283f) \equiv (3.271b)$ has characteristic polynomial $(3.273b) \equiv (3.303b)$ with double roots only for position (3.275a–d) with the condition (3.303a) leading to (3.303b, c):

$$a = \pm \frac{c^2}{2\,q}, \qquad \left[b_\pm (a) \right]^2 = \frac{c^2}{2\,q^2}, \qquad b_\pm (a) = \pm \frac{c}{q\,\sqrt{2}}. \tag{3.303a–c}$$

In the case (3.303a) the characteristic polynomial (3.302e) for the beam equation becomes (3.304a)

$$P(b) = \frac{c^4}{4\,q^2} - b^2 c^2 + q^2 b^4 = q^2 \left(b^2 - \frac{c^2}{2\,q^2} \right)^2, \tag{3.304a}$$

and its first-order derivative with regard to b becomes (3.304b)

$$P'(b) = 2\,q^2 b \left(b^2 - \frac{c^2}{2\,q^2} \right), \qquad P\!\left(\frac{c}{q\,\sqrt{2}} \right) = 0 = P'\!\left(\frac{c}{q\,\sqrt{2}} \right), \tag{3.304b–d}$$

confirming that (3.303c) is a root of both (3.304c, d) and hence a double root. It is not a triple root because the second-order derivative (3.304e) does not vanish (3.304f)

$$P''(b) = 12\,q^2\,b^2 - 2\,c^2, \qquad P''\!\left(\frac{c}{q\sqrt{2}} \right) = 4\,c^2 \equiv E, \tag{3.304e–g}$$

and appears in the double resonant (3.282b) particular integral (3.305a, b):

$$\Phi(x,t) = \frac{B\,x^2}{E}\,\exp(a\,x + b\,t) = \frac{B\,x^2}{4\,c^2}\,\exp\!\left(a\,x \pm \frac{c\,t}{q\,\sqrt{2}} \right), \tag{3.305a, b}$$

There is no double temporal resonance corresponding to the double spatial resonance (3.305a–c) because the characteristic polynomial $(3.273a) \equiv (3.302f)$ has no double root for a in (3.273c) except zero.

Thus, *there are four cases for the particular integral of the beam equation with exponential forcing (3.283f), namely:*

$$P_{2,4}(a,b) \begin{cases} \neq 0: & non-resonant & (3.306a) \\[2mm] = 0 \neq \begin{array}{c} P'_{2,4}(a,b) \\ simple \\ resonance \end{array} \begin{cases} a = a_\pm (b): & resonance\ in\ t & (3.306b) \\ b = b_\pm (a),\ b_+ \neq b_- : & resonance\ in\ x & (3.306c) \end{cases} \\[4mm] = P'_{2,4}(a,b) = 0 \neq P''_{2,4}(a,b): & double\ resonance\ in\ x & (3.306d) \end{cases}$$

for: (i) non-zero (3.306a) characteristic polynomial (3.273b), the non-resonant solution (3.289a, b); (ii) for a single root (3.306b) [(3.306c)] with regard to a (b) of the characteristic polynomial, a singly resonant solution in time (3.295a–f) [position (3.301a–d)]; and (iii) for a double root (3.306d) of the characteristic polynomial (3.303a–c; 3.304a–g), a doubly resonant solution in position (3.305a–c). The cases of no resonance/single (double) resonance for six equations of mathematical physics [Subsection 3.4.17/3.4.18–3.4.19 (3.4.20)] are compared next (Subsection 3.4.21).

3.4.21 COMPARISON OF NON-RESONANT WITH SINGLE/DOUBLE RESONANT SOLUTIONS

The preceding results on non-resonant/singly/doubly resonant exponential forcing of six linear partial differential equations with constant coefficients are summarised next and in Table 3.4. *The two-dimensional Cartesian harmonic equation with exponential forcing (3.283a) has: (i) non-resonant (3.284a) solution (3.284b); and (ii/iii) a singly resonant (3.290a) [(3.296a)] solution (3.290f) [(3.296f)] in time (position) corresponding to (3.307a, b) [(3.307c, d)]:*

$$\frac{\partial^2 \Phi}{\partial x^2} + \frac{\partial^2 \Phi}{\partial t} = Re\{B\ e^{b(x\pm it)}\} = B\ e^{bx}\cos(b\ t),$$

$$= Re\{B\ e^{a(t\pm ix)}\} = B\ e^{at}\cos(a\ x) \qquad (3.307a\text{–}d)$$

The one-dimensional Cartesian wave equation with exponential forcing (3.283b) has: (i) non-resonant (3.285a) solution (3.285b); and (ii/iii) a singly resonant (3.291a) [(3.297a)] solution in time (3.291e) [position (3.297e)] corresponding to (3.308a) [(3.308b)]:

$$\frac{\partial^2 \Phi}{\partial x^2} - \frac{1}{c^2}\frac{\partial^2 \Phi}{\partial t^2} = B\ \exp[b\ (x\pm c\ t)], \quad B\ \exp\left[a\left(t\pm \frac{x}{c}\right)\right]. \qquad (3.308a,\ b)$$

The one-dimensional Cartesian diffusion equation with exponential forcing (3.283c) has: (i) non-resonant (3.286a) solution (3.286b); and (ii/iii) a singly resonant (3.292a) [(3.298a)] solution in time (3.292d) [space (3.298e)] corresponding to (3.309a) [(3.309b)]:

$$\chi\frac{\partial^2 \Phi}{\partial x^2} - \frac{\partial \Phi}{\partial t} = B\ \exp[b\ (x+\chi\ b\ t)], \quad B\ \exp\left(a\ t\pm x\ \sqrt{\frac{a}{\chi}}\ \right). \qquad (3.309a,\ b)$$

The one-dimensional Cartesian wave-diffusion equation with exponential forcing (3.283d) has: (i) non-resonant (3.287a) solution (3.287b); and (ii/iii) a singly resonant (3.293a) [(3.299a)] solution in time (3.293c) [position (3.299d)] corresponding to (3.310a) [(3.310b)]:

$$\frac{\partial^2 \Phi}{\partial x^2} - \frac{1}{c^2}\frac{\partial^2 \Phi}{\partial t^2} - \frac{1}{\chi}\frac{\partial \Phi}{\partial t} = B\ \exp\left[b\ x \pm \frac{c^2\ t}{2\ \chi}\left(1 \pm \sqrt{1 + \frac{4\ b^2\ \chi^2}{c^2}}\ \right)\right]$$

$$= B\ \exp\left(a\ t \pm x\ \sqrt{\frac{a}{\chi} + \frac{a^2}{c^2}}\ \right). \qquad (3.310a, b)$$

The elastic bar equation forced by an exponential (3.283e) has: (i) non-resonant (3.288a) solution (3.288b); and (ii-iii) a singly resonant solution (3.294a) [(3.300a, b)] in time (3.294e) [position (3.300h)] corresponding to (3.311a, b) [(3.311c, d)]:

$$\frac{\partial^2 \Phi}{\partial t^2} + q^2\ \frac{\partial^4 \Phi}{\partial x^4} = Re\left\{B\ \exp\left(b\ x \pm i\ q\ b^2\ t\right)\right\} = B\ e^{bx}\ \cos\left(q\ b^2\ t\right)$$

$$= B\ Re\left\{\exp\left[a\ t + 2\ x\ (\pm 1 \pm i)\ \sqrt{2\ a^3\ q}\ \right]\right\}$$

$$= B\ \exp\left(a\ t \pm 2\ x\ \sqrt{2\ a^3\ q}\ \right)\cos\left(2\ x\ \sqrt{2\ a^3\ q}\ \right). \qquad (3.311a\text{--}d)$$

The elastic beam equation with exponential forcing (3.283f) has: (i) non-resonant (3.289a) solution (3.289b); (ii/iii) a singly resonant (3.295a) [(3.301a)] solution in time (3.295e) [position (3.301d)] corresponding to (3.312a, b) [(3.312c)]:

$$\frac{\partial^2 \Phi}{\partial t^2} - c^2\ \frac{\partial^2 \Phi}{\partial x^2} + q^2\ \frac{\partial^4 \Phi}{\partial x^4} = Re\left\{B\ \exp\left(b\ x \pm i\ b\ t\ \sqrt{q^2\ b^2 - c^2}\ \right)\right\}$$

$$= B\ e^{bx}\ \left(\cos b\ t\ \sqrt{q^2\ b^2 - c^2}\ \right)$$

$$= Re\left\{B\ \exp\left[a\ t \pm x\ b_\pm\ (a)\right]\right\}, \qquad (3.312a\text{--}c)$$

and *(iv) double resonance (3.303a–c) in space (3.305b) corresponding to (3.313):*

$$\frac{\partial^2 \Phi}{\partial t^2} - c^2\ \frac{\partial^2 \Phi}{\partial x^2} + q^2\ \frac{\partial^4 \Phi}{\partial x^4} = B\ \exp\left(a\ x \pm \frac{c\ t}{2\ q}\right), \qquad (3.313)$$

The combination of exponential forcings leads to the forcing by circular and hyperbolic cosines and sines (Subsection 3.4.22), and their products, that are considered for the harmonic/wave/diffusion equation (Subsections 3.4.23/3.4.24/3.4.25).

3.4.22   FORCING BY CIRCULAR/HYPERBOLIC COSINES/SINES

The forcing of a linear partial differential equation with constant coefficients (3.277a, b) can be extended from an exponential (3.278a; 3.280a, b) to: (i) the product of an exponential by a circular cosine or sine (3.314a, b) leading to the particular integral (3.314c):

$$B(x,y) = B \ e^{ax+by} \cos, \sin(r \ x+s \ y) = B \ Re, Im\{\exp[(a+i \ r) \ x+(b+i \ s) \ y]\},$$

$$(3.314a, b)$$

$$\Phi_*(x,y) = B \ Re, Im \left\{ \frac{e^{(a+ir)x \ + \ (b+is)y}}{P_{N,M}(a+ir, \ b+is)} \right\};$$

$$(3.314c)$$

(ii) the product of an exponential by a hyperbolic cosine or sine (3.315a, b) leading to the particular integral (3.315c):

$$B(x,y) = B \ e^{ax+by} \cosh, \sinh(\varphi \ x+\psi \ y)$$

$$= \frac{B}{2} \left\{ \exp[(a+\varphi) \ x+(b+\psi) \ y] \pm \exp[(a-\varphi) \ x+(b-\psi) \ y] \right\}, \quad (3.315a, b)$$

$$\Phi_*(x,y) = \frac{B}{2} \left[\frac{e^{(a+\varphi)x \ + \ (b+\psi)y}}{P_{N,M}(a+\varphi, \ b+\psi)} \pm \frac{e^{(a-\varphi)x \ + \ (b-\psi)y}}{P_{N,M}(a-\varphi, \ b-\psi)} \right];$$

$$(3.315c)$$

and (iii) the product of an exponential by circular and hyperbolic cosines and sines (3.316a, b) leading to the particular integral (3.316c):

$$B(x,y) = B \ e^{ax+by} \cos, \sin(r \ x+s \ y)\cosh, \sinh(\varphi \ x+\psi \ y)$$

$$= \frac{B}{2} \ Re, Im \left\{ \left[\exp(a+i \ r+\varphi) \ x+(b+i \ s+\psi) \ y \right] \right.$$

$$\left. \pm \exp[(a+i \ r-\varphi) \ x+(b+i \ s-\psi) \ y] \right\}, \quad (3.316a, b)$$

$$\Phi_*(x,y) = \frac{B}{2} \ Re, Im \left\{ \frac{e^{(a+ir+\varphi)x \ + \ (b+is+\psi)y}}{P_{N,M}(a+ir+\varphi, \ b+is+\psi)} \pm \frac{e^{(a+ir-\varphi)x \ + \ (b+is-\psi)y}}{P_{N,M}(a+ir-\varphi, \ b+is-\psi)} \right\}.$$

$$(3.316c)$$

The particular integrals (3.314c; 3.315c; 3.316c) apply in the non-resonant case (3.280a, b) and are replaced in the resonant case by (3.282a, b). The forcing by an exponential multiplied by circular (3.314a–c)/hyperbolic(3.315a–c)/both (3.316a–c) functions is exemplified next (Subsections 3.4.23/3.4.24/3.4.25) by applications to the harmonic/wave/diffusion equations, respectively.

3.4.23 FORCING OF THE HARMONIC EQUATION BY THE CIRCULAR COSINE

The forcing (3.277a, b) of the two-dimensional Cartesian harmonic equation (3.283a) by the product of an exponential and a circular cosine (3.317a, b):

$$\frac{\partial^2 \Phi}{\partial x^2} + \frac{\partial^2 \Phi}{\partial y^2} = B \ e^{ax+by} \cos(r \ x+s \ y) = B \ Re\{\exp[(a+i \ r) \ x+(b+i \ s) \ y]\},$$

$$(3.317a, b)$$

involves (3.302a) the characteristic polynomial (3.318a–c):

$$P_{2,2}(a+i\ r,\ b+i\ s)=(a+i\ r)^2+(b+i\ s)^2=a^2+b^2-r^2-s^2+2\ i\ (a\ r+b\ s)$$
$$\equiv \lambda + i\ \mu.\tag{3.318a–c}$$

In the non-resonant case (3.319a) when the characteristic polynomial (3.318a) is not zero, the particular integral (3.282b) is (3.319b–d):

$$a+i\ r \neq \pm\ i\ (b+i\ s):$$

$$\Phi_*(x,y)=B\ Re\left\{\frac{e^{(a+ir)x\ +\ (b+is)y}}{\lambda+i\ \mu}\right\}=B\ \frac{e^{ax+by}}{\lambda^2+\mu^2}\ Re\left\{(\lambda-i\ \mu)\ e^{i(rx+sy)}\right\}$$

$$=B\ \frac{e^{ax+by}}{\lambda^2+\mu^2}\left[\lambda\cos(r\ x+s\ y)+\mu\sin(r\ x+s\ y)\right].\tag{3.319a–d}$$

The resonant case when the characteristic polynomial (3.318c) vanishes corresponds to (3.320a, b) ≡ (3.320c, d):

$$a+i\ r=\pm\ i\ (b+i\ s)=\mp\ s\pm i\ b:\qquad s=\mp\ a,\qquad r=\pm\ b,\tag{3.320a–d}$$

and taking the single root as (3.321a, b) the particular integral (3.282b) is (3.321c–f):

$$s=\mp\ a,\qquad r=\pm\ b:$$

$$\Phi_*(x,y)=\lim_{\substack{r\to\pm b \\ s\to\mp a}}\frac{B\ x}{2}\ Re\left\{\frac{e^{(a+ir)x\ +\ (b+is)y}}{a+i\ r}\right\}=\frac{B\ x}{2}\ e^{ax\ +\ by}\ Re\left\{\frac{e^{\pm\ i\ (bx-ay)}}{a\pm i\ b}\right\}$$

$$=\frac{B\ x}{2}\ \frac{e^{ax\ +\ by}}{a^2+b^2}\ Re\left\{(a\mp i\ b)\ e^{\pm\ i(bx\ -\ ay)}\right\}$$

$$=\frac{B\ x}{2}\ \frac{e^{ax\ +\ by}}{a^2+b^2}\left[a\cos(b\ x-a\ y)+b\sin(b\ x-a\ y)\right],\tag{3.321a–f}$$

corresponding to the forced harmonic equation (3.317b; 3.321a, b) ≡ (3.321g):

$$\frac{\partial^2\Phi}{\partial x^2}+\frac{\partial^2\Phi}{\partial y^2}=B\ e^{ax\ +\ by}\cos(b\ x-a\ y).\tag{3.321g}$$

Thus, *the simplest particular integral of the two-dimensional Cartesian harmonic equation forced by the product of an exponential and a circular cosine (3.317a, b) is (3.319d; 3.318c) in the non-resonant case (3.319a); in the resonant case (3.320a, b) ≡ (3.320c, d) ≡ (3.321a, b), the particular integral (3.321f) corresponds to the forcing (3.321g).* Next, the harmonic is replaced by the wave equation with forcing by the circular cosine replaced by the hyperbolic sine (Subsection 3.4.24).

3.4.24   FORCING OF THE WAVE EQUATION BY THE HYPERBOLIC SINE

The forcing (3.277a, b) of the one-dimensional Cartesian wave equation (3.283b) by the product of an exponential and a hyperbolic sine (3.322a, b):

$$\frac{\partial^2 \Phi}{\partial t^2} - c^2 \frac{\partial^2 \Phi}{\partial x^2} = B\ e^{at+bx} \sinh(\varphi\ t + \psi\ x)$$

$$= \frac{B}{2} \left[e^{(a+\varphi)t\ +\ (b+\psi)x} - e^{(a-\varphi)t\ +\ (b-\psi)x} \right], \quad (3.322a, b)$$

involves (3.302b) the characteristic polynomial (3.323a–c):

$$P_{2,2}(a\pm\varphi,\ b\pm\psi) = (a\pm\varphi)^2 - c^2 (b\pm\psi)^2$$

$$= a^2 + \varphi^2 - c^2 (b^2 + \psi^2) \pm 2 (a\ \varphi - c^2\ b\ \psi)$$

$$= \lambda \pm \mu. \quad (3.323a\text{–}c)$$

In the non-resonant case (3.280b) when the characteristic polynomial (3.323a) is not zero (3.324a), the particular integral is (3.324b–e):

$a \pm \varphi \neq \pm c (b \pm \psi)$:

$$\Phi_*(x,t) = \frac{B}{2} \left[\frac{e^{(a+\varphi)t\ +\ (b+\psi)x}}{(a+\varphi)^2 - c^2 (b+\psi)^2} - \frac{e^{(a-\varphi)t\ +\ (b-\psi)x}}{(a-\varphi)^2 - c^2 (b-\psi)^2} \right]$$

$$= \frac{B}{2} e^{at+bx} \left(\frac{e^{\varphi t + \psi x}}{\lambda + \mu} - \frac{e^{-\varphi t - \psi x}}{\lambda - \mu} \right)$$

$$= \frac{B}{2} \frac{e^{at+bx}}{\lambda^2 - \mu^2} \left[(\lambda - \mu)\ e^{\varphi t + \psi x} - (\lambda + \mu)\ e^{-\varphi t - \psi x} \right]$$

$$= B\ \frac{e^{at+bx}}{\lambda^2 - \mu^2} \left[\lambda \sinh(\varphi\ t + \psi\ x) - \mu \cosh(\varphi\ t + \psi\ x) \right].$$
$$(3.324a\text{–}e)$$

The roots of (3.323a) correspond to four resonant cases (3.325a, b):

$$a \pm \varphi = \pm c (b \pm \psi) = \pm (c\ b \pm c\ \psi); \qquad a = c\ b, \qquad \varphi = c\ \psi, \qquad (3.325a\text{-}d)$$

choosing the resonant case (3.325c, d) for the wave speed $+ c$ corresponding to propagation in the positive x-direction, and the upper sign (3.326a) in (3.325a) leads to (3.326b, c):

$$a + \varphi = c (b + \psi): \qquad a - \varphi = a - \left[c (b+\psi) - a \right] = 2\ a - c (b+\psi). \quad (3.326a\text{-}c)$$

The condition (3.326a) [(3.326b)] implies that the denominator of the first (second) term on the r.h.s. of (3.324b) is zero [is not zero (3.327a, b)]

$$(a-\varphi)^2 - c^2 (b-\psi)^2 = \left[2\, a - c\, (b+\psi) \right]^2 - c^2\, (b-\psi)^2$$

$$= 4\, a^2 - 4\, a\, c\, (b+\psi) + 4\, c^2\, b\, \psi, \qquad (3.327\text{a, b})$$

and, thus, the corresponding term is resonant (3.282b) [non-resonant (3.280b)] in the first (second) term on the r.h.s. of the particular integral (3.328b, c) for (3.326a) $\equiv$ (3.328a):

$$a + \varphi = c\, (b+\psi):$$

$$\Phi_* (x,t) = \lim_{a+\varphi\,\to\,c(b+\psi)} \left\{ \frac{B\,t}{4}\, \frac{e^{(a\,+\,\varphi)t\,+\,(b\,+\,\psi)x}}{a+\varphi} - \frac{B}{2}\, \frac{e^{(a\,-\,\varphi)t\,+\,(b\,-\,\psi)x}}{(a-\varphi)^2 - c^2\,(b-\psi)^2} \right\}$$

$$= \frac{B\,t}{4\,c}\, \frac{e^{(b\,+\,\psi)\,(x\,+\,ct)}}{b+\psi} - \frac{B}{8}\, \frac{e^{2at\,+\,b(x\,-\,ct)\,-\,\psi(x\,+\,ct)}}{a^2 - a\,c\,(b+\psi) + c^2\,b\,\psi}, \qquad (3.328\text{a–c})$$

that involves waves propagating in opposite directions $x \pm c\,t$; this is the simplest particular integral of the wave equation with forcing (3.322a; 3.326a) $\equiv$ (3.328d):

$$\frac{\partial^2 \Phi}{\partial t^2} - c^2\, \frac{\partial^2 \Phi}{\partial x^2} = B\, e^{at+bx}\, \sinh\left\{ \psi\, x + \left[c\, (b+\psi) - a \right] t \right\}. \qquad (3.328\text{d})$$

The other three resonance cases in (3.325a, b) can be treated like (3.325c, d) with suitable changes of (3.326a–c; 3.327a, b; 3.328a–d). Thus, *the one-dimensional Cartesian wave equation forced by the product of an exponential and a hyperbolic sine (3.322a, b) has the simplest particular integral (3.324e) in the non-resonant case (3.324a, b); in the resonant case (3.328a), the particular integral is (3.328c), corresponding to the forced wave equation (3.328d).* The product of a circular sine by a hyperbolic cosine by an exponential is considered next (Subsection 3.4.25) forcing a diffusion equation.

3.4.25   DIFFUSION EQUATION FORCED BY THE PRODUCT OF CIRCULAR AND HYPERBOLIC FUNCTIONS

The forcing (3.277a, b) of the one-dimensional Cartesian diffusion equation (3.283c) by the (3.329a, b) product of an exponential by a circular sine and a hyperbolic cosine:

$$\frac{\partial \Phi}{\partial t} - \chi\, \frac{\partial^2 \Phi}{\partial x^2} = B\, e^{at+bx}\, \sin(r\,t + s\,x)\cosh(\varphi\,t + \psi\,x)$$

$$= \frac{B}{2}\, Im\left\{ e^{(a+ir+\varphi)t\,+\,(b+is+\psi)x} + e^{(a+ir-\varphi)t\,+\,(b+is-\psi)x} \right\}, \qquad (3.329\text{a, b})$$

involves (3.302c) the characteristic polynomial (3.330a–c):

$$P_{2,2}\left(a+i\ r\pm\varphi\ ,\ b+i\ s\pm\psi\right)=\chi\left(b+i\ s\pm\psi\right)^2-\left(a+i\ r\pm\varphi\right)$$

$$=\chi\left[\left(b\pm\psi\right)^2-s^2\right]-\left(a\pm\varphi\right)+i\left[2\ \chi\ s\left(b\pm\psi\right)-r\right]\equiv\lambda_\pm+i\ \mu_\pm.$$

$$(3.330\text{a–c})$$

In the non-resonant (3.280b) case (3.331a) when (3.330a) is not zero, the simplest particular integral is (3.331b–d):

$$a+i\ r+\varphi\neq\chi\left(b+i\ s+\psi\right)^2:$$

$$\Phi_*\left(x,t\right)=\frac{B}{2}\ Im\left\{\frac{e^{(a+ir+\varphi)t\ +\ (b+is+\psi)x}}{\lambda_+ +i\ \mu_+}+\frac{e^{(a+ir-\varphi)t\ +\ (b+is-\psi)x}}{\lambda_- +i\ \mu_-}\right\}$$

$$=\frac{B}{2}\ e^{at+bx}\left[\frac{e^{\varphi t\ +\ \psi x}}{\lambda_+^2 +\mu_+^2}\ Im\left\{\left(\lambda_+ -i\ \mu_+\right)e^{i(rt+sx)}\right\}\right.$$

$$\left.+\frac{e^{-\varphi t\ -\psi x}}{\lambda_-^2 +\mu_-^2}\ Im\left\{\left(\lambda_- -i\ \mu_-\right)e^{i(rt+sx)}\right\}\right]$$

$$=\frac{B}{2}\ e^{at+bx}\left\{\frac{e^{\varphi t\ +\ \psi x}}{\lambda_+^2 +\mu_+^2}\left[\lambda_+ \sin\left(r\ t+s\ x\right)-\mu_+ \cos\left(r\ t+s\ x\right)\right]\right.$$

$$\left.+\frac{e^{-\varphi t\ -\psi x}}{\lambda_-^2 -\mu_-^2}\left[\lambda_- \sin\left(r\ t+s\ x\right)-\mu_- \cos\left(r\ t+s\ x\right)\right]\right\}.$$

$$(3.331\text{a–d})$$

The zero of the characteristic polynomial (3.330a–c) leads to two resonant cases (3.332a) ≡ (3.332a, b):

$$a+i\ r\pm\varphi=\chi\left(b+i\ s\pm\psi\right)^2:$$

$$a\pm\varphi=\chi\left[\left(b\pm\psi\right)^2-s^2\right],\qquad r=2\ \chi\ s\left(b\pm\psi\right).\qquad(3.332\text{a–c})$$

Choosing the resonant case with the upper sign in (3.332a) ≡ (3.333a) leads to (3.333b–e):

$$a+i\ r+\varphi=\chi\left(b+i\ s+\psi\right)^2:$$

$$a+i\ r-\varphi=a+i\ r-\left[\chi\left(b+i\ s+\psi\right)^2-\left(a+i\ r\right)\right]$$

$$=2\left(a+i\ r\right)-\chi\left(b+i\ s+\psi\right)^2$$

$$=2\ a-\chi\left[\left(b+\psi\right)^2-s^2\right]+2\ i\left[r-\chi\ s\left(b+\psi\right)\right]\equiv\xi+i\ \eta$$

$$(3.333\text{a–e})$$

The condition (3.333a) [(3.333c)] implies that the denominator of the first (second) term on the r.h.s. of (3.331b) vanishes [does not vanish (3.334a–e)]:

$$\alpha + i\ r - \varphi - \chi\ (b + i\ s - \psi)^2 = 2\ (a + i\ r) - \chi\left[(b + i\ s + \psi)^2 + (b + i\ s - \psi)^2\right]$$

$$= 2\ (a + i\ r) - 2\ \chi\ (b^2 + \psi^2 - s^2 + 2\ i\ s\ b)$$

$$= 2\ a - 2\ \chi\ (b^2 + \psi^2 - s^2) + 2\ i\ (r - 2\ \chi\ s\ b)$$

$$\equiv \lambda + i\ \mu. \tag{3.334a–e}$$

This leads to a resonant (3.282b) [non-resonant (3.280b)] term first (second) on the r.h.s. of the particular integral (3.335c–e) in the case (3.332b, c) ≡ (3.335a, b):

$$a + \varphi = \chi\left[(b + \psi)^2 - s^2\right], \qquad r = 2\ \chi\ s\ (b + \psi):$$

$$\Phi_*(x,t) = \frac{B\ t}{2}\ Im\left\{\exp\left[(b + i\ s + \psi)\ x + \chi\ (b + i\ s + \psi)^2\ t\right]\right\}$$

$$+ \frac{B}{2}\ Im\left\{\frac{\exp\left[(b + i\ s - \psi)\ x + (\xi + i\ \eta)\ t\right]}{\lambda + i\ \mu}\right\}$$

$$= \frac{B\ t}{2}\ \exp\left\{(b + \psi)\ x + \chi\left[(b + \psi)^2 - s^2\right]t\right\}$$

$$\times Im\left\{\exp\left[i\ s\ x + 2\ i\ \chi\ s\ (b + \psi)\ t\right]\right\}$$

$$+ \frac{B}{2}\ \frac{\exp\left[(b - \psi)\ x + \xi\ t\right]}{\lambda^2 + \mu^2}\ Im\left\{(\lambda - i\ \mu)\ e^{i(sx + \eta t)}\right\}$$

$$= \frac{B\ t}{2}\ \exp\left\{(b + \psi)x + \left[(b + \psi)^2 - s^2\right]\chi\ t\right\}\ \sin\left[sx + 2\ s\ \chi(b + \psi)\right]$$

$$+ \frac{B}{2}\ \frac{\exp\left[(b - \psi)x + \xi\ t\right]}{\lambda^2 + \mu^2}\left[\lambda \sin(s\ x + \eta\ t) - \mu \cos(s\ x + \eta\ t)\right], \tag{3.335a–e}$$

which satisfies the forced diffusion equation (3.329a; 3.335a, b) ≡ (3.335f):

$$\frac{\partial \Phi}{\partial t} - \chi\ \frac{\partial^2 \Phi}{\partial x^2} = B \exp\left\{b\ x - \varphi\ t + \chi\left[(b + \psi)^2 - s^2\right]t\right\}$$

$$\sin\left[s\ x + 2\ \chi\ s\ (b + \psi)\ t\right]\cosh(\varphi\ t + \psi\ x). \tag{3.335f}$$

The second resonance case with lower − sign in (3.332a–c) would be solved similarly with suitable changes in (3.333a–e; 3.334a–e; 3.335a–f). Thus, *the simplest particular integral of the one-dimensional Cartesian diffusion equation forced by the product of an exponential by a circular sine and a hyperbolic cosine (3.329a, b) [(3.335f)] has simplest non-resonant (resonant) particular integral (3.331a–d; 3.330a–c) [(3.335a–e; 3.334a–e; 3.333a–e)].* Whereas the forced diffusion equation has only non-resonant and singly resonant solutions (Subsection 3.4.25), the forced harmonic (wave) equation [Subsection 3.4.23 (3.4.24)] also has doubly resonant solutions (Subsection 3.4.26).

3.4.26   DOUBLY RESONANT SOLUTIONS OF FORCED HARMONIC AND WAVE EQUATIONS

The case (3.336a, b) leads in the first term of the r.h.s. of (3.328a, c) to the doubly resonant particular integral (3.336c, d):

$$\varphi = -a, \qquad \psi = -b:$$

$$\Phi_*(x,t) = \lim_{\varphi \to -b} \left\{ \frac{B\,t^2}{4}\, e^{(b+\psi)x} - \frac{B}{2}\, \frac{e^{2at+(b-\psi)x}}{4\,a^2 - c^2\left(b-\psi\right)^2} \right\}$$

$$= \frac{B\,t^2}{4} - \frac{B}{8}\, \frac{e^{2(at+bx)}}{a^2 - c^2\,b^2}, \tag{3.336a–d}$$

of the forced wave equation (3.322a; 3.336a, b) ≡ (3.336e):

$$\frac{\partial^2 \Phi}{\partial t^2} - c^2\, \frac{\partial^2 \Phi}{\partial x^2} = -B\, e^{at+bx}\, \sinh\!\left(a\,t + b\,x\right). \tag{3.336e}$$

The second non-resonant term on the r.h.s. of (3.336d) becomes resonant for (3.337a), leading to the particular integral (3.337b):

$$a = \pm\, b\, c: \qquad\qquad \Phi_*^{\pm}(x,t) = \frac{B\,t^2}{4} \mp \frac{B\,t}{8\,b\,c}\, e^{2b(x\,\pm\,ct)}, \tag{3.337a, b}$$

corresponding to the forced wave equation (3.336e; 3.337a) = (3.337c, d):

$$\frac{\partial^2 \Phi}{\partial t^2} - c^2\, \frac{\partial^2 \Phi}{\partial x^2} = -B\, e^{b(x\,\pm\,ct)}\, \sinh\!\left[b\,(x \pm c\,t)\right] = \frac{B}{2}\left[1 - e^{2b(x\,\pm\,ct)}\right]. \tag{3.337c, d}$$

The limit (3.338a) in the r.h.s. of (3.321d) leads to the singly resonant particular integral (3.338b–d):

$$b \to -i\,a: \qquad \Phi_*(x,t) = \frac{B\,x}{2}\, \lim_{b \to -ia} e^{ax}\, Re\left\{ \frac{e^{by}}{a+i\,b}\, e^{i(bx-ay)} \right\}$$

$$= \frac{B\,x}{4\,a}\, e^{2ax}\, Re\{e^{-2iay}\} = \frac{B\,x}{4\,a}\, e^{2ax}\, \cos\left(2ay\right), \tag{3.338a–d}$$

corresponding to the forced harmonic equation (3.321g; 3.338a) ≡ (3.338e–g):

$$\frac{\partial^2 \Phi}{\partial x^2} + \frac{\partial^2 \Phi}{\partial y^2} = B \lim_{b \to -ia} Re\left\{e^{a(x-iy)+b(y+ix)}\right\}$$

$$= B\ Re\left\{e^{2a(x-iy)}\right\} = B\ e^{2ax}\cos(2ay). \qquad (3.338e\text{–}g)$$

Thus, *the forced one-dimensional Cartesian wave equation (3.336e) has particular integral (3.336d) with a doubly resonant (non-resonant) first (second) term; the second term becomes singly resonant for (3.337a), leading to the particular integral (3.337b) of the forced wave equation (3.337c, d). The two-dimensional Cartesian forced harmonic equation (3.338g) has the particular singly resonant integral (3.338d).* The two simplest particular integrals (3.338d) [(3.337b)] of the forced harmonic (3.338g) [wave (3.337c, d)] equation are obtained next [Subsection(s) 3.4.30 (3.4.27–3.4.29)] by the alternative method of phase (complex) variables, providing an independent check of the results.

3.4.27 COMPLETE INTEGRAL OF THE FORCED WAVE EQUATION

The general integral (3.228b) of the unforced one-dimensional Cartesian wave equation (3.222b) is the sum of two arbitrary differentiable functions (3.227b, c) of the **phase variables** (3.339a) [(3.339b)] that are constant (3.339c) [(3.339d)] for propagation at constant speed c in the positive (3.339e) [negative (3.339f)] x-direction:

$$\xi = t - \frac{x}{c}, \qquad \eta = t + \frac{x}{c}, \qquad \xi, \eta = const: \qquad \frac{dx}{dt} = \{+\,c\,,-\,c\}. \qquad (3.339a\text{–}f)$$

Inverting (3.339a, b) specifies time (3.340a) and position (3.340b) as a function of the phase variables, leading to (3.340c–f):

$$t = \frac{\xi + \eta}{2}, \qquad x = c\,\frac{\eta - \xi}{2}: \qquad\qquad\qquad (3.340a, b)$$

$$\frac{\partial}{\partial \xi} = \frac{\partial t}{\partial \xi}\frac{\partial}{\partial t} + \frac{\partial x}{\partial \xi}\frac{\partial}{\partial x} = \frac{1}{2}\left(\frac{\partial}{\partial t} - c\,\frac{\partial}{\partial x}\right), \qquad (3.340c, d)$$

$$\frac{\partial}{\partial \eta} = \frac{\partial t}{\partial \eta}\frac{\partial}{\partial t} + \frac{\partial x}{\partial \eta}\frac{\partial}{\partial x} = \frac{1}{2}\left(\frac{\partial}{\partial t} + c\,\frac{\partial}{\partial x}\right). \qquad (3.340e, f)$$

The Cartesian one-dimensional wave operator (3.222b) ≡ (3.341b) is given by (3.341c) in terms of phase variables for the wave function (3.341a)

$$\Phi(x,t) = \Psi(\xi,\eta): \qquad \frac{\partial^2 \Phi}{\partial t^2} - c^2\,\frac{\partial^2 \Phi}{\partial x^2} = \left\{\left(\frac{\partial}{\partial t} - c\,\frac{\partial}{\partial x}\right)\left(\frac{\partial}{\partial t} + c\,\frac{\partial}{\partial x}\right)\right\}\Phi(x,t)$$

$$= \frac{\partial^2}{\partial \xi\,\partial \eta}\left[\Psi(\xi,\eta)\right]. \qquad (3.341a\text{–}c)$$

Thus, the similarity variables for the one-dimensional Cartesian wave equation (3.222b) ≡ (3.341b) are the phase variables (3.339a) [(3.339b)] for propagation in the positive (negative) x-direction; the one-dimensional Cartesian wave equation takes a simpler form (3.341c) in terms of phase variables, leading to the complete integral with arbitrary forcing, as shown next. *The change from the independent variable position x and time t to the phase variables (3.339a) [(3.339b)] for propagation in the positive (3.339c, e) [negative (3.339d, f)] x-direction in the wave (3.342a, b) and forcing (3.342c, d) functions:*

$$\Phi(x,t) = \Psi(\xi,\eta) = \Psi\left(t - \frac{x}{c}, \ t + \frac{x}{c}\right), \qquad B(x,t) = F(\xi,\eta) = F\left(t - \frac{x}{c}, \ t + \frac{x}{c}\right)$$

$$(3.342a\text{--}d)$$

transforms the forced one-dimensional Cartesian wave equation (3.343a) to (3.343b):

$$\frac{\partial^2 \Phi}{\partial t^2} - c^2 \frac{\partial^2 \Phi}{\partial x^2} = B(x,t) \quad \Leftrightarrow \quad 4 \frac{\partial^2 \Psi}{\partial \xi \, \partial \eta} = F(\xi,\eta); \qquad (3.343a, b)$$

the latter (3.343b) has complete integral (3.344c) involving two arbitrary twice differentiable functions (3.344a, b):

$$f^{\pm} \in D^{2}(\mathbb{R}): \qquad \Psi(\xi,\eta) = f^{+}(\xi) + f^{-}(\eta) + \int^{\xi} d\alpha \int^{\eta} d\beta \ F(\alpha,\beta). \qquad (3.344a\text{--}c)$$

and specifies the complete integral (3.345) of the former (3.343a):

$$\Phi(x,t) = f^{+}\left(t - \frac{x}{c}\right) + f^{-}\left(t + \frac{x}{c}\right) + \frac{1}{4} \int^{t-x/c} d\xi \int^{t+x/c} d\eta \ B\left(c \, \frac{\eta - \xi}{2}, \ \frac{\eta + \xi}{2}\right).$$

$$(3.345)$$

In the unforced case $B = 0$, (3.345) simplifies to the general integral (3.228b) of the unforced wave equation (3.222b). There is (is not) an analogous result for the Laplace (diffusion) equation, because (Subsection 3.4.28) it has (has not) all derivatives of the same order and, thus, has (has no) similarity solutions.

3.4.28 PHASE VARIABLES FOR THE WAVE EQUATION

There is an analogy between the complete integral (3.345) [(1.293a, d) ≡ (3.346)]:

$$\Phi(x,y) = f_{+}(x + i\,y) + f_{-}(x - i\,y) + \frac{1}{4} \int^{x+iy} dz \int^{x-iy} dz^{*} \ B\left(\frac{z + z^{*}}{2}, \ \frac{z - z^{*}}{2\,i}\right), \qquad (3.346)$$

of the forced one (two)-dimensional Cartesian wave (3.343a) [harmonic (3.347c)] equation using as similarity variables the phases for propagation at speed c in

opposite directions (3.339a, b) [the complex conjugate variables (3.216a, b) ≡
(3.347a, b)]:

$$z \equiv x+i\ y, \qquad z^* = x-i\ y: \qquad \frac{\partial^2 \Phi}{\partial x^2} + \frac{\partial^2 \Phi}{\partial y^2} = B(x,y). \qquad (3.347a\text{--}c)$$

There is no comparable result for the forced one-dimensional Cartesian diffusion equation (3.348a) because it involves partial derivatives with regard to time (position) of different orders one (two) and, thus, there is no similarity solution involving only one variable ζ combining x,t:

$$\frac{\partial \Phi}{\partial t} - \chi \frac{\partial^2 \Phi}{\partial x^2} = B(x,t); \qquad G(x,t) = \frac{1}{2\ \sqrt{\pi\ \chi\ t}}\ \exp\left(-\frac{x^2}{4\ \chi\ t}\right), \qquad (3.348a, b)$$

if a similarity solution of (3.348a) existed then, in particular the Green function (3.241b) ≡ (3.348b) would involve only one similarity variable. That is not the case because (3.348b) cannot be written in terms of a single variable ζ instead of x,t; for example, the choice (3.349a) would lead to (3.349b) [(3.349c)] that still involves t (x)

$$\zeta = \frac{x}{2\ \sqrt{\chi\ t}}: \qquad G(x,t) = \frac{1}{2\ \sqrt{\pi\ \chi\ t}}\ \exp\left(-\ \zeta^2\right) = \frac{\zeta}{x\ \sqrt{\pi}}\ \exp\left(-\ \zeta^2\right). \ (3.349a\text{--}c)$$

The Green function (3.348b) can be used (Subsection 3.4.10) to obtain a particular integral (3.243d) of the diffusion equation with integrable forcing (3.243a) using the convolution integral (3.243b, c). The method of solution of the one-dimensional Cartesian forced wave equation by means of phase variables is illustrated by next obtaining an alternative solution of (3.337c, d).

The forced wave equation (3.337d) with upper sign in space-time becomes (3.350) in terms (3.341c) of phase variables (3.340a, b):

$$4\ \frac{\partial^2 \Psi}{\partial \xi\ \partial \eta} = \frac{B}{2} - \frac{B}{2}\ e^{2bc\eta}; \qquad (3.350)$$

the simplest particular integral of (3.350) is (3.351a)

$$\Psi(\xi,\eta) = \frac{B\ \xi\ \eta}{8} - \frac{B\ \xi}{16\ b\ c}\ e^{2bc\eta} = \frac{B}{8}\left(t^2 - \frac{x^2}{c^2}\right) - \frac{B}{16\ b\ c}\left(t - \frac{x}{c}\right)e^{2b(x+ct)} = \hat{\Phi}(x,t),$$

$$(3.351a\text{--}c)$$

which is equivalent (3.339a, b; 3.341a) to (3.351b, c). The two particular integrals (3.337b) ≠ (3.351c) of the same forced wave equation (3.337d) need not coincide, and must only differ by a solution of the unforced equation, as shown next (Subsection 3.4.29).

3.4.29 GENERAL, PARTICULAR, AND COMPLETE INTEGRALS

Consider a linear forced partial differential equation with variable coefficients (3.352a) corresponding to the linear differential operator (3.352b):

$$B(x,y) = \sum_{n=0}^{N} \sum_{m=0}^{M} A_{n,m}(x,y) \frac{\partial^{n+m} \Phi}{\partial x^n \partial y^m} = \left\{ L\left(\frac{\partial}{\partial x}, \frac{\partial}{\partial y} \right) \right\} \Phi(x,y). \quad \text{(3.352a, b)}$$

The **complete (particular) integral** *of the forced equation involving (not involving) arbitrary functions (3.353a, b) has as its difference the* **general integral** *of the unforced equation involving arbitrary functions (3.353c):*

$$\left\{ L\left(\frac{\partial}{\partial x}, \frac{\partial}{\partial y} \right) \right\} \Phi_{1,2}(x,y) = B(x,y):$$

$$\left\{ L\left(\frac{\partial}{\partial x}, \frac{\partial}{\partial y} \right) \right\} [\Phi_1(x,y) - \Phi_2(x,y)] = 0.$$

$$\text{(3.353a–c)}$$

Thus, the difference between the two particular integrals (3.337b) and (3.351c) of the same forced wave equation (3.337d) must be of the form (3.354a, b):

$$\Phi_*^+(x,t) - \hat{\Phi}(x,t) = f^+\left(t - \frac{x}{c} \right) + f^-\left(t + \frac{x}{c} \right) = f^+(\xi) + f^-(\eta), \quad \text{(3.354a, b)}$$

which is a solution (3.328b) ≡ (3.354a) of the unforced wave equation (3.222b), with a particular choice determined next.

Thus, *the forced one-dimensional Cartesian wave equation (3.337c, d) has two particular integrals (3.337b) [≠ (3.351c)] that: (i) need not coincide since they were obtained by different methods of characteristic polynomial (phase variables); and (ii) must differ by a solution (3.328b) of the unforced wave equation. It can be checked that their difference (3.355a):*

$$\Phi_*^+(x,t) - \hat{\Phi}(x,t) = \frac{B}{8}\left(t^2 + \frac{x^2}{c^2} \right) - \frac{B}{16\,b\,c}\left(t + \frac{x}{c} \right) e^{2bc\eta}$$

$$= \frac{B}{32} \left[(\xi + \eta)^2 + (\eta - \xi)^2 \right] - \frac{B\,\eta}{16\,b\,c} e^{2bc\eta}$$

$$= \frac{B}{16} (\xi^2 + \eta^2) - \frac{B\,\eta}{16\,b\,c} e^{2bc\eta} = f^+(\xi) + f^-(\eta), \quad \text{(3.335a–d)}$$

is (3.355b, c) of the form (3.355d) ≡ (3.354b) involving the functions (3.356a, b):

$$f^+(\xi) = \frac{B\,\xi^2}{16}, \qquad f^-(\eta) = \frac{B\,\eta^2}{16} - \frac{B\,\eta}{16\,b\,c} e^{2bc\eta}. \quad \text{(3.356a, b)}$$

The methods of phase (complex conjugate) variables are analogous [Sections 1.6 and 1.9 (Subsections 3.4.27–3.4.29)] in providing solutions of the two (one)-dimensional Cartesian harmonic and biharmonic (wave) equation and an example is the derivation (Subsection 3.4.30) of the solution (3.338g) using an alternative method to (3.338d).

3.4.30 METHOD OF COMPLEX VARIABLES FOR THE HARMONIC EQUATION

The general integral (3.217b) of the unforced two-dimensional Cartesian harmonic equation (3.212b) is the sum of two arbitrary twice differentiable functions of the **complex variable** (3.357a) and its **conjugate** (3.357b) leading by inversion to (3.357c, d):

$$z = x + i\ y, \qquad z^* = x - i\ y: \qquad x = \frac{z + z^*}{2}, \qquad y = \frac{z - z^*}{2\ i}. \qquad (3.357a\text{–}d)$$

Also, the two-dimensional Cartesian Laplacian operator becomes (1.239b) $\equiv$ (3.358b) using a dependent variable (3.358a) in terms of complex conjugate variables

$$\Phi(x, y) = \Psi(z, z^*): \qquad \frac{\partial^2 \Phi}{\partial x^2} + \frac{\partial^2 \Phi}{\partial y^2} = 4\ \frac{\partial^2 \Psi}{\partial z\ \partial z^*}. \qquad (3.358a, b)$$

The forced harmonic equation (3.338f) in terms of complex conjugate variables (3.357c, d; 3.358a, b) becomes (3.359a, b):

$$4\ \frac{\partial^2 \Psi}{\partial z\ \partial z^*} = B\ Re\left\{ e^{2a(x - iy)} \right\} = B\ Re\left\{ e^{2az^*} \right\}. \qquad (3.359a, b)$$

The integral of (3.359b) is (3.360a–d):

$$\Psi(z, z^*) = Re\left\{ \frac{B\ z}{4}\ \frac{e^{2az^*}}{2\ a} \right\} = Re\left\{ \frac{B}{8\ a}\ (x + i\ y)\ e^{2a(x - iy)} \right\}$$

$$= \frac{B}{8\ a}\ e^{2ax}\ \left\{ x \cos(2ay) + y \sin(2ay) \right\} \equiv \hat{\Phi}(x, y). \qquad (3.360a\text{–}d)$$

Thus, *the two-dimensional forced Cartesian harmonic equation (3.338g) has particular integrals (3.338d) [$\neq$ (3.360d)] that: (i) are distinct because they were obtained by different methods of characteristic polynomial (complex conjugate variables); and (ii) must differ by a solution (3.217b) of the unforced harmonic equation. It can be checked that their difference (3.361a–d):*

$$\Phi_*(x, y) - \hat{\Phi}(x, y) = \frac{B}{8\ a}\ e^{2ax}\ \left[x \cos(2ay) - y \sin(2ay) \right]$$

$$= \frac{B}{8\ a}\ e^{2ax}\ Re\left\{ (x + i\ y)\ e^{i2ay} \right\} = \frac{B}{8\ a}\ Re\left\{ (x + iy)\ e^{2a(x+iy)} \right\}$$

$$= \frac{B}{8\ a}\ Re\left\{ z\ e^{2az} \right\} = Re\left\{ f_+(z) + f_-(z^*) \right\}, \qquad (3.361a\text{–}e)$$

equals (3.361e) $\equiv$ (3.217b) with the functions (3.362a, b):

$$f_+(z) = \frac{B\ z}{8\ a}\ e^{2az}, \qquad f_-(z^*) = 0. \qquad (3.362a, b)$$

The characteristic polynomial has been used to obtain the general integral of a linear unforced partial differential equation with constant coefficients (Subsections 3.4.1–3.4.15) and also applies to forcing by exponentials and circular and hyperbolic cosines and sines (Subsections 3.4.16–3.4.30). It is used next (Subsections 3.4.31–3.4.39) in the method of the inverse characteristic polynomial of derivatives.

3.4.31 INVERSE CHARACTERISTIC POLYNOMIAL OF PARTIAL DERIVATIVES

In the sequel, the short-hand notation (3.363a) [(3.363b)] is used for the operator partial derivative with regard to the independent variable x (y):

$$\{\partial_x , \partial_y\} \, \Phi(x,y) \equiv \left\{\frac{\partial\Phi}{\partial x} , \frac{\partial\Phi}{\partial y}\right\}: \qquad \{\partial_x^{-1} , \partial_y^{-1}\} \, \Phi \equiv \int \Phi(x,y) \, \{dx, dy\}$$

(3.363a–d)

and (3.363c) [(3.363d)] for the inverse operator of indefinite integration with regard to x (y). A particular integral of the linear forced partial differential equation with constant coefficients (3.277a, b) $\equiv$ (3.364a-c):

$$B(x,y) = \sum_{n=0}^{N} \sum_{m=0}^{M} A_{n,m} \frac{\partial^{n+m}\Phi}{\partial x^n \, \partial y^m} \equiv \left\{\sum_{n=0}^{N} \sum_{m=0}^{M} A_{n,m} \, \partial_x^n \, \partial_y^m\right\} \Phi(x,y)$$

$$= \left\{P_{N,M}(\partial_x , \partial_y)\right\} \Phi(x,y):$$

$$\Phi(x,y) = \left\{P_{N,M}(\partial_x , \partial_y)\right\}^{-1} B(x,y),$$

(3.364a–d)

is (3.364d), provided that a suitable interpretation is given to the **inverse characteristic polynomial of partial derivatives** appearing in (3.364d). There are three cases of which the first two (third) concern derivatives all (not all) of the same order.

If in the forced linear partial differential equation with constant coefficients (3.364a) all derivatives are of the same order (3.365a-c):

$$B(x,y) = \sum_{n=0}^{N} A_n \frac{\partial^N \Phi}{\partial x^n \, \partial y^{N-n}} = \left\{\sum_{n=0}^{N} A_n \, \partial_x^n \, \partial_y^{N-n}\right\} \Phi(x,y)$$

$$= \left\{P_{N,N}(\partial_x , \partial_y)\right\} \Phi(x,y),$$

(3.365a–c)

and the characteristic polynomial can be factorised (case I) with distinct roots (3.366a), its inverse is (3.366b) where (Section I.38.1 and Subsection IV.1.5.3) the coefficients are the residues:

$$P_{N,N}(\partial_x , \partial_y) = A_N \prod_{\ell=1}^{N} (\partial_x - a_\ell \, \partial_y):$$

$$\left\{P_{N,N}(\partial_x , \partial_y)\right\}^{-1} = \partial_y^{1-N} \sum_{\ell=1}^{N} E_\ell \, (\partial_x - a_\ell \, \partial_y)^{-1}.$$

(3.366a, b)

This result follows from (3.367a–c):

$$\left\{A_N \prod_{\ell=1}^{N} \left(\partial_x - a_\ell \, \partial_y\right)\right\}^{-1} = \partial_y^{-N} \left\{A_N \prod_{\ell=1}^{N} \left(\partial_y^{-1} \, \partial_x - a_\ell\right)\right\}^{-1}$$

$$= \partial_y^{-N} \sum_{\ell=1}^{N} \frac{E_\ell}{\partial_y^{-1} \, \partial_x - a_\ell}$$

$$= \partial_y^{1-N} \sum_{\ell=1}^{N} E_\ell \left(\partial_x - a_\ell \, \partial_y\right)^{-1}; \qquad (3.367a–c)$$

where the coefficients are specified by the residues at simple poles (IV.1.203a–d) ≡ (3.367d):

$$A_N \, E_\ell = \sum_{\substack{m=1 \\ m \neq l}}^{L} \frac{1}{a_\ell - a_m}. \qquad (3.367d)$$

The operators in (3.366b) can be interpreted (I.21.62a) as an arithmetic series (3.368a, b):

$$\left(\partial_x - a_\ell \, \partial_y\right)^{-1} = \partial_x^{-1} \left(1 - a_\ell \, \partial_x^{-1} \, \partial_y\right)^{-1} = \sum_{r=0}^{\infty} \left(a_\ell\right)^r \partial_x^{-r-1} \, \partial_y^r. \qquad (3.368a, b)$$

Similar factorisations and interpretations are given (Subsection 3.4.32) in the cases II and III.

3.4.32   FACTORISATION AND INTERPRETATION OF THE INVERSE POLYNOMIAL OF PARTIAL DERIVATIVES

Again, for all derivatives of the same order (3.365a,b) but in the case of multiple roots (3.369a, b), the inverse characteristic polynomial is given by the extended residue rule (Section I.31.9 and Subsection IV.1.5.4) by (3.369c):

$$\sum_{\ell=1}^{L} \alpha_\ell = N, \qquad P_{N,N}\left(\partial_x, \partial_y\right) = A_N \prod_{\ell=1}^{L} \left(\partial_x - a_\ell \, \partial_y\right)^{\alpha_\ell}:$$

$$\left\{P_{N,N}\left(\partial_x, \partial_y\right)\right\}^{-1} = \sum_{\ell=1}^{L} \sum_{k=1}^{\alpha_\ell} E_{\ell,k} \, \partial_y^{k-N} \left(\partial_x - a_\ell \, \partial_y\right)^{-k}, \qquad (3.369a–c)$$

using a similar derivation to (3.367a–d) for multiple poles, where the coefficients are specified by the extended residue rule (IV.1.209a–g) ≡ (3.369d):

$$A_N \, E_{\ell,k} = \frac{(-)^{\alpha_\ell - k}}{(\alpha_\ell - k)!} \prod_{\substack{m=1 \\ m \neq l}}^{L} \frac{(\alpha_\ell - \alpha_m - k)!}{(\alpha_m - 1)!} \left(a_\ell - a_m\right)^{k - \alpha_m - \alpha_\ell}. \qquad (3.369d)$$

The operators in (3.368c) may be interpreted (I.25.37a–c) as a binomial series (3.370a,b)

$$\left(\partial_x - a_\ell\, \partial_y\right)^{-k} = \partial_x^{-k}\left(1 - a_\ell\, \partial_x^{-1}\, \partial_y\right)^{-k} = \partial_x^{-k} \sum_{r=0}^{\infty} \binom{-k}{r}\left(-a_\ell\, \partial_x^{-1}\, \partial_y\right)^r, \qquad (3.370a, b)$$

where the permutations (3.370c-f) appear:

$$\binom{-k}{r} = \frac{(-k)(-k-1)...(-k-r+1)}{r!} = \frac{(-)^r}{r!}\, k\,(k+1)...(k+r-1)$$

$$= \frac{(-)^r}{r!}\frac{(k+r-1)!}{(k-1)!} \equiv (-)^r\binom{k+r-1}{r}. \qquad (3.370c\text{–}f)$$

Thus, substitution of (3.370f) in (3.370b) leads in the case II to (3.371):

$$\left(\partial_x - a_\ell\, \partial_y\right)^{-k} = \sum_{r=0}^{\infty}\binom{k+r-1}{r}\left(a_\ell\right)^r\, \partial_x^{-k-r}\, \partial_y^r. \qquad (3.371)$$

In the general case III of derivatives of unequal order in (3.364a, b) ≡ (3.372a–c):

$$\sum_{\ell=1}^{L}\alpha_\ell = N, \qquad \sum_{\ell=1}^{L}\beta_\ell = M:$$

$$P_{N,M}\left(\partial_x\,,\partial_y\right) = A_{N,M}\prod_{\ell=1}^{L}\left(\partial_x^{\alpha_\ell} - a_\ell\, \partial_y^{\beta_\ell}\right), \qquad (3.372a\text{–}c)$$

the inverse characteristic polynomial of partial derivatives is a sum of terms of the form (3.373a, b):

$$\left(\partial_x^\alpha - a\, \partial_y^\beta\right)^{-k} = \partial_x^{-\alpha}\left(1 - a\, \partial_x^{-\alpha}\, \partial_y^{-\beta}\right)^{-k} = \sum_{r=0}^{\infty}\binom{r+k-1}{r}a^k\, \partial_x^{-\alpha(k+1)}\, \partial_y^{\beta\,k}.$$

$$(3.373a, b)$$

Thus, *the linear forced partial differential equation with constant coefficients (3.277a) ≡ (3.364a–c) has a particular integral (3.364d) where the inverse characteristic polynomial of partial derivatives is interpreted as a series (3.366a, b; 3.368a, b)/(3.369a–c; 3.371)/(3.372a–c; 3.373a, b) of partial differentiation (3.363a, b) and indefinite integration (3.363c, d) operators. It is assumed that the forcing function B(x,y) is smooth, that is infinitely differentiable, and that the resulting series converge. If the forcing function is a polynomial of degree r (s) in x (y), then the derivatives (3.363a) [(3.363b)] of an order higher than r (s) vanish, and the series solutions terminate, so there is no convergence issue.* Thus, the simplest application of the method of inverse characteristic polynomial of partial derivatives is forcing by polynomials, illustrated next for the harmonic/wave/diffusion equations (Subsections 3.4.33–3.4.34).

3.4.33   POLYNOMIAL FORCING OF HARMONIC/WAVE/DIFFUSION EQUATIONS

The forcing of a linear partial differential equation by a polynomial is the superposition of a sum of powers and, thus, one term may serve as illustration. The forcing by a first (second) power is considered for the harmonic (3.212b) $\equiv$ (3.374a), wave (3.222b) $\equiv$ (3.374b), and diffusion (3.235b) $\equiv$ (3.374c) equations:

$$b\,t\,x^2 = \left\{ \frac{\partial^2}{\partial t^2} + \frac{\partial^2}{\partial x^2} \;,\; \frac{\partial^2}{\partial t^2} - c^2 \frac{\partial^2}{\partial x^2} \;,\; \frac{\partial}{\partial t} - \chi \frac{\partial^2}{\partial x^2} \right\} \Phi(x,t). \qquad (3.374a\text{–}c)$$

Since the lowest power is t, the simplest particular integral is obtained expanding in powers of $\partial/\partial t$ in the forced: (i) harmonic equation (3.374a) leading to the particular integral (3.375a–d):

$$\tilde{\Phi}_1(x,t) = \left(\partial_t^2 + \partial_x^2 \right)^{-1} b\,t\,x^2 = b\,\partial_x^{-2} \left(1 + \partial_t^2\,\partial_x^{-2} \right)^{-1} t\,x^2$$

$$= b\,\partial_x^{-2} \left(t\,x^2 \right) = \frac{b\,t\,x^4}{12}; \qquad (3.375a\text{–}d)$$

(ii) wave equation (3.374b) leading to the particular integral (3.376a–d):

$$\tilde{\Phi}_2(x,t) = \left(\partial_t^2 - c^2\,\partial_x^2 \right)^{-1} b\,t\,x^2 = -\,b\,c^{-2}\,\partial_x^{-2} \left(1 - c^{-2}\,\partial_x^{-2}\,\partial_t^2 \right)^{-1} t\,x^2$$

$$= -\,b\,c^{-2}\,\partial_x^{-2} \left(t\,x^2 \right) = -\frac{b\,t\,x^4}{12\,c^2}; \qquad (3.376a\text{–}d)$$

and (iii) diffusion equation (3.374c) leading to the particular integral (3.377a–e):

$$\tilde{\Phi}_3(x,t) = \left(\partial_t - \chi\,\partial_x^2 \right) b\,t\,x^2 = -\,b\,\chi^{-1}\,\partial_x^{-2} \left(1 - \chi^{-1}\,\partial_x^{-2}\,\partial_t \right)^{-1} t\,x^2$$

$$= -\,b\,\chi^{-1}\,\partial_x^{-2} \left(1 + \chi^{-1}\,\partial_x^{-2}\,\partial_t \right) t\,x^2$$

$$= -\,b\,\chi^{-1}\,\partial_x^{-2} \left(t\,x^2 \right) - b\,\chi^{-2}\,\partial_x^{-4} \left(x^2 \right)$$

$$= -\frac{b\,t\,x^4}{12\,\chi} - \frac{b\,x^6}{360\,\chi^2}. \qquad (3.377a\text{–}e)$$

In (3.375a–d; 3.3.76a–d) [(3.377a–e)], it is sufficient to expand the series to zero (first) order in ∂_t because the higher-order terms ∂_t^n with $n = 2$, 3 ,... give zero when applied to $t\,x^2$.

The particular integrals are less simple, expanding in powers of ∂_x since x appears to the square in the forced: (i) harmonic equation (3.374a) leading to the particular integral (3.378a–d):

$$\hat{\Phi}_1(x,t) = \left(\partial_t^2 + \partial_x^2 \right)^{-1} b\,t\,x^2 = b\,\partial_t^{-2} \left(1 + \partial_t^{-2}\,\partial_x^2 \right)^{-1} t\,x^2$$

$$= b\,\partial_t^{-2} \left(t\,x^2 \right) - b\,\partial_t^{-4}\,\partial_x^2 \left(t\,x^2 \right) = \frac{b\,t^3\,x^2}{6} - \frac{b\,t^5}{60}; \qquad (3.378a\text{–}d)$$

(ii) wave equation (3.374b) leading to the particular integral:

$$\hat{\Phi}_2(x,t) = \left(\partial_t^2 - c^2 \, \partial_x^2\right)^{-1} b \, t \, x^2 = b \, \partial_t^{-2} \left(1 - c^2 \, \partial_t^{-2} \, \partial_x^2\right)^{-1} t \, x^2$$

$$= b \left(\partial_t^{-2} + c^2 \, \partial_t^{-4} \, \partial_x^2\right) t \, x^2 = \frac{b \, t^3 \, x^2}{6} + \frac{b \, c^2 \, t^5}{60}; \qquad (3.379a\text{--}d)$$

and (iii) diffusion equation (3.374c) leading to the particular integral (3.380a–d):

$$\hat{\Phi}_3(x,t) = \left(\partial_t - \chi \, \partial_x^2\right) b \, t \, x^2 = b \, \partial_t^{-1} \left(1 - \chi \, \partial_t^{-1} \, \partial_x^2\right) t \, x^2$$

$$= b \left(\partial_t^{-1} + \chi \, \partial_t^{-2} \, \partial_x^2\right) t \, x^2 = \frac{b \, t^2 \, x^2}{2} + \frac{\chi \, b \, t^3}{3}. \qquad (3.380a\text{--}d)$$

In (3.378a–d; 3.379a–d; 3.380a–d), only the first two terms of the power series are used, because the higher-order terms ∂_x^n with $n = 3$, 4 , ... give zero when applied to $t \, x^2$. The differences of forced solutions (Subsection 3.4.33) specify unforced solutions (Subsection 3.4.34).

3.4.34 POLYNOMIAL SOLUTIONS OF THE HARMONIC/WAVE/ DIFFUSION EQUATIONS

The particular integrals do not coincide and their difference satisfy the unforced partial differential equation as can be checked: (i) from (3.375d; 3.378d) for (3.374a) the harmonic equation (3.381a–c):

$$\Phi_1(x,t) = \hat{\Phi}_1(x,t) - \tilde{\Phi}_1(x,t) = \frac{b \, t \, x^2}{12} \left(2 \, t^2 - x^2\right) - \frac{b \, t^5}{60}:$$

$$\left(\partial_t^2 + \partial_x^2\right) \Phi_1 = 0; \qquad (3.381a\text{--}c)$$

(ii) from (3.376d; 3.379d) for (3.374b) the wave equation (3.382a–c):

$$\Phi_2(x,t) = \hat{\Phi}_2(x,t) - \tilde{\Phi}_2(x,t) = \frac{b \, t^3 \, x^2}{6} + \frac{b \, c^2 \, t^5}{60} + \frac{b \, t \, x^4}{12 \, c^2}:$$

$$\left(\partial_t^2 - c^2 \, \partial_x^2\right) \Phi_2 = 0; \qquad (3.382a\text{--}c)$$

and (iii) from (3.377e; 3.380d) for (3.374c) the diffusion equation (3.383a–c):

$$\Phi_3(x,t) = \hat{\Phi}_3(x,t) - \tilde{\Phi}_3(x,t) = \frac{b \, t^2 \, x^2}{2} + \frac{\chi \, b \, t^3}{3} + \frac{b \, t \, x^4}{12 \, \chi} + \frac{b \, x^6}{360 \, \chi^2}:$$

$$\left(\partial_t - \chi \, \partial_x^2\right) \Phi_3 = 0. \qquad (3.383a\text{--}c)$$

The method of the inverse characteristic polynomial of partial derivatives can be extended to the product of a smooth function by exponential and/or circular or hyperbolic cosines and sines (Subsections 3.4.35–3.4.39).

3.4.35 FORCING BY THE PRODUCT OF A SMOOTH FUNCTION
BY AN EXPONENTIAL

The method of inverse characteristic polynomial of partial derivatives can be applied to the product of a smooth or infinitely differentiable function (3.384a) by an exponential as follows: (i) the starting point is the identity (3.384a):

$$\Theta \in \mathcal{D}^\infty (|R^2): \qquad \left\{\partial_x, \partial_y\right\} e^{ax+by} \Theta = e^{ax+by} \left\{a+\partial_x, b+\partial_y\right\} \Theta; \qquad \text{(3.384a, b)}$$

(ii) this extends to polynomials of derivatives:

$$P\!\left(\partial_x, \partial_y\right) e^{ax+by} \Theta = e^{ax+by} P\!\left(a+\partial_x, b+\partial_y\right) \Theta, \qquad \text{(3.385a)}$$

and with the substitution (3.385b) leads to (3.385c)

$$\Phi \equiv e^{ax+by} \Theta: \qquad \left\{P\!\left(\partial_x, \partial_y\right)\right\} \Phi = e^{ax+by} \left\{P\!\left(a+\partial_x, b+\partial_y\right)\right\} e^{-ax-by} \Phi; \qquad \text{(3.385b, c)}$$

(iii) using the property (3.385c), the differential equation (3.386a) can be written (3.386b):

$$\Psi\, e^{ax+by} = \left\{P\!\left(\partial_x, \partial_y\right)\right\} \Phi = e^{ax+by} \left\{P\!\left(a+\partial_x, b+\partial_y\right)\right\} e^{-ax-by} \Phi; \qquad \text{(3.386a, b)}$$

(iv) the relation (3.386b) is equivalent to (3.386c):

$$\Psi = \left\{P\!\left(a+\partial_x, b+\partial_y\right)\right\} e^{-ax-by} \Phi, \qquad \text{(3.386c)}$$

and may be inverted (3.386c) ≡ (3.387a):

$$\left\{P\!\left(a+\partial_x, b+\partial_y\right)\right\}^{-1} \Psi = e^{-ax-by} \Phi = e^{-ax-by} \left\{P\!\left(\partial_x, \partial_y\right)\right\}^{-1} \Psi\, e^{ax+by}; \qquad \text{(3.387a, b)}$$

the substitution of (3.386a) leads to (3.387b); and (v) the differential equation (3.386a) has solution (3.388a):

$$\Phi = \left\{P\!\left(\partial_x, \partial_y\right)\right\}^{-1} \Psi\, e^{ax+by} = e^{ax+by} \left\{P\!\left(a+\partial_x, b+\partial_y\right)\right\}^{-1} \Psi, \qquad \text{(3.388a, b)}$$

and substitution of (3.387b) leads to (3.388b). Thus, *the linear forced partial differential equation (3.386a) ≡ (3.389a) has particular integral (3.388b) ≡ (3.389b):*

$$\left\{P\!\left(\partial_x, \partial_y\right)\right\} \Phi = e^{ax+by} \Psi:$$

$$\Phi = e^{ax+by} \left\{P\!\left(a+\partial_x, b+\partial_y\right)\right\}^{-1} \Psi, \qquad \text{(3.389a, b)}$$

where the inverse of the characteristic polynomial of partial derivatives is interpreted as before (Subsections 3.4.31–3.4.34). The method of the inverse characteristic polynomial of partial derivatives can be extended from the forcing by a smooth function (Subsections 3.4.31–3.4.34) to include the product by an exponential (Subsection 3.4.35) and circular and/or hyperbolic cosines and sines (Subsections 3.4.36–3.4.39).

3.4.36 Forcing by Product of Smooth, Exponential, Circular, and Hyperbolic Functions

The result (3.389a, b) can be extended as follows: *a linear partial differential equation with constant coefficients (3.365a–c) forced by the product of a smooth function (3.384a) by an exponential (3.389a, b) and by: (i) a circular cosine or sine (3.390a, b) has particular integral (3.390c):*

$$\left\{ P\left(\partial_x, \partial_y \right) \right\} \Phi(x,y) = \Psi(x,y) \, e^{ax+by} \cos, \sin(r\ x + s\ y)$$

$$= Re,\ Im\ \left\{ \Psi(x,y) \exp\left[(a+i\ r)\ x + (b+i\ s)\ y \right] \right\}, \qquad (3.390a, b)$$

$$\Phi(x,y) = Re, Im\ \left\{ \frac{\exp\left[(a+i\ r)\ x + (b+i\ s)\ y \right]}{P\left(a+i\ r+\partial_x,\ b+i\ s+\partial_y \right)} \Psi(x,y) \right\}; \qquad (3.390c)$$

(ii) a hyperbolic cosine or sine (3.391a, b) has a particular integral (3.391c):

$$\left\{ P\left(\partial_x, \partial_y \right) \right\} \Phi(x,y) = \Psi(x,y) \, e^{ax+by} \cosh, \sinh(\varphi\ x + \psi\ y) \qquad (3.391a)$$

$$= \frac{\Psi(x,y)}{2} \left\{ \exp\left[(a+\varphi)\ x + (b+\psi)\ y \right] \right.$$

$$\left. \pm \exp\left[(a-\varphi)\ x + (b-\psi)\ y \right] \right\}, \qquad (3.391b)$$

$$\Phi(x,y) = \left\{ \frac{\exp\left[(a+\varphi)\ x + (b+\psi)\ y \right]}{P\left(a+\varphi+\partial_x,\ b+\psi+\partial_y \right)} \pm \frac{\exp\left[(a-\Phi)\ x + (b-\psi)\ y \right]}{P\left(a-\varphi+\partial_x,\ b-\psi+\partial_y \right)} \right\} \frac{\Psi(x,y)}{2};$$

$$(3.391c)$$

(iii) circular and hyperbolic cosines or sines (3.392a, b) has a particular integral (3.392c):

$$\left\{ P\left(\partial_x, \partial_y \right) \right\} \Phi(x,y) = \Psi(x,y) \, e^{ax+by} \cos, \sin(r\ x + s\ y) \cosh(\varphi\ x + \psi\ y)$$

$$= \frac{\Psi(x,y)}{2} Re, Im\ \left\{ \exp\left[(a+i\ r+\varphi)\ x + (b+i\ s+\psi)\ y \right] \right.$$

$$\left. \pm \exp\left[(a+i\ r-\varphi)\ x + (b+i\ s-\psi)\ y \right] \right\}, \qquad (3.392a, b)$$

$$\Phi(x,y) = Re, Im \left[\left\{ \frac{\exp\left[(a + i\,r + \varphi)\,x + (b + i\,s + \psi)\,y\right]}{P\left(a + i\,r + \varphi + \partial_x\,,\ b + i\,s + \psi + \partial_y\right)} \right. \right.$$

$$\left. \left. \pm\, \frac{\exp\left[(a + i\,r - \varphi)\,x + (b + i\,s - \psi)\,y\right]}{P\left(a + i\,r - \varphi + \partial_x\,,\ b + i\,s - \psi + \partial_y\right)} \right\} \frac{\Psi(x,y)}{2} \right].$$

(3.392c)

As examples (Subsections 3.4.37/3.4.38/3.4.39) will be considered the forcing of the harmonic/wave/diffusion equation by the product of a power and, respectively, an exponential/and a circular sine/and a hyperbolic cosine.

3.4.37   HARMONIC EQUATION FORCED BY THE PRODUCT OF POWER AND EXPONENTIAL

The two-dimensional Cartesian harmonic equation (3.393a) ≡ (3.212b) forced by the product of powers and exponentials

$$\frac{\partial^2 \Phi}{\partial x^2} + \frac{\partial^2 \Phi}{\partial y^2} = b\,x\,y\,\exp(x + 2\,y),$$

(3.393a)

has (3.389b) ≡ (3.393b) particular integral (3.393c–g):

$$\Phi_*(x,y) = \left\{ \frac{b}{\partial_x^2 + \partial_y^2} \right\} b\,x\,y\,e^{x+2y} = b\,e^{x+2y} \left\{ \frac{1}{(1 + \partial_x)^2 + (2 + \partial_y)^2} \right\} x\,y$$

$$= b\,e^{x+2y} \left\{ \frac{1}{5 + 2\,\partial_x + 4\,\partial_y + \partial_x^2 + \partial_y^2} \right\} x\,y$$

$$= b\,\frac{e^{x+2y}}{5} \left\{ 1 + \frac{2}{5}\,\partial_x + \frac{4}{5}\,\partial_y + 0\left(\partial_x^2\,,\partial_y^2\right) \right\}^{-1} x\,y$$

$$= b\,\frac{e^{x+2y}}{5} \left(1 - \frac{2}{5}\,\partial_x - \frac{4}{5}\,\partial_y \right) x\,y = b\,\frac{e^{x+2y}}{5} \left(x\,y - \frac{2\,y}{5} - \frac{4\,x}{5} \right). \text{(3.393b–g)}$$

Thus, *(3.393g) is a particular integral of the two-dimensional Cartesian harmonic equation (3.393a) forced by a product of powers and exponentials.* A particular integral of (3.393a) could be sought in the form (3.394a):

$$\Phi_*(x,y) = b\,e^{x+2y}\,(A\,x\,y + B\,x + C\,y),$$

(3.394a)

and substitution in (3.393a) would lead to (3.394b–d):

$$\{A\,,\,B\,,\,C\} = \left\{ \frac{1}{5}\,,\, -\frac{4}{25}\,,\, -\frac{2}{25} \right\},$$

(3.394b–d)

in agreement with (3.393g), to within an added function of the form (3.217a) of the solution of the unforced Laplace equation (3.212b). The method (3.393b–g) is more direct than (3.394a–d). Next (Subsection 3.4.38), the harmonic is replaced by the wave equation and a circular sine is included as factor in the forcing term.

3.4.38   WAVE EQUATION FORCED BY PRODUCT OF POWER, EXPONENTIAL, AND CIRCULAR SINE

The one-dimensional Cartesian wave equation (3.222b) forced by the product of power by an exponential and circular sine (3.395a, b) ≡ (3.390a, b):

$$\frac{\partial^2 \Phi}{\partial t^2} - c^2 \frac{\partial^2 \Phi}{\partial x^2} = b \, x \, t \, e^{a(x+2ct)} \sin[a \, (2 \, x + c \, t)]$$

$$= Im \left\{ b \, x \, t \exp\big[(1+2 \, i) \, a \, x + (2+i) \, a \, c \, t\big] \right\}, \qquad (3.395a, b)$$

has (3.390c) ≡ (3.395c) the particular integral (3.396d–h):

$$\Phi_*(x,t) = Im \left\{ \frac{b}{\partial_t^2 - c^2 \, \partial_x^2} \exp\big[(1+2 \, i) \, a \, x + (2+i) \, a \, c \, t\big] \right\} x \, t$$

$$= b \, Im \left\{ \frac{\exp\big[(1 + 2 \, i) \, a \, x + (2 + i) \, a \, c \, t\big]}{\big[(2 + i) \, a \, c + \partial_t \, \big]^2 - c^2 \big[(1 + 2 \, i) \, a + \partial_x \, \big]^2} \, x \, t \right\}$$

$$= b \, Im \left\{ \frac{\exp\big[(1+ 2 \, i) \, a \, x + (2+ i) \, a \, c \, t\big]}{6 \, a^2 \, c^2 + 2 \, (2+i) \, a \, c \, \partial_t - 2 \, (1+2 \, i) \, a \, c^2 \, \partial_x + \partial_t^2 - c^2 \, \partial_x^2} \, x \, t \right\}$$

$$= \frac{b}{6 \, a^2 \, c^2} \, e^{a(x+2ct)} \, Im \left\{ \exp\big[i \, a \, (2 \, x + c \, t)\big] \right.$$

$$\left. \times \left(1 - \frac{2+i}{3 \, a \, c} \, \partial_t + \frac{1+2 \, i}{3 \, a} \, \partial_x \right) x \, t \right\}$$

$$= \frac{b}{6 \, a^2 \, c^2} \, e^{a(x+2ct)} \, Im \left\{ \left[\left(x \, t - \frac{2 \, x}{3 \, a \, c} + \frac{t}{3 \, a}\right) - i \left(\frac{x}{3 \, a \, c} - \frac{2 \, t}{3 \, a}\right)\right] e^{ia(x+ct)} \right\}$$

$$= \frac{b}{6 \, a^2 \, c^2} \, e^{a(x+2ct)} \left\{ \left(x \, t - \frac{2 \, x}{3 \, a \, c} + \frac{t}{3 \, a}\right) \sin\big[(a \, (x+c \, t)\big] \right.$$

$$\left. - \left(\frac{x}{3 \, a \, c} - \frac{2 \, t}{3 \, a}\right) \cos\big[(a \, (x+c \, t)\big] \right\}. \qquad (3.395c–h)$$

Thus, *(3.396h) is a particular integral of the one-dimensional Cartesian wave equation (3.396a) forced by a product of powers, exponentials, and a circular sine.* A particular integral of (3.395a) could be sought in the form (3.396a)

$$\Phi(x,t) = e^{a(x\,+\,2ct)}\{(A_1\ x\ t + A_2\ x + A_3\ t)\cos[a(x+ct)]$$

$$+ (B_1\ x\ t + B_2\ x + B_3\ t)\sin[a(x+ct)]\}, \qquad (3.396a)$$

and (3.395a) would be satisfied by (3.396b–g):

$$\{A_1\ ,\ A_2\ ,\ A_3\ ; B_1\ ,\ B_2\ ,\ B_3\ \} = \frac{b}{6\ a^2\ c^2}\left\{1\ ,\ -\frac{2}{3\ a\ c}\ ,\ \frac{1}{3\ a}\ ; 0\ ,\ -\frac{1}{3\ a\ c}\ ,\ \frac{2}{3\ a}\right\},$$
$$(3.396b\text{–}g)$$

in agreement with (3.395h), to within an added function of the form of the general integral (3.228a) of the unforced wave equation (3.222b). The method (3.395c–h) of the solution of (3.395a, b) is less cumbersome than the substitution of (3.396a–g). In the forcing terms in (3.395a, b) and the particular integral (3.396a–f), the constant a has the dimensions of inverse length so that $a\,x$ is dimensionless, and distance x is added to multiples of $c\,t$ that have the same dimension and, thus, $a\,x$, $a\,c\,t$ are dimensionless and appear in the arguments of exponential and circular cosines and sines; also $a\ (c\ t)$ has the same dimensions as $\partial_x\ (\partial_t)$ in the inverse characteristic polynomial. Next (Subsection 3.4.39), the wave is replaced by the diffusion equation and in the forcing term is included a hyperbolic cosine factor.

3.4.39   DIFFUSION EQUATION FORCED BY THE PRODUCT OF POWER, EXPONENTIAL AND CIRCULAR AND HYPERBOLIC FUNCTIONS

The one-dimensional Cartesian diffusion equation forced by the product of powers by exponentials by circular sine and hyperbolic cosine (3.397a, b) $\equiv$ (3.392a, b):

$$\frac{\partial \Phi}{\partial t} - \chi \frac{\partial^2 \Phi}{\partial x^2} = b\ x\ t\ e^{a(x+2a\chi t)}\sin\big[a\ (x+a\ \chi\ t)\big]\cosh\big[a\ (x-a\ \chi\ t)\big]$$

$$= \frac{b\ x\ t}{2}\ Im\ \Big\{\exp\big[(2+i)\ a\ x+(1+i)\ a^2\ \chi\ t\big]$$

$$+\exp\big[i\ x\ a+(3+i)\ a^2\ \chi\ t\big]\Big\}, \qquad (3.397a,\ b)$$

has (3.392c) $\equiv$ (3.397b), the particular integral (3.398a–g) using (3.398c):

$$\xi \equiv a\ (x+a\ \chi\ t):$$

$$\Phi_*(x,t) = \frac{b}{2}\ Im\ \Big\{(\partial_t - \chi\ \partial_x^2)^{-1}\ \Big[\big\{\exp\big[a\ (2+i)\ x+a^2\ \chi\ (1+i)\ t\big]$$

$$+\exp\big[i\ a\ x+a^2\ \chi\ (3+i)\ t\big]\big\}x\ t\Big]$$

$$= \frac{b}{2} \, Im \left[\left\{ \frac{\exp\left[a\,(2+i)\,x + a^2\chi\,(1+i)\,t\right]}{a^2\,\chi\,(1+i) + \partial_t - \chi\left[a\,(2+i) + \partial_x\right]^2} \right. \right.$$

$$\left. \left. + \frac{\exp\left[i\,a\,x + a^2\,\chi\,(3+i)t\right]}{a^2\,\chi\,(3+i) + \partial_t - \chi\left(i\,a + \partial_x\right)^2} \right\} x\,t \right]$$

$$= \frac{b}{2} \, e^{a(2x\,+\,a\chi t)} \, Im \left\{ \frac{\exp\left[i\,a\,(x + a\,\chi\,t)\right]}{-(2+3\,i)\,a^2\,\chi + \partial_t - 2\,a\,\chi\,(2+i)\,\partial_x - \chi\,\partial_x^2} x\,t \right\}$$

$$+ \frac{b}{2} \, e^{3a^2\chi t} \, Im \left\{ \frac{\exp\left[i\,a\,(x + a\,\chi\,t)\right]}{a^2\,\chi\,(4+i) + \partial_t - 2\,i\,a\,\chi\,\partial_x - \chi\,\partial_x^2} x\,t \right\}$$

$$= -\frac{b}{2\,a^2\,\chi} \, e^{a(2x\,+\,a\chi t)} \, Im \left[\frac{e^{i\xi}}{2+3i} \times \left\{ 1 + \frac{1}{(2+3i)a^2\,\chi}\,\partial_t - \frac{2(2+i)}{a(2+3i)}\,\partial_x \right\} x\,t \right]$$

$$+ \frac{b}{2\,a^2\,\chi} \, e^{3a^2\chi t} \, Im \left[\frac{e^{i\xi}}{4+i} \times \left\{ 1 - \frac{1}{a^2\,\chi\,(4+i)}\,\partial_t + \frac{2\,i}{a\,(4+i)}\,\partial_x \right\} x\,t \right]$$

$$= -\frac{b}{26} \, \frac{e^{a(2x\,+\,a\chi t)}}{a^2\,\chi} \, Im \left\{ e^{i\xi}\,(2-3i) \times \left[x\,t + \frac{2-3i}{13}\,\frac{x}{a^2\,\chi} - \frac{(4+2i)(2-3i)}{13}\,\frac{t}{a} \right] \right\}$$

$$+ \frac{b}{34} \, \frac{e^{3a^2\chi t}}{a^2\,\chi} \, Im \left\{ e^{i\xi}\,(4-i) \left[x\,t - \frac{4-i}{17}\,\frac{x}{a^2\,\chi} + \frac{2+8i}{17}\,\frac{t}{a} \right] \right\}$$

$$= -\frac{b}{26} \, \frac{e^{a(2x\,+\,a\chi t)}}{a^2\,\chi} \left\{ \left(2\,x\,t - \frac{5\,x}{13\,a^2\,\chi} + \frac{4\,t}{13\,a} \right) \sin\left[a\,(x + a\,\chi\,t)\right] \right.$$

$$\left. -\left(3\,x\,t + \frac{12}{13}\,\frac{x}{a^2\,\chi} - \frac{58\,t}{13\,a} \right) \cos\left[a\,(x + a\,\chi\,t)\right] \right\}$$

$$+ \frac{b}{34} \, \frac{e^{3a^2\chi t}}{a^2\,\chi} \left\{ \left(4\,x\,t - \frac{15\,x}{17\,a^2\,\chi} + \frac{16\,t}{17\,a} \right) \sin\left[a\,(x + a\,\chi\,t)\right] \right.$$

$$\left. -\left(x\,t - \frac{8\,x}{17\,a^2\,\chi} - \frac{30\,t}{17\,a} \right) \cos\left[a\,(x + a\,\chi\,t)\right] \right\}. \qquad (3.398a\text{--}g)$$

Thus, *(3.398i)* is a particular integral of the one-dimensional Cartesian diffusion equation *(3.398a)* forced by the product of powers by exponentials by a circular sine and hyperbolic cosine.

A particular integral of (3.398a, b) could be sought in the form (3.399a):

$$\Phi_*(x,t) = b\ e^{a(2x\,+\,a\chi t)} \left\{\left(A_1\ x\ t + A_2 x + A_3\ t\right)\sinh\left[a\left(x + a\ \chi\ t\right)\right]\right.$$

$$+ \left(B_1\ x\ t + B_2\ x + B_3\ t\right)\cosh\left[a\left(x + a\ \chi\ t\right)\right]\right\}$$

$$+ b\ e^{3a^2\chi t} \left\{\left(C_1\ x\ t + C_2\ x + C_3\ t\right)\sinh\left[a\left(x + a\ \chi\ t\right)\right]\right.$$

$$+ \left(D_1\ x\ t + D_2\ x + D_3\ t\right)\cosh\left[a\left(x + a\ \chi\ t\right)\right]\right\}, \quad (3.399a)$$

leading, on substitution in (3.398a), to (3.399b–m):

$$\left\{A_1\,,\ A_2\,,\ A_3\ ; B_1\,,\ B_2\,,\ B_3\right\} = -\frac{b}{26\ a^2\ \chi}\left\{2\ ,\ -\frac{5}{13\ a^2\ \chi}\ ,\ \frac{4}{13\ a}\ ; -\ 3\ ,\ -\frac{12}{13\ a^2\ \chi}\ ,\ \frac{58}{13\ a}\right\},$$
$$(3.399\text{b--g})$$

$$\left\{C_1\,,\ C_2\,,\ C_3\ ; D_1\,,\ D_2\,,\ D_3\right\} = \frac{b}{34\ a^2\ \chi}\left\{4\ ,\ -\frac{15}{17\ a^2\ \chi}\ ,\ \frac{16}{17\ a}\ ; -\ 1\ ,\ \frac{8}{17\ a^2\ \chi}\ ,\ \frac{30}{17\ a}\right\},$$
$$(3.399\text{h--m})$$

which agrees with (3.398i) to within an added function that is a solution of the unforced diffusion equation (3.235b). In the forcing terms in (3.398a, b) and in the particular integrals (3.398c–i), the constant a has the dimensions of inverse length so that $a^2\ \chi$ has the dimensions of inverse time; it follows that $a\ x$, $a^2\ \chi\ t$ are dimensionless as they appear in the arguments of exponentials and circular or hyperbolic cosines and sines. Also $a\ (a^2\ \chi)$ has the same dimensions as $\partial_x\ (\partial_t)$ in the inverse characteristic polynomial. The method of the inverse characteristic polynomial of partial derivatives (Subsections 3.4.31–3.4.39) applies (Table 3.5) to a linear forced partial differential equation with constant coefficients (3.364a–c), and if the derivatives are all of the same order, (3.365a, b) can be combined with the method of factorisation (Sections 1.6–1.9), distinguishing the cases of single (multiple) roots of the characteristic polynomial [Subsections 3.4.40–3.4.42 (3.4.43–3.4.44)].

3.4.40   COMBINATION OF THE METHODS OF FACTORISATION AND INVERSION OF THE POLYNOMIAL OF PARTIAL DERIVATIVES

Consider a forced linear partial differential equation with constant coefficients (3.364a–d) in the cases of (i) all derivatives of the same order (3.365a–c) allowing factorisation of the characteristic polynomial (3.366a) with (ii) distinct roots (3.366b), leading to the particular integral (3.400a):

$$\Phi_*(x,y) = \left\{\partial_y^{1-N}\sum_{m=1}^{N}\frac{E_m}{\partial_x - a_m\ \partial_y}\right\}B(x,y) = \partial_y^{1-N}\sum_{m=1}^{N}E_m\ \Psi_m(x,y), \quad (3.400\text{a, b})$$

that is a sum (3.400b) of **intermediate functions** satisfying (3.401a), whose inverses are (3.401b):

$$\Psi_m(x,y) = \left(\partial_x - a_m \, \partial_y\right)^{-1} B(x,y) \quad \Leftrightarrow \quad B(x,y) = \left(\partial_x - a_m \, \partial_y\right) \Psi_m(x,y).$$

$$\text{(3.401a, b)}$$

The latter (3.401b) is a linear first-order differential equation with constant coefficients that can be converted from partial to ordinary derivatives (3.402b):

$$\lambda_m \equiv a_m \, \frac{d}{dy} : \qquad\qquad \frac{d\Psi_m}{dx} - \lambda_m \, \Psi_m = B(x,y). \qquad \text{(3.402a, b)}$$

The forced linear first-order ordinary differential equation (3.402b) has a solution (Section IV.3.3) in the form of (3.403a), with the coefficient function $C(x)$ determined by substitution into (3.402b), leading to (3.403b, c):

$$\Psi_m(x,y) = C_m(x) \, e^{\lambda_m x}:$$

$$B(x,y) = \frac{d}{dx}\left[C_m(x) \, e^{\lambda_m x}\right] - C_m(x) \, e^{\lambda_m x} = e^{\lambda_m x} \, \frac{dC_m}{dx}. \qquad \text{(3.403a–c)}$$

Thus, the coefficient function is given by (3.404a):

$$C_m(x) = \int^x B(\xi,y) \, e^{-\lambda_m \xi} \, d\xi:$$

$$\Psi_m(x,y) = e^{a_m x \, d/dy} \int^x e^{-a_m \xi \, d/dy} \, B(\xi,y) \, d\xi, \qquad \text{(3.404a, b)}$$

and the particular integral of (3.402b) is (3.403a; 3.402a) given by (3.404b). Assuming that the forcing function is analytic in y in (3.405a) with the Taylor series (I.23.32b) $\equiv$ (3.101a–d) $\equiv$ (3.405b, c):

$$B(\ldots,y) \in \mathcal{A}\,(|R): \qquad e^{\zeta \, d/dy} \, B(\xi,y) = \sum_{n=0}^{\infty} \frac{\zeta^n}{n!} \frac{d^n}{dy^n}\left[B(\xi,y)\right] = B(\xi, y + \zeta \, x),$$

$$\text{(3.405a–c)}$$

and applying (3.405c) to (3.404b) with (3.406a) leads to (3.406b):

$$\zeta \equiv a_m \, (x-\xi): \qquad \Psi_m(x,y) = \int^x B(\,\xi\,, y + a_m \, x - a_m \, \xi)\, d\xi. \qquad \text{(3.406a, b)}$$

Substituting (3.406b) in (3.400b), it follows that *the linear partial differential equation with constant coefficients (3.364a–c) with: (i) all derivatives of the same order (3.365a–c); (ii) characteristic polynomial factorised by (3.366a) with single roots and inverse partial fraction decomposition (3.366b; 3.367d); and (iii) forcing function analytic (3.405a) ≡ (3.407a) in y and (N − 1)-times integrable (3.407b) in x:*

$$B(\ldots,y)\in\mathcal{A}\,(\mathbb{R}),\qquad B(x,\ldots)\in E^{\,N-1}\,(\mathbb{R}):$$

$$\Phi_*(x,y)=\int^{y}dy_1\int^{y_1}dy_2\ldots\int^{y_{N-2}}dy_{N-1}\sum_{m=1}^{N}E_m\int^{x}B(\xi,y+a_m\,x-a_m\,\xi)\,d\xi,\qquad (3.407\text{a–c})$$

has a particular integral given by (3.407c). An example of forcing by inverse powers is given (Subsection 3.4.41–3.4.42) before relaxing the condition (ii) to multiple roots (Subsections 3.4.43–3.4.44).

3.4.41   FORCING OF A LINEAR EQUATION WITH SECOND-ORDER PARTIAL DERIVATIVES

Consider the forcing of the linear partial differential equation with constant coefficients and all derivatives of the second-order (3.408):

$$B(x,y)=\frac{\partial^2\Phi}{\partial x^2}-2\,\alpha\,\frac{\partial^2\Phi}{\partial x\,\partial y}+\beta\,\frac{\partial^2\Phi}{\partial y^2},\qquad (3.408)$$

where the coefficient of $\partial^2\Phi/\partial x^2$ is unity without loss of generality because it can be incorporated in B in the l.h.s. of (3.408). The differential operator in (3.408) can be factorised (3.409a, b):

$$\frac{\partial^2}{\partial x^2}-2\,\alpha\,\frac{\partial^2}{\partial x\,\partial y}+\beta\,\frac{\partial^2}{\partial y^2}=\left(\frac{\partial}{\partial x}-a_+\frac{\partial}{\partial y}\right)\left(\frac{\partial}{\partial x}-a_-\frac{\partial}{\partial y}\right),$$

$$=\frac{\partial^2}{\partial x^2}-\left(a_++a_-\right)\frac{\partial^2}{\partial x\,\partial y}+a_+\,a_-\frac{\partial^2}{\partial y^2},\qquad (3.409\text{a, b})$$

with roots satisfying (3.410a, b), hence also (3.410c–e) and, thus, given by (3.410f, g):

$$a_++a_-=-\,2\,\alpha,\qquad a_+\,a_-=\beta:$$

$$0=(a-a_+)(a-a_-)=a^2-(a_++a_-)\,a+a_+\,a_-=a^2-2\,\alpha\,a+\beta,$$

$$a_\pm=\alpha\pm\sqrt{\alpha^2-\beta}.\qquad (3.410\text{a–g})$$

A particular integral of (3.408) is given by (3.411):

$$\Phi_*(x,y)=\frac{1}{\left(\partial_x-a_+\,\partial_y\right)\left(\partial_x-a_-\,\partial_y\right)}\,B(x,y),\qquad (3.411)$$

with $a_\pm$ specified by (3.410g).

In the case of distinct roots, (3.412a) can be used the decomposition into partial fractions (3.412b):

$$a_+ \neq a_- : \qquad \Phi_*(x,y) = \frac{1}{(a_+ - a_-)\,\partial_y} \left\{ \frac{1}{\partial_x - a_+\,\partial_y} - \frac{1}{\partial_x - a_-\,\partial_y} \right\} B(x,y)$$

$$= \frac{\partial_y^{-1}}{a_+ - a_-} \left[\Psi_+(x,y) - \Psi_-(x,y) \right], \qquad (3.412a\text{–}c)$$

where (3.401a; 3.406b) the two terms on the r.h.s. of (3.412c) involve (3.413a, b):

$$\Psi_\pm(x,y) = \left(\partial_x - a_\pm\,\partial_y\right)^{-1} B(x,y) = \int^x B\!\left(\xi, y + a_\pm\, x - a_\pm\, \xi\right) d\xi. \qquad (3.413a, b)$$

Substituting (3.413b) in (3.412c) follows that *the linear partial differential equation with constant coefficients and all derivatives of second-order (3.408) in the case (3.414a) of distinct (3.412a) roots (3.410g) with forcing function analytic (3.414b) in x and integrable (3.414c) in y has particular integral (3.414d):*

$$\alpha^2 \neq \beta; \qquad B(x,\ldots) \in \mathcal{A}\,(\mathbb{R}); \qquad B(\ldots,y) \in E\,(\mathbb{R}):$$

$$\Phi_*(x,y) = \frac{1}{2\sqrt{\alpha^2 - \beta}} \int^y d\eta \int^x d\xi \left[B(\xi + a_+\, x - a_+\xi) - B(\xi + a_-\, x - a_-\, \xi) \right].$$
$$(3.414a\text{–}d)$$

The result (3.414d) is a particular case of (3.407c) with (3.415a–f):

$$N = 2, \quad a_1 = a_+, \quad a_2 = a_-, \quad E_1 = -E_2 = \frac{1}{a_+ - a_-} = \frac{1}{2\sqrt{\alpha^2 - \beta}}. \qquad (3.415a\text{–}f)$$

Next, forcing by inverse independent variables (Subsection 3.4.42) is considered, including the harmonic and wave equations.

3.4.42 FORCING OF THE HARMONIC AND WAVE EQUATIONS BY INVERSE VARIABLES

In the case of (3.408) with forcing by inverse independent variables (3.416a), a particular integral is obtained in two steps: (i) a first integration for the intermediate functions (3.413b) using (3.416a) leading to (3.416b–e):

$$B(x,y) = \frac{b}{x\,y}:$$

$$\Psi_\pm(x,y) = b \int^x \frac{d\xi}{\xi\,(y + a_\pm\, x - a_\pm\, \xi)} = \frac{b}{y + a_\pm\, x} \int^x \left(\frac{1}{\xi} + \frac{a_\pm}{y + a_\pm\, x - a_\pm\, \xi} \right) d\xi$$

$$= \frac{b}{y + a_\pm\, x} \left[\log\xi - \log(y + a_\pm\, x - a_\pm\, \xi) \right]_{\xi = x} = b\,\frac{\log(x/y)}{y + a_\pm\, x};$$
$$(3.416a\text{–}e)$$

and (ii) a second integration after substitution in (3.414a–d) ≡ (3.417a–f):

$$\alpha^2 \neq \beta: \qquad \Phi_*(x,y) = \frac{b}{2\sqrt{\alpha^2 - \beta}} \int^y [\Psi_+(x,\eta) - \Psi_-(x,\eta)]\, d\eta$$

$$= \frac{b}{a_+ - a_-} \int^y (\log x - \log \eta)\left(\frac{1}{\eta + a_+ \, x} - \frac{1}{\eta + a_- \, x}\right) d\eta$$

$$= - b\, x \int^y \frac{\log x - \log \eta}{(\eta + a_+ \, x)(\eta + a_- \, x)}\, d\eta$$

$$= b\, x \int^y \frac{\log \eta - \log x}{\eta^2 + (a_+ + a_-)\,\eta\, x + a_+ \, a_- \, x^2}\, d\eta$$

$$= b\, x \int^y \frac{\log \eta - \log x}{\eta^2 - 2\,\alpha\,\eta\, x + \beta\, x^2}\, d\eta, \qquad (3.417\text{a–f})$$

with (3.410f, g) used in the passage from (3.417e) to (3.417f).

Thus, *the linear partial differential equation with constant coefficients, all derivatives of second-order, distinct roots (3.418a), and forcing by inverse independent variables (3.408; 3.416a) ≡ (3.418b):*

$$\alpha \neq \beta^2, \qquad \frac{b}{x\, y} = \frac{\partial^2 \Phi}{\partial x^2} - 2\,\alpha\, \frac{\partial^2 \Phi}{\partial x\, \partial y} + \beta\, \frac{\partial^2 \Phi}{\partial y^2}:$$

$$\Phi_*(x,y) = b\, x \int^y \frac{\log(\eta/x)}{\eta^2 - 2\,\alpha\,\eta\, x^2 + \beta\, x^2}\, d\eta \qquad (3.418\text{a–c})$$

has a particular integral (3.418c), including: (i) for (3.419a, b) leading (3.410g) to conjugate imaginary roots (3.419c, d), the particular integral (3.419f) of the two-dimensional Cartesian harmonic equation forced by inverse independent variables (3.419e):

$$\alpha = 0, \qquad \beta = 1; \qquad a_\pm = \pm\, i; \qquad \frac{\partial^2 \Phi}{\partial x^2} + \frac{\partial^2 \Phi}{\partial y^2} = \frac{b}{x\, y}:$$

$$\Phi_*(x,y) = b\, x \int^y \frac{\log(\eta/x)}{\eta^2 + x^2}\, d\eta, \qquad (3.419\text{a–f})$$

corresponding to (3.419h) via the change of integration variable (3.419g):

$$z \equiv \frac{\eta}{x}: \qquad \Phi_*(x,y) = b \int^{y/x} \frac{\log z}{1 + z^2}\, dz = f\left(\frac{y}{x}\right), \qquad (3.419\text{g–i})$$

*and showing that the particular integral (3.419i) depends only on the ratio of inde-
pendent variables y/x; and (ii) for (3.420c, d) leading (3.410g) to real symmetric
roots (3.420e, f), the particular integral (3.420h) of the one-dimensional Cartesian
wave equation forced (3.420g) by inverse independent variables (3.420a, b):*

$$x \to t , \qquad y \to x , \qquad \alpha = 0 , \qquad \beta = - c^2 , \qquad a_\pm = \pm\, c:$$

$$\frac{\partial^2 \Phi}{\partial t^2} - c^2 \frac{\partial^2 \Phi}{\partial x^2} = \frac{b}{x\, t}, \qquad \Phi_*\left(x,t\right) = b\, t \int^x \frac{\log\left(\eta/x\right)}{\eta^2 - c^2\, t^2}\, d\eta , \qquad \text{(3.420a-h)}$$

corresponding via the change of variable (3.420i) to (3.420j):

$$z \equiv \frac{\eta}{t}: \qquad \Phi_*\left(x,t\right) = b \int^{x/t} \frac{\log z}{z^2 - c^2}\, dz = g\left(\frac{x}{t}\right), \qquad \text{(3.420i–k)}$$

*which shows that the particular integral (3.420k) depends only on the ratio of inde-
pendent variables x/t with the dimensions of velocity.* The general case (3.407a–c)
[particular example (3.418b)] of a particular integral of a forced linear partial differential equation with constant coefficients with all derivatives of the same order
(order two) assumes distinct roots in (3.401a, b) [(3.418a)] and is extended next to
multiple roots [Subsection 3.4.43 (3.4.44)].

3.4.43 MULTIPLE ROOTS OF THE FACTORISED CHARACTERISTIC POLYNOMIAL OF PARTIAL DERIVATIVES

Returning to the forced linear partial differential equation with constant coefficients
(3.364a–d) and all derivatives of the same order (3.365a–c), and allowing for multiple
roots of the characteristic polynomial (3.369a–c), a particular integral is (3.421a, b):

$$\Phi_*\left(x,y\right) = \left\{ \sum_{m=1}^{S} \sum_{k=1}^{\alpha_m} E_{m,k}\, \partial_y^{k-N} \left(\partial_x - a_m\, \partial_y\right)^{-k} \right\} B\left(x,y\right)$$

$$= \sum_{m=1}^{S} \sum_{k=1}^{\alpha_m} E_{m,k}\, \partial_y^{k-N}\, \Psi_{m,k}\left(x,y\right), \qquad \text{(3.421a, b)}$$

a sum of **intermediate functions** satisfying (3.422a) whose inverses are (3.422b):

$$\left\{\left(\partial_x - a_m\, \partial_y\right)^{-k}\right\} B\left(x,y\right) = \Psi_{m,k}\left(x,y\right) \qquad \Leftrightarrow \qquad B\left(x,y\right) = \left(\partial_x - a_m\, \partial_y\right)^{k}\, \Psi_{m,k}\left(x,y\right). \qquad \text{(3.422a, b)}$$

Using again the symbolic operator (3.402a) leads from (3.422b) to (3.423a):

$$\left(\frac{d}{dx} - \lambda_m\right)^{k}\, \Psi_{m,k}\left(x,y\right) = B\left(x,y\right): \qquad \Psi_{m,k}\left(x,y\right) = C_{m,k}\left(x\right) e^{\lambda_m x}, \qquad \text{(3.423a, b)}$$

whose solution is similar (3.423b) $\equiv$ (3.403a) with the coefficient function satisfying (3.424a, b) as follows, applying k times (3.403b, c):

$$B(x,y) = \left(\frac{d}{dx} - \lambda_m\right)^k \left[C_{m,k}(x)\, e^{\lambda_m x}\right] = e^{\lambda_m x}\, \frac{d^k}{dx^k}\left[C_{m,k}(x)\right]. \qquad (3.424a, b)$$

The coefficient function in (3.424b) involves k integrations before substitution in (3.423b), leading to (3.425b):

$$B(\ldots,y) \in \mathcal{A}\,(|R): \qquad \Psi_{m,k}(x,y) = e^{x a_m\, d/dy}\, \partial_x^{-k+1} \int_x^\xi e^{\xi a_m\, d/dy}\, B(\xi,y)\, d\xi$$

$$= e^{x a_m\, d/dy}\, \partial_x^{1-k} \int^x B(\xi, y - a_m\, \xi)\, d\xi, \qquad (3.425a\text{--}c)$$

and hence (3.425c) for forcing by an analytic function (3.425a) in y. Substitution of (3.425b) in (3.421b) shows that *the linear partial differential equation with constant coefficients (3.364a–d) with: (i) all derivatives of the same order (3.365a–c); (ii) characteristic polynomial with multiple roots (3.369a–c); and (iii) forcing function analytic (3.426a) in y and (3.426b) times integrable (3.426c) in x:*

$$B(\ldots,y) \in \mathcal{A}\,(|R); \qquad \alpha \equiv \max(\alpha_m): \qquad B(x,\ldots) \in E^{\,\alpha}\,(|R):$$

$$\Phi_*(x,y) = \sum_{m=1}^{S} \sum_{k=1}^{\alpha_m} E_{m,k}\, \partial_y^{k-N}\, F_m(x, y + a_m\, x),$$

$$F_m(x,y) \equiv \partial_x^{1-k} \int^x B(\xi, y - a_m\, \xi)\, d\xi \qquad (3.426a\text{--}e)$$

has particular integral (3.426d, e) involving up to $N - \alpha$ (α) integrations with regard to $y(x)$. A concluding example (Subsection 3.4.44) is considered (3.408) with a double root, in general, and in the particular case of forcing by inverse independent variables.

3.4.44   PARTIAL DIFFERENTIAL EQUATION WITH SECOND-ORDER DERIVATIVES AND DOUBLE ROOT

The linear differential equation with constant coefficients and all derivatives of second-order (3.408) has a double root (3.410f, g) $\equiv$ (3.427b, c) for (3.427a) leading to (3.427d, e):

$$\beta = \alpha^2; \qquad a_+ = a_- = \alpha:$$

$$B(x,y) = \frac{\partial^2 \Phi}{\partial x^2} - 2\,\alpha\,\frac{\partial^2 \Phi}{\partial x\, \partial y} + \alpha^2\, \frac{\partial^2 \Phi}{\partial y^2} = \left\{\left(\frac{\partial}{\partial x} - \alpha\,\frac{\partial}{\partial y}\right)^2\right\} \Phi(x,y).$$

$$(3.427a\text{--}e)$$

Comparing (3.427e) ≡ (3.422a, b) follows (3.428a–c) and the particular integral (3.426a, b) ≡ (3.428d–f):

$$a_m = \alpha, \qquad k = 2 = N:$$

$$F(x,y) = \partial_x^{-1} \int^x B(\xi, y - \alpha \, \xi) \, d\xi = \int^x d\eta \int^\eta d\xi \, B(\xi, y - \alpha \, \xi),$$

$$\Phi_*(x,y) = F(x, y + \alpha \, x), \tag{3.428a–f}$$

where it is assumed that the forcing function is analytic in y and twice integrable in x. Thus, *the linear partial differential equation (3.429c) with: (i) constant coefficients; (ii) all derivatives of second-order; (iii) double root (3.427d, e) ≡ (3.429c); and (iv) forcing function analytic (3.429a) in y and twice integrable (3.429b) in x:*

$$B(\ldots, y) \in \mathcal{A}\,(\mathbb{R}), \qquad B(x, \ldots) \in E^2\,(\mathbb{R}):$$

$$\frac{\partial^2 \Phi}{\partial x^2} - 2\,\alpha\,\frac{\partial^2 \Phi}{\partial x\,\partial y} + \alpha^2\,\frac{\partial^2 \Phi}{\partial y^2} = B(x, y), \tag{3.429a–c}$$

has particular integral (3.428e, f).

In the case of forcing by the inverse of the independent variables (3.430a) ≡ (3.416a), the first integration in (3.428e) is similar to (3.417b–f) leading to (3.430b–e):

$$B(x,y) = \frac{b}{x\,y}:$$

$$F(x,y) = \int^x d\eta \int^\eta d\xi \, \frac{b}{\xi\,(y - \alpha\,\xi)} = \frac{b}{y} \int^x d\eta \int^\eta \left(\frac{1}{\xi} + \frac{\alpha}{y - \alpha\,\xi} \right) d\xi$$

$$= \frac{b}{y} \int^x [\log \eta - \log(y - \alpha\,\eta)] \, d\eta$$

$$= \frac{b}{y} (x \log x - x) + \frac{b}{\alpha\,y}[(y - \alpha\,x)\log(y - \alpha\,x) - (y - \alpha\,x)]$$

$$= -\frac{b}{\alpha} + \frac{b}{y}\left[x \log x + \left(\frac{y}{\alpha} - x \right)\log(y - \alpha\,x) \right], \tag{3.430a–f}$$

where was used (3.431a, b):

$$\int^x \log \zeta \, d\zeta = x \log x - x \qquad \Leftrightarrow \qquad \frac{d}{dx}(x \log x - x) = \log x. \tag{3.431a, b}$$

Omitting the first term on the r.h.s. of (3.430f) that vanishes under differentiation in (3.427d, e) and substituting the second term in (3.428e) shows that *the linear partial differential equation with (i) constant coefficients; (ii) all derivatives of second-order; (iii) double root; and (iv) forcing by the inverse of the independent variables (3.427d; 3.430a) ≡ (3.432a):*

$$\frac{b}{x\,y} = \frac{\partial^2 \Phi}{\partial x^2} - 2\,\alpha\,\frac{\partial^2 \Phi}{\partial x\,\partial y} + \alpha^2\,\frac{\partial^2 \Phi}{\partial y^2}: \qquad \Phi(x,y) = \frac{b}{y + \alpha\,x}\left(x\log x + \frac{y}{\alpha}\log y\right),$$

$$(3.432a, b)$$

has a particular integral (3.432b). The examples of the solution of linear partial differential equations with constant coefficients by method V of characteristic polynomials (Subsections 3.4.1–3.4.44) are summarised (Tables 3.5 and 3.6) next (Subsection 3.4.45) before indicating (Subsection 3.4.46) other equations and simultaneous systems that also have characteristic polynomials (Section 3.5–3.9).

3.4.45   COMPLETE INTEGRALS BY THE METHOD OF CHARACTERISTIC POLYNOMIALS

The complete integral of a forced linear partial differential equation (3.352a, b) with constant or variable coefficients (Subsection 3.4.29) consists of the sum of: (i) the general integral of the unforced equation involving arbitrary functions; and (ii) any particular integral of the forced equation, the simpler the better. The method of characteristic polynomials applies in the case of constant coefficients to unforced (forced) cases [Subsections 3.4.1–3.4.15 (3.4.16–3.4.44)]; the most numerous examples concern the harmonic, wave and diffusion equations, and the telegraph, bar, and beam equations are also considered (Diagram 3.1).

For each of these, linear partial differential equations or simultaneous systems (Diagram 3.2) can be obtained: (i) in the unforced case, the natural integrals that added together specify the general integral and do not include the special or singular integrals in the case when the latter solutions exist; and (ii) the resonant and non-resonant particular integrals associated with forcing by exponentials and circular and hyperbolic cosines and sines, and their products, and other particular integrals for other types of forcing.

The general integral of the unforced two-dimensional Cartesian harmonic equation (3.212b) is (3.217a–d). Particular integrals were obtained for forcing by: (i) exponentials (3.283a) in non-resonant (3.284a, b), singly resonant [(3.307a, b; 3.290a–f) and (3.307c, d; 3.296a–f)] cases; (ii) the product of the exponential by a circular cosine in non-resonant (3.317a, b; 3.318a–c; 3.319a–d), singly resonant (3.321a–g), and doubly resonant (3.338a–g; 3.360a–d) cases; (iii) by powers (3.374a; 3.375a–d; 3.378a–d); (iv) by the product of powers and exponentials (3.393a–g; 3.394a–d); and (v) by inverses of the independent variables (3.419e, f).

The general integral of the unforced one-dimensional Cartesian (3.222b) [spherical (3.230a, b)] wave equation is (3.228a, b) [(3.232c)]. Particular integrals were obtained for forcing by: (i) exponentials (3.283b) in non-resonant (3.285a, b) and singly resonant time (3.308a; 3.291a–e) and space (3.308b; 3.297a-e) cases; (ii) product of an exponential by an hyperbolic sine in non-resonant (3.322a, b; 3.323a–c; 3.324a–e), singly resonant (3.328a-d), and doubly resonant [(3.336a–e) and (3.337a–d;

3.351a–c)] cases; (iii) by powers (3.374b; 3.376a–d; 3.379a–d); (iv) by the product of power by exponentials by a circular sine (3.395a–h; 3.396a–g); and (v) by the inverses of the independent variables (3.420a–k).

The general integral of the unforced one-dimensional Cartesian diffusion equation (3.235b) is (3.238a; 3.236d) or (3.238b; 3.237b, c). Particular integrals were obtained for forcing (3.243a) by: (i) an integrable function (3.243b–d), including the influence (3.242b) [error (3.249b; 3.248a)] function for forcing by a unit impulse (3.242a) [jump (3.245a–d)]; (ii) an exponential (3.283c) in non-resonant case (3.286a, b) and singly resonant case in time (3.309a; 3.292a–d) or space (3.309b; 3.298a–e); (iii) the product of powers by exponentials by a circular sine and a hyperbolic cosine in non-resonant (3.329a, b; 3.330a–c; 3.331a–d) and resonant (3.333a–e; 3.334a–e; 3.335a–f) cases; (iv) by powers (3.374c; 3.377a–e; 3.380a–d); and (v) by the product of powers by exponentials by a circular sine and a hyperbolic cosine (3.397a, b;3.398a–g; 3.399a–m).

The unforced telegraph equation (3.257b) has general integrals (3.261a; 3.259a, b) or (3.261b; 3.260a–c); the particular integral for forcing by exponentials is (3.283d; 3.287a, b) in the non-resonant case and (3.310a; 3.293a–e) [(3.310b; 3.299a–d)] for single resonance in space (time).

The unforced bar equation (3.264b) has general integrals (3.270a; 3.267a, b) or (3.270b; 3.269a–d); the particular integral with exponential forcing is (3.283e; 3.288a, b) in the non-resonant case and (3.311a, b; 3.294a–e) [(3.311c, d; 3.300a–j)] for single resonance in time (space).

The unforced beam equation (3.271b) has general integrals (3.270a; 3.274a, b) or (3.270b; 3.276a–d; 3.275a–d); the particular integral for exponential forcing is (3.283f; 3.289a, b) in the non-resonant case, (3.312a, b; 3.295a–f) [(3.312c; 3.301a–d)] for single resonance in time (space), and (3.313; 3.305a, b) for double resonance.

Method V of the characteristic polynomial is (Section 3.4.46) compared with the four preceding methods, I–IV, to show that it is the simplest and most efficient and (Table 3.1), thus: (i) developed in more detail (Section 3.4) than the other methods I–IV (Sections 1.6–1.9 and 3.1–3.3); and (ii) adopted in the sequel for other equations (Sections 3.5–3.9).

3.4.46 OTHER EQUATIONS WITH CHARACTERISTIC POLYNOMIALS

The five methods of solution of linear partial differential equations with constant coefficients are indicated in Table 3.2: (I) factorisation of the differential operator (Sections 1.6–1.9); (II) separation of variables in the solution (Section 3.1); (III) solution by similarity or modified similarity functions (Section 3.2); (IV) use of a symbolic operator of partial derivatives (Section 3.3); and (V) exponential solutions or characteristic polynomial of several variables (Section 3.4). These five methods, when possible, are compared in three cases, by applying first (Table 3.3) to the Cartesian (i) two-dimensional harmonic and [ii (iii)] one-dimensional wave (diffusion) equations because: (a) they are the simplest forms of linear partial differential equations of second-order of the respectively elliptic/hyperbolic/parabolic type, illustrating the mathematical properties; and (b) they apply to a substantial number of problems of physical and engineering interest associated with potential fields/propagation/dissipation.

Method V of exponential solutions or characteristic polynomial of several variables is the simplest and is, thus, also applied (Table 3.4) to the telegraph, bar, and beam equations and to a generalised equation of mathematical physics. The simplicity of method V of exponential solutions is that it relies on the existence of a characteristic polynomial that specifies: (i) the general integral of the unforced equation (Table 3.5); and (ii) a particular integral for forcing by smooth functions (Table 3.6). In the case of a linear partial differential equation with constant coefficients and all derivatives of the same order, the characteristic polynomial can be factorised (Table 3.7).

The method of the characteristic polynomial applies (Table 3.8) to linear partial differential equations with: (α) constant coefficients (Section 3.4); and (β) power coefficients leading to homogeneous derivatives (Section 3.6). It extends to simultaneous systems of linear partial differential equations with constant coefficients and ordinary (homogeneous) derivatives [Section 3.5 (3.7)]. An analogue method of characteristic polynomial applies to single (simultaneous) linear finite difference equations with constant coefficients [Sections 3.8 (3.9)]. The method applies to unforced and some forced partial differential equations of any order. For example, it applies to general fourth-order partial differential equation of mathematical physics (notes 3.1–3.12).

TABLE 3.2

Comparison of Five Methods of Solution of Linear Partial Differential Equations

Number	Method	Allows Derivatives of Different Order?	Allows Variable Coefficients?	Level of Complexity of Application
I	Factorisation of differential operator	No	No	Low
II	Separation of variables	Yes	Yes	Medium
III	Similarity functions	Yes	No	Medium
IV	Symbolic operator of derivatives	Yes	Yes	High
V	Exponential solution or characteristic polynomial	Yes	No	Low

Note: Five methods of solution of linear partial differential equations are compared on three criteria: (i) allowing derivatives of different order; (ii) allowing for variable coefficients; (iii) level of complexity.

TABLE 3.3
Solutions of Harmonic, Wave, and Diffusion Equations

Type	Elliptic	Hyperbolic	Parabolic
Partial differential equation	Two-dimensional Cartesian harmonic	One-dimensional Cartesian wave	One-dimensional Cartesian diffusion
Constant coefficients	$\dfrac{\partial^2 \Phi}{\partial x^2} + \dfrac{\partial^2 \Phi}{\partial y^2} = 0$	$\dfrac{\partial^2 \Phi}{\partial t^2} = c^2 \dfrac{\partial^2 \Phi}{\partial x^2}$	$\dfrac{\partial \Phi}{\partial t} = \chi \dfrac{\partial^2 \Phi}{\partial x^2}$
Method of separation of variables	3.1.1	3.1.2–3.1.3	3.1.4–3.1.5
Method of similarity solutions	3.2.1	3.2.2	3.2.3
Method of symbolic differentiation operators	3.3.1	3.3.2–3.3.4	3.3.5–3.3.8
Method of exponential of several variables or characteristic polynomial	3.4.6, 3.4.17–3.4.19, 3.4.21, 3.4.23, 3.4.26, 3.4.28–3.4.30, 3.4.33–3.4.34, 3.4.37–3.4.42	3.4.7, 3.4.17– 3.4.19, 3.4.24, 3.4.26–3.4.30, 3.4.33–3.4.34, 3.4.38, 3.4.42	3.4.8–3.4.12, 3.4.17–3.4.19, 3.4.21, 3.4.25, 3.4.33–3.4.34, 3.4.39

Note: Comparison of the four methods of solution of linear partial differential equations (Table 3.1) applied to the simplest second-order equations of elliptic/hyperbolic/parabolic type.

TABLE 3.4
Partial Differential Equations of Physics and Engineering

Name	Equation	Effects
Harmonic	$\dfrac{\partial^2 \Phi}{\partial x^2} + \dfrac{\partial^2 \Phi}{\partial y^2} = 0$	Potential field
Wave	$\dfrac{\partial^2 \Phi}{\partial t^2} - c^2 \dfrac{\partial^2 \Phi}{\partial x^2} = 0$	Propagation at speed c
Diffusion	$\dfrac{\partial^2 \Phi}{\partial t} = \chi \dfrac{\partial^2 \Phi}{\partial x^2}$	Dissipation with diffusivity χ
Telegraph or wave-diffusion	$\dfrac{\partial^2 \Phi}{\partial x^2} - \dfrac{1}{c^2} \dfrac{\partial^2 \Phi}{\partial y^2} - \dfrac{1}{\chi} \dfrac{\partial \Phi}{\partial t} = 0$	Wave-diffusion or dissipative propagation
Bar	$\dfrac{\partial^2 \Phi}{\partial t^2} + q^2 \dfrac{\partial^4 \Phi}{\partial x^4} = 0$	Bending waves stiffness with parameter q
Beam	$\dfrac{\partial^2 \Phi}{\partial t^2} - c^2 \dfrac{\partial^2 \Phi}{\partial x^2} + q^2 \dfrac{\partial^4 \Phi}{\partial x^4} = 0$	Bending waves with axial traction with speed c and stiffness q

Note: The examples of linear partial differential equations with constant coefficients include, besides the three in Table 3.3, three more, namely, the telegraph, bar, and beam equations in Table 3.4.

TABLE 3.5

Solutions of Partial Differential Equations of Mathematical Physics

			Exponential Forcing: Particular Integral			
				Resonant		
		Unforced: General/	**Non-**		**Single**	
Case	**Equation**	**Special Integral**	**resonant**	**Time**	**Position**	**Double**
1	Harmonic (3.212b)	(3.215b–d)	(3.283a; 3.284a, b)	(3.307a, b; 3.290a–f)	(3.307c, d; 3.296a–f)	–
2	Wave (3.222b)	(3.228a, b)	(3.283b; 3.285a, b)	(3.308a; 3.291a–e)	(3.308b; 3.297a–e)	–
3	Diffusion (3.235b)	(3.238a, b; 3.237b–c; 3.236d)	(3.283c; 3.286a, b)	(3.309a; 3.292a–d)	(3.309b; 3.298a–e)	–
4	Telegraph (3.257b)	(3.261a, b; 3.260a–c; 3.259a, b)	(3.283d; 3.287a, b)	(3.310a; 3.293a–e)	(3.310b; 3.299a–d)	–
5	Bar (3.264b)	(3.270a, b; 3.269a–d; 3.267a, b)	(3.283e; 3.288a, b)	(3.311a, b; 3.294a–e)	(3.311c, d; 3.300a–j)	–
6	Beam (3.271b)	(3.270a, b; 3.274a, b; 3.275a–d; 3.276a–d)	(3.283f; 3.289a, b)	(3.312a, b; 3.295a–f)	(3.312c; 3.301a–d)	(3.313; 3.305a, b)

Note: The six equations in Table 3.4 are considered for: (i) general and special integrals in the unforced case; and (ii) non-resonant and resonant solutions due to exponential forcing.

TABLE 3.6

Forcing by Products of Powers, Exponentials, and Circular and Hyperbolic Cosines and Sines

			Equation		
Case	**Forcing**	**General**	**Harmonic**	**Wave**	**Diffusion**
Exponential (3.279a, b; 3.280a, b; 3.282a, b)	× circular	(3.314a–c)	(3.317a, b; 3.318a–c; 3.319a–d)	–	(3.329a, b; 3.330a–c; 3.331a–d)
	× hyperbolic	(3.315a–c)	–	–	–
	× circular × hyperbolic	(3.316a–c)	–	(3.322a, b; 3.323a–c; 3.324a–c)	–
	Resonant	–	(3.321a–g)	(3.326a–c; 3.327a, b; 3.328a–d)	(3.333a–e; 3.334a–e; 3.335a–f)
SmoothI: (3.365a–c; 3.366a, b; 3.368a, b) II:(3.365a–c; 3.369a–d; 3.371) III:(3.364a–d; 3.372a–c; 3.373a, b)	× exponential	(3.389a, b)	(3.393a–g) ≡ (3.394a–d)	-	–
	× exponential × circular	(3.390a–c)	–	(3.395a–h) ≡ (3.396a–g)	–
	× exponential × hyperbolic	(3.391a–c)	–	–	–
	× exponential × circular × hyperbolic	(3.392a–c)	–	–	(3.397a, b; 3.398a–g) ≡ (3.399a–m)

(Continued)

TABLE 3.6 (*Continued*)

Forcing by Products of Powers, Exponentials, and Circular and Hyperbolic Cosines and Sines

| Case | Forcing | General | Equation | | |
			Harmonic	Wave	Diffusion
Rational	Power	–	(3.374a; 3.375a–d; 3.378a–d; 3.381a–c)	(3.374b; 3.376a–d; 3.379a–d; 3.382a–c)	(3.374c; 3.377a–e; 3.380a–d; 3.383a–c)
	Inverse variable	(3.365a–c; 3.366a; 3.467a–c)	(3.419e–i)	(3.420g–k)	–

Note: The exponential forcing considered in the Table 3.5 is extended to products involving powers, polynomials, circular and hyperbolic cosines and sines and smooth functions in general, and applied to the harmonic, wave, and diffusion equations.

TABLE 3.7

Factorisation and Inverse Characteristic Polynomial of Partial Derivatives

Case	Single Roots	Multiple Roots
General forcing	(3.365a–c; 3.366a, b; 3.367a–d; 3.407a–c)	(3.365a–c; 3.369a–d; 3.426a–e)
Second-order equation	(3.408; 3.410g; 3.414a–d)	(3.429a–c; 3.428d–f)
Forcing by inverse variables	(3.418a–c)	(3.432a, b)

Note: Two methods in Table 3.2, namely (I) factorisation of the differential operator and (V) inverse characteristic polynomial of derivatives can be combined to find particular integrals of a forced linear p.d.e. with constant coefficients and all the derivatives of the same order.

TABLE 3.8

Method V of Characteristic Polynomials

Equation	Single	Simultaneous System
A	Section 3.1–3.4	Section 3.6
B	Section 3.5	Section 3.7
C	Section 3.8	Section 3.9

Note: The method of characteristic polynomials applies to six classes, consisting of three pairs of single (simultaneous systems) of linear equations: (i/ii) partial differential with constant (homogeneous power) coefficients; and (iii) multiple finite differences with constant coefficients. There are analogies in the method of characteristic polynomials applied to the six classes of equations in Table 3.8, for example the similarity (analogous) solutions of an unforced linear partial differential equation with all derivatives of the same order and constant (homogeneous power) coefficients.
A(B) = linear partial differential equation with constant coefficients (homogeneous derivatives); C = linear multiple finite difference equation with constant coefficients.

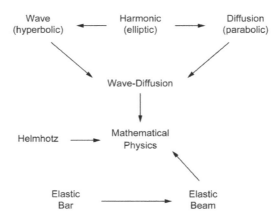

*Partial Differential Equations
of Physics and Engineering*

DIAGRAM 3.1   The six examples of linear partial differential equations in Tables 3.3 and 3.4, plus the Helmholtz equation, are combined as particular cases of the one-dimensional equations of mathematical physics in Table 3.14.

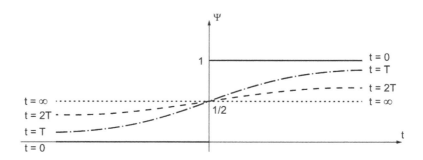

FIGURE 3.3   In the case of an initial finite jump at zero time, the diffusion equation leads to a more gradual evolution as time increases and a monotonic decay to the uniform average at infinite time.

Integrals of Partial Differential Equations

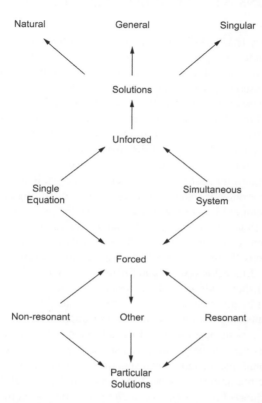

DIAGRAM 3.2 Solutions of linear partial differential equations and finite difference equations and simultaneous systems that can be obtained by the method of characteristic polynomials, separating unforced (forced) cases relating to natural, general, and special (non-resonant, resonant, and other particular) integrals.

3.5 SIMULTANEOUS SYSTEMS OF PARTIAL DIFFERENTIAL EQUATIONS WITH CONSTANT COEFFICIENTS

The linear ordinary (partial) differential equation with constant coefficients corresponds to a characteristic polynomial in one (several) variables [Sections IV.1.3–IV.1.5 (V.3.4)], and a simultaneous system of linear ordinary (partial) differential equations with constant coefficients involves a matrix of polynomials of derivatives with regard to one (several) variables [Sections IV.7.4–IV.7.5 (V.3.5)]. The characteristic polynomial of the simultaneous system is the determinant of the matrix of polynomials of derivatives, and its degree in each variable determines the order of the system (Subsection 3.5.1). The roots of the characteristic polynomial specify the natural integrals of the unforced linear simultaneous system of partial differential equations (p.d.e.) with constant coefficients; the general integral is specified by the sum of natural integrals for one dependent variable, and the remaining dependent variables are specified by compatibility of the system. Two examples are the Cauchy-Riemann (alternate) system [Subsection 3.5.2 (3.5.3)] that is equivalent to the two-dimensional Cartesian harmonic equation, which lead to distinct solutions even for the same boundary conditions [Subsection 3.5.4 (3.5.5)]. A third example is the coupling of the Cartesian one-dimensional wave and diffusion equations through gradient terms (Subsection 3.5.6) that leads to a characteristic polynomial with single or double roots. The forcing by an exponential of a simultaneous linear system of p.d.e. with constant coefficients leads to non-resonant and resonant solutions (Subsection 3.5.7). In the case of the exponential forcing of the alternate harmonic system (Subsection 3.5.3), whose characteristic polynomial has single roots, there are non-resonant and singly resonant solutions, the latter corresponding to the forcing by the product of an exponential by a circular cosine (Subsection 3.5.8). In the case of the wave-diffusion coupled system (Subsection 3.5.4), whose characteristic polynomial has four single (or two double) roots, there are non-resonant solutions, and four (two) singly (doubly) resonant solutions (Subsection 3.5.9).

3.5.1 SIMULTANEOUS LINEAR SYSTEMS WITH CONSTANT COEFFICIENTS

A system of simultaneous linear p.d.e. with constant coefficients is (3.433b):

$$i, j = 1,\ldots,L: \qquad \sum_{j=1}^{L} \left\{ P_{ij}\left(\frac{\partial}{\partial x}, \frac{\partial}{\partial y} \right) \right\} \Phi_j(x,y) = B_i(x,y), \qquad \text{(3.433a, b)}$$

where there are: (i) two independent variables x, y ; (ii) L dependent (3.433a) variables Φ_j ; (iii) to which are applied L^2 polynomials of partial derivatives with regard to x and y, each, respectively, with arbitrary order N_{ij} and M_{ij}:

$$P_{ij}\left(\frac{\partial}{\partial x}, \frac{\partial}{\partial y} \right) = \sum_{n=0}^{N_{ij}} \sum_{m=0}^{M_{ij}} A_{n,m}^{i,j} \frac{\partial^{n+m}}{\partial x^n \, \partial y^m}; \qquad \text{(3.434)}$$

and (iv) the L forcing functions B_i of the two independent variables x, y. The system (3.433a, b) may be written in matrix form (3.435):

$$
\begin{bmatrix}
P_{11}\left(\dfrac{\partial}{\partial x}, \dfrac{\partial}{\partial y}\right) & \cdots & P_{1L}\left(\dfrac{\partial}{\partial x}, \dfrac{\partial}{\partial y}\right) \\
\vdots & \ddots & \vdots \\
P_{L1}\left(\dfrac{\partial}{\partial x}, \dfrac{\partial}{\partial y}\right) & \cdots & P_{LL}\left(\dfrac{\partial}{\partial x}, \dfrac{\partial}{\partial y}\right)
\end{bmatrix}
\begin{bmatrix}
\Phi_1(x,y) \\
\vdots \\
\Phi_L(x,y)
\end{bmatrix}
=
\begin{bmatrix}
B_1(x,y) \\
\vdots \\
B_L(x,y)
\end{bmatrix}.
\tag{3.435}
$$

In the case of an unforced system (3.436a), a solution may be sought in the form of a common exponential with a distinct coefficient for each dependent variable:

$$
B_i(x,y) = 0: \qquad\qquad \Phi_j(x,y) = C_j \exp(a\,x + b\,y); \tag{3.436a, b}
$$

substitution of (3.436b) in (3.433a, b; 3.436a) and use of (3.198a–d) leads to (3.437a, b):

$$
0 = e^{ax+by} \sum_{j=1}^{L} P_{ij}(a,b)\, C_j = \sum_{j=1}^{L} P_{ij}(a,b)\, \Phi_j(x,y); \tag{3.437a, b}
$$

the linear unforced system (3.437b) [(3.437a)] has non-trivial solution (3.438a) [(3.438b)], with at least one $C_j \neq 0 \left(\Phi_j \neq 0\right)$ iff the determinant of coefficients is zero (3.438c):

$$
\{C_1, \ldots, C_L\} \neq \{0, \ldots, 0\} \neq \{\Phi_1, \ldots, \Phi_L\}: \qquad 0 = Det\left\{P_{ij}(a,b)\right\} \equiv P_{N,M}(a,b),
\tag{3.438a–d}
$$

specifying the characteristic polynomial (3.438d).

*The **characteristic polynomial** of the simultaneous system of linear p.d.e. with constant coefficients (3.433a, b) is (3.438d) the determinant of the matrix (3.435) of polynomials of partial derivatives (3.434) and its degree in $\partial/\partial x$ $(\partial/\partial y)$ is the order N (M) of the system in the independent variable x (y). The single (3.203a–c) [multiple (3.206a–f)] roots of the characteristic polynomial (3.438b, c) specify the* **natural integrals** *(3.204a, b) [(3.207; 3.208)]. All dependent variables are linear combinations of natural integrals (3.204a) [(3.204b)] in the case of single roots (3.439c) [(3.439d)]:*

$$
E_{1,n} = 1 = E_1^m: \qquad \Phi_j(x,y) = \sum_{n=1}^{N} E_{j,n}\, \Phi_n(x,y) = \sum_{m=1}^{M} E_j^m\, \Phi^m(x,y). \tag{3.439a–d}
$$

Since the first dependent variable $\Phi_1(x,y)$ already involves N (M) arbitrary functions, its coefficients can be set to unity (3.439a) [(3.439b)]. For all other dependent variables $\Phi_2, \ldots, \Phi_L$, the coefficients $E_{j,n}\left(E_n^j\right)$ with $j = 2, \ldots, L$ of the natural integrals are determined from compatibility of the system (3.437a, b). In the case of *multiple roots (3.206a–c) [(3.206d–f)] of the characteristic polynomial (3.438b, c), the natural integrals (3.207) [(3.208)] are (3.440a, b) [(3.441a, b)]*

$$p_r = 1,\ldots,\alpha_r: \qquad \Phi_{r,p_r}(x,y) = \int e^{by}\, x^{p_r-1} \exp\left[x\, a_r(b)\right] db, \qquad (3.440a, b)$$

$$q_s = 1,\ldots,\beta_s: \qquad \Phi^{s,q_s}(x,y) = \int e^{ax}\, y^{q_s-1} \exp\left[y\, b_s(a)\right] da; \qquad (3.441a, b)$$

the general integral of the unforced system (3.433a, b; 3.436a) is (3.442b) [(3.443b)]:

$$E_{1,r,p_r} = 1: \qquad \Phi(x,y) = \sum_{j=1}^{L} \sum_{r=1}^{R} \sum_{p_r=1}^{\alpha_r} E_{j,r,p_r}\, \Phi_{r,p_r}(x,y), \qquad (3.442a, b)$$

$$E_{s,q_s}^1 = 1: \qquad \Phi(x,y) = \sum_{j=1}^{L} \sum_{s=1}^{S} \sum_{q_s=1}^{\beta_s} E_{s,q_s}^{j}\, \Phi_{s,q_s}(x,y), \qquad (3.443a, b)$$

where: (i) the coefficients of the first dependent variable may be set to unity (3.432a) [(3.433a)] for j = 1; and (ii) all other coefficients j = 2,…,L are determined by compatibility of the unforced system (3.437a, b). In the case of single roots $\alpha_r = 1\,(\beta_r = 1)$, *there are no powers in (3.440a, b) [(3.441a, b)], and (3.442a, b) [(3.443a, b)] simplify to (3.439a, c) [(3.439b, d)]. The general integral is obtained next [Subsections 3.5.2–3.5.3 (3.5.4)] for a linear unforced simultaneous system of p.d.e. with constant coefficients equivalent to the harmonic equation (coupling the wave and diffusion equations through gradients).*

3.5.2   Conditions for a Complex Holomorphic Function (Cauchy 1821, Riemann 1851)

A complex function of a complex variable that is holomorphic (Chapter I.11) leads to the Cauchy-Riemann conditions and to the two-dimensional Cartesian harmonic equation as shown next. A complex (3.444b) function (3.444a) of a complex variable (3.444c):

$$w = f(z): \qquad w = \Phi + i\,\Psi, \qquad z = x + i\,y, \qquad (3.444a\text{–}c)$$

corresponds to two real functions of two real variables (3.445):

$$f(x + i\,y) = \Phi(x,y) + i\,\Psi(x,y). \qquad (3.445)$$

   The complex function is **holomorphic** (3.446a) iff (Section I.11.3) it has a unique derivative (3.446b) independent of direction; thus, the derivative can in particular be calculated (3.446c–e) [(3.446f–h)] in the x (y)-direction:

$$f \in \mathcal{D}\,(|C); \qquad f'(z) \equiv \frac{df}{dz}: \qquad (3.446a, b)$$

$$f'(z) = \frac{\partial f}{\partial x} = \frac{\partial(\Phi + i\,\Psi)}{\partial x} = \frac{\partial \Phi}{\partial x} + i\,\frac{\partial \Psi}{\partial x} \tag{3.446c-e}$$

$$f'(z) = \frac{\partial f}{\partial(i\,y)} = -i\,\frac{\partial(\Phi + i\,\Psi)}{\partial y} = \frac{\partial \Psi}{\partial y} - i\,\frac{\partial \Phi}{\partial y}. \tag{3.446f-h}$$

The equality of the real and imaginary parts of the derivatives (3.446e) ≡ (3.446h) leads to *the **Cauchy-Riemann conditions** (3.447b, c) that are necessary for the function (3.445) to be holomorphic (3.446a, b) ≡ (3.447a):*

$$\Phi + i\,\Psi \in \mathcal{D}\,(\mathbb{C}): \qquad \frac{\partial \Phi}{\partial x} = \frac{\partial \Psi}{\partial y}, \qquad \frac{\partial \Phi}{\partial y} = -\frac{\partial \Psi}{\partial x}. \tag{3.447a-c}$$

If the real and imaginary parts of the complex function have continuous second-order derivatives with regard to the real variables (3.448a), then the equality of cross-derivatives (3.448c) [(3.448e)] implies (3.448b) [(3.448f)] that they satisfy the two-dimensional harmonic equation (3.448d) [(3.448g)]:

$$\Phi,\,\Psi \in C^{2}\,(\mathbb{R}^{2}): \qquad \frac{\partial^2 \Phi}{\partial x^2} = \frac{\partial^2 \Psi}{\partial x\,\partial y} = \frac{\partial^2 \Psi}{\partial y\,\partial x} = -\frac{\partial^2 \Phi}{\partial y^2}, \tag{3.448a-d}$$

$$\frac{\partial^2 \Psi}{\partial x^2} = -\frac{\partial^2 \Phi}{\partial x\,\partial y} = -\frac{\partial^2 \Phi}{\partial y\,\partial x} = -\frac{\partial^2 \Psi}{\partial y^2}. \tag{3.448e-g}$$

It follows (3.217a) ≡ (3.449a) that the real part is the sum of two arbitrary functions of the complex variable (3.444c) and its conjugate:

$$\Phi(x,y) = f(x + i\,y) + g(x - i\,y); \tag{3.449a}$$

$$\frac{\partial \Psi}{\partial x} = -\frac{\partial \Phi}{\partial y} = -i\,f'(x + i\,y) + i\,g'(x - i\,y). \tag{3.449b, c}$$

Substitution of (3.449a) in the second Cauchy-Riemann condition (3.447c) ≡ (3.449b) leads to (3.449c), where the prime denotes a derivative with regard to the argument. Integrating (3.449c) leads to (3.450):

$$\Psi(x,y) = -i\,f(x + i\,y) + i\,g(x - i\,y), \tag{3.450}$$

showing that (3.449a; 3.450) are solutions of the Cauchy-Riemann conditions (3.447a-c). It can be checked that the first Cauchy-Riemann condition (3.447b) is also satisfied by (3.449a; 3.450) because substituting (3.450) in (3.447b) leads to (3.451a-c):

$$\frac{\partial \Psi}{\partial y} = -i^2\,f'(x + i\,y) - i^2\,g(x - i\,y) = f'(x + i\,y) + g'(x - i\,y) = \frac{\partial \Phi}{\partial x}, \tag{3.451a-c}$$

whose primitive is (3.449a). Thus, *the Cauchy-Riemann conditions (3.447a–c) are a pair of coupled linear first-order unforced p.d.e. with constant coefficients whose general integral is (3.449a; 3.450); both functions satisfy the two-dimensional Cartesian harmonic equation (3.448d, g), and involve the same functions $f\,(g)$ of the complex (3.216a) $\equiv$ (3.444c) [conjugate (3.216b)] variable with distinct coefficients to ensure compatibility with the Cauchy-Riemann conditions (3.447b, c).* The same equation may be obtained by elimination of distinct simultaneous systems; as an example, a simultaneous system distinct from the Cauchy-Riemann system that also leads to the harmonic equation (Subsection 3.5.3) is considered next.

3.5.3 Two Simultaneous Partial Differential Equations Reducible to the Harmonic Equation

Consider the simultaneous system of two linear unforced coupled p.d.e. with constant coefficients:

$$\frac{\partial \Phi}{\partial x} + \frac{\partial \Phi}{\partial y} + \frac{\partial \Psi}{\partial x} = 0 = 2\frac{\partial \Phi}{\partial y} + \frac{\partial \Psi}{\partial x} + \frac{\partial \Psi}{\partial y}. \qquad (3.452\text{a, b})$$

The simultaneous system (3.452a, b) can be written in matrix form (3.453):

$$\begin{bmatrix} \dfrac{\partial}{\partial x} + \dfrac{\partial}{\partial y} & \dfrac{\partial}{\partial x} \\[2ex] 2\dfrac{\partial}{\partial y} & \dfrac{\partial}{\partial x} + \dfrac{\partial}{\partial y} \end{bmatrix} \begin{bmatrix} \Phi \\ \Psi \end{bmatrix} = 0. \qquad (3.453)$$

The characteristic polynomial is the determinant (3.454a) of the matrix in (3.453):

$$P_{2,2}\left(\frac{\partial}{\partial x},\ \frac{\partial}{\partial y}\right) = \left(\frac{\partial}{\partial x} + \frac{\partial}{\partial y}\right)^2 - 2\frac{\partial^2}{\partial x\,\partial y} = \frac{\partial^2}{\partial x^2} + \frac{\partial^2}{\partial y^2}, \qquad (3.454\text{a, b})$$

leading to the two-dimensional Cartesian harmonic equation (3.454b). This can be confirmed by starting with the equation (3.452a) [(3.452b)] and substituting the other (3.452b) [(3.452a)], leading to (3.455a–d) [(3.456a–d)]:

$$2\frac{\partial^2 \Psi}{\partial y\,\partial x} = -2\frac{\partial^2 \Phi}{\partial x\,\partial y} - 2\frac{\partial^2 \Phi}{\partial y^2} = -2\left(\frac{\partial}{\partial x} + \frac{\partial}{\partial y}\right)\frac{\partial \Phi}{\partial y}$$

$$= \left(\frac{\partial}{\partial x} + \frac{\partial}{\partial y}\right)\left(\frac{\partial \Psi}{\partial x} + \frac{\partial \Psi}{\partial y}\right) = \frac{\partial^2 \Psi}{\partial x^2} + \frac{\partial^2 \Psi}{\partial y^2} + 2\frac{\partial^2 \Psi}{\partial y\,\partial x}, \qquad (3.455\text{a–d})$$

$$2\frac{\partial^2 \Phi}{\partial x\,\partial y} = -\frac{\partial^2 \Psi}{\partial x^2} - \frac{\partial^2 \Psi}{\partial x\,\partial y} = -\left(\frac{\partial}{\partial x} + \frac{\partial}{\partial y}\right)\frac{\partial \Psi}{\partial x}$$

$$= \left(\frac{\partial}{\partial x} + \frac{\partial}{\partial y}\right)\left(\frac{\partial \Phi}{\partial x} + \frac{\partial \Phi}{\partial y}\right) = \frac{\partial^2 \Phi}{\partial x^2} + \frac{\partial^2 \Phi}{\partial y^2} + 2\frac{\partial^2 \Phi}{\partial x\,\partial y}. \qquad (3.456\text{a–d})$$

Both equations (3.455d) [(3.456d)] are equivalent to the two-dimensional Cartesian harmonic equation (3.457c, d) [(3.457a, b)]:

$$\frac{\partial^2 \Phi}{\partial x^2} + \frac{\partial^2 \Phi}{\partial y^2} = \nabla^2\Phi = 0 = \nabla^2\Psi = \frac{\partial^2 \Psi}{\partial x^2} + \frac{\partial^2 \Psi}{\partial y^2}, \qquad (3.457\text{a–d})$$

in agreement with (3.448d) [(3.448g)].

It follows that the general integral for one of the dependent variables is (3.449a) and the other dependent variable satisfies (3.452a) ≡ (3.458a) leading to (3.458b):

$$\frac{\partial \Psi}{\partial x} = -\frac{\partial \Phi}{\partial x} - \frac{\partial \Phi}{\partial y} = -(1+i)f'(x+i\ y) - (1-i)\ g'(x-i\ y), \qquad (3.458\text{a, b})$$

where the prime denotes a derivative with regard to the argument. Integrating (3.458b) leads to (3.458c):

$$\Psi(x,y) = -(i+1)\ f(x+i\ y) + (i-1)\ g(x-i\ y). \qquad (3.458\text{c})$$

It can be checked that the general integral (3.449a; 3.458c) also satisfies the second equation (3.452b) ≡ (3.459a), since substitution of (3.458c) leads to (3.459b) ≡ (3.459c):

$$2\frac{\partial \Phi}{\partial y} = -\frac{\partial \Psi}{\partial x} - \frac{\partial \Psi}{\partial y}$$
$$= \left[1 + i + i\ (1+i)\right] f'(x+i\ y) + \left[1 - i - i\ (1-i)\right] g'(x-i\ y)$$
$$= 2\ i\ f'(x+i\ y) - 2\ i\ g'(x-i\ y), \qquad (3.459\text{a–c})$$

whose primitive is (3.449a). Thus, *the Cauchy-Rieman-coupled (3.447a–c) [alternate (3.452a, b)] linear unforced system of first-order p.d.e. with constant coefficients: (i) is of the second order in both independent variables because it can be eliminated for the two-dimensional Cartesian harmonic equation (3.448a–g) [(3.454a, b)]; and (ii) has general integral (3.449a; 3.450) [(3.449a; 3.458c)] that coincides (3.449a) for the first dependent variable Φ and differs (3.450) [≠(3.459)] for the second dependent variable Ψ because the compatibility conditions (3.447b, c) [(3.452a, b)] are different.*

3.5.4 Unicity of Solutions and Boundary Conditions

The general integral (3.449a; 3.450) of the Cauchy-Riemann system (3.447a–c) in the whole complex plane becomes unique, specifying the functions f, g through boundary conditions on one axis, for example, the x-axis in (3.460a, b):

$$\Phi(\xi,0) = f(\xi) + g(\xi), \qquad \Psi(\xi,0) = -i\ f(\xi) + i\ g(\xi). \qquad (3.460\text{a, b})$$

Solving (3.460a, b) for f, g yields (3.461a, b):

$$2\left\{f(\xi),\ g(\xi)\right\} = \left\{\Phi(\xi,0) + i\ \Psi(\xi,0),\ \Phi(\xi,0) - i\ \Psi(\xi,0)\right\}, \qquad (3.461\text{a, b})$$

which imply (3.462a, b):

$$2 f(x+i\ y) = \Phi(x+i\ y,\ 0) + i\ \Psi(x+i\ y,\ 0), \tag{3.462a}$$

$$2 g(x-i\ y) = \Phi(x-i\ y,\ 0) - i\ \Psi(x-i\ y,\ 0). \tag{3.462b}$$

Substituting (3.462a, b) in (3.449a; 3.450) specifies the general integral:

$$2\ \Phi(x,y) = \Phi(x+i\ y,\ 0) + \Phi(x-i\ y,\ 0) + i\left[\Psi(x+i\ y,\ 0) - \Psi(x-i\ y,\ 0)\right], \tag{3.463a}$$

$$2\ \Psi(x,y) = \Psi(x+i\ y,\ 0) + \Psi(x-i\ y,\ 0) + i\left[\Phi(x-i\ y,\ 0) - \Phi(x+i\ y,\ 0)\right]. \tag{3.463b}$$

It can be checked that (3.463a, b) reduce to identities for $y = 0$ and, thus, satisfy the boundary conditions (3.460a, b).

In particular, the boundary conditions (3.464a, b)

$$\Phi(x,0) = e^x, \qquad\qquad \Psi(x,0) = e^{-x}, \tag{3.464a, b}$$

substituted in (3.463a, b) lead to the general integrals (3.465a–d):

$$\Phi(x,y) = \frac{1}{2}\ e^x\ (e^{iy} + e^{-iy}) + \frac{i}{2}\ e^{-x}\ (e^{-iy} - e^{iy}) = e^x \cos y + e^{-x} \sin y, \quad (3.465a, b)$$

$$\Psi(x,y) = \frac{1}{2}\ e^{-x}\ (e^{-iy} + e^{iy}) + \frac{i}{2}\ e^x\ (e^{-iy} - e^{iy}) = e^{-x} \cos y + e^x \sin y. \quad (3.465c, d)$$

It can be checked that the functions (3.465b, d) satisfy: (i) the boundary conditions (3.464a, b); (ii) the harmonic equation (3.448d, g); and (iii) the Cauchy-Riemann conditions (3.447a, b). The same boundary conditions applied to the Cauchy-Riemann (alternate harmonic) system [Subsection 3.5.2 (3.5.3)] lead to different results [Subsection 3.5.3 (3.5.4)].

3.5.5   COMPARISON OF CAUCHY-RIEMANN AND ALTERNATE HARMONIC SYSTEM

The same boundary conditions (3.460a, b) on the x-axis applied to the general integral (3.449a; 3.458c) of the alternate harmonic system (3.452a, b) lead to (3.466a,b):

$$\Phi(\xi,0) = f(\xi) + g(\xi), \qquad \Psi(\xi,0) = -(i+1)\ f(\xi) + (i-1)\ g(\xi), \quad (3.466a, b)$$

so that compared with (3.460a, b), the first equation coincides (3.460a) ≡ (3.466a) but not the second (3.460b) ≠ (3.466b). Thus, solving (3.466a, b) for f, g gives a result (3.467a, b) distinct from (3.461a, b):

$$2\left\{f(\xi),\ g(\xi)\right\} = \left\{(1+i)\ \Phi(\xi,0) + i\ \Psi(\xi,0)\ ,\ (1-i)\ \Phi(\xi,0) - i\ \Psi(\xi,0)\right\}. \tag{3.467a, b}$$

From (3.467a, b) follows (3.468a, b):

$$2\,f(x+i\,y)=(1+i)\,\Phi(x+i\,y\,,\,0)+i\,\Psi(x+i\,y\,,\,0), \qquad (3.468a)$$

$$2\,g(x-i\,y)=(1-i)\,\Phi(x-i\,y\,,\,0)-i\,\Psi(x-i\,y\,,\,0), \qquad (3.468b)$$

and substitution in (3.449a; 3.458c) yields (3.469a, b)

$$2\,\Phi(x,y)=\Phi(x+i\,y,\,0)+\Phi(x-i\,y,\,0)$$
$$+\,i\left[\Phi(x+i\,y,\,0)+\Psi(x+i\,y,\,0)-\Phi(x-i\,y,\,0)-\Psi(x-i\,y,\,0)\right].$$
$$(3.469a)$$

$$2\,\Psi(x,y)=\Psi(x+i\,y,\,0)+\Psi(x-i\,y,\,0)$$
$$+\,i\left[\Psi(x-i\,y,\,0)-\Psi(x+i\,y,\,0)+2\,\Phi(x-i\,y,0)-2\,\Phi(x+i\,y,\,0)\right].$$
$$(3.469b)$$

The general integral (3.469a, b) of the alternate harmonic system (3.452a, b) reduces to an identity for $y=0$ and, thus, satisfies the boundary conditions (3.460a, b).

The same particular boundary conditions (3.464a, b) applied to the general integral (3.469a, b) of the alternate harmonic system (3.452a, b) lead to (3.470a–f):

$$\Phi(x,y)=\frac{e^x}{2}\left(e^{iy}+e^{-iy}\right)+\frac{i}{2}\,e^x\left(e^{iy}-e^{-iy}\right)+\frac{i}{2}\,e^{-x}\left(e^{-iy}-e^{iy}\right)$$
$$=e^x\cos y-e^x\sin y+e^{-x}\sin y=e^x\cos y-2\sinh x\sin y, \qquad (3.470a\text{–}c)$$

$$\Psi(x,y)=\frac{e^{-x}}{2}\left(e^{-iy}+e^{iy}\right)+\frac{i}{2}\,e^{-x}\left(e^{iy}-e^{-iy}\right)+i\,e^x\left(e^{-iy}-e^{iy}\right)$$
$$=e^{-x}\cos y-e^{-x}\sin y+2\,e^x\sin y=e^{-x}\cos y+e^x\sin y+2\sinh x\sin y.$$
$$(3.470d\text{–}f)$$

It can be checked that the general integral (3.470c, e) satisfies: (i) the boundary conditions (3.464a, b); (ii) the harmonic equation (3.457a, d); and (iii) the coupled system (3.452a, b). Thus, *the Cauchy-Riemann (3.447a–c) [alternate harmonic (3.452a, b)] coupled linear unforced first-order p.d.e. with constant coefficients, with boundary conditions specifying* $\left[\Phi(x,0),\Psi(x,0)\right]$ *on the x-axis, has general integral (3.463a, b) [(3.469a, b)]; the particular choice of boundary conditions (3.460a, b) leads to the general integral (3.465b, d) [(3.470c, f)].* Next is considered a higher order system coupling the wave and diffusion equations through gradients (Subsection 3.5.6).

3.5.6 COUPLING OF THE WAVE AND DIFFUSION EQUATIONS THROUGH GRADIENTS

The coupling of the one-dimensional wave (3.222b) [diffusion (3.235b)] equations through gradients (3.471a, b):

$$\frac{\partial^2 \Phi}{\partial t^2} - c^2 \frac{\partial^2 \Phi}{\partial x^2} - \mu \frac{\partial \Psi}{\partial x} = 0 = \frac{\partial \Psi}{\partial t} - \chi \frac{\partial^2 \Psi}{\partial x^2} - v \frac{\partial \Phi}{\partial x}, \qquad (3.471a, b)$$

leads to a simultaneous linear unforced system of p.d.e. with constant coefficients that can be put in matrix form (3.471a, b) = (3.472):

$$\begin{bmatrix} \dfrac{\partial^2}{\partial t^2} - c^2 \dfrac{\partial^2}{\partial x^2} & - \mu \dfrac{\partial}{\partial x} \\[2ex] - v \dfrac{\partial}{\partial x} & \dfrac{\partial}{\partial t} - \chi \dfrac{\partial^2}{\partial x^2} \end{bmatrix} \begin{bmatrix} \Phi \\ \Psi \end{bmatrix} = 0 . \qquad (3.472)$$

The determinant (3.473c, d) of the matrix in (3.472) shows that the system (3.471a, b) is (3.473a, b) of the third (fourth) order in time (position):

$$0 = \left\{ P_{3,4}\left(\frac{\partial}{\partial t}, \frac{\partial}{\partial x} \right) \right\} \Phi , \Psi: \qquad (3.473a, b)$$

$$P_{3,4}\left(\frac{\partial}{\partial t}, \frac{\partial}{\partial x} \right) \equiv \left(\frac{\partial^2}{\partial t^2} - c^2 \frac{\partial^2}{\partial x^2} \right)\left(\frac{\partial}{\partial t} - \chi \frac{\partial^2}{\partial x^2} \right) - \mu \, v \, \frac{\partial^2}{\partial x^2} \qquad (3.473c)$$

$$= \frac{\partial^3}{\partial t^3} - \chi \frac{\partial^4}{\partial t^2 \, \partial x^2} - c^2 \frac{\partial^3}{\partial t \, \partial x^2} + c^2 \, \chi \, \frac{\partial^4}{\partial x^4} - \mu \, v \, \frac{\partial^2}{\partial x^2} \qquad (3.473d)$$

and both dependent variables (3.473a, b) satisfy (3.473c, d). A unique solution requires three initial (four boundary) conditions.

The parameters in (3.471a, b) or (3.473c, d) have the following dimensions: (i/ii) length (length square) divided by time for the wave speed (3.474a) [diffusivity (3.474b)]; and (iii/iv) length divided by time (time squared) for the second (3.474d) [first (3.474c)] coupling parameter:

$$\{c, \chi, \mu, v, \mu\, v\} = \{L\, T^{-1}, L^2\, T^{-1}, L\, T^{-2}, L\, T^{-1}, L^2\, T^{-3}\}; \qquad (3.474a\text{--}e)$$

the coupling parameters appear together only as a product (3.474e) with the dimensions of length squared divided by time cubed. The characteristic polynomial (3.473d) ≡ (3.475a) is bi-quadratic (3.475b) in position:

$$P_{3,4}(a,b) = c^2 \, \chi \, b^4 - \left(\mu \, v + c^2 \, a + \chi \, a^2 \right) b^2 + a^3$$

$$= c^2 \, \chi \, [b - b_+(a)] \, [b + b_+(a)] \, [b - b_-(a)] \, [b + b_-(a)], \qquad (3.475a, b)$$

with roots (3.476a) involving (3.476b):

$$2 c^2 \chi \left[b_\pm(a) \right]^2 = \mu v + c^2 a + \chi a^2 \pm \lambda, \tag{3.476a}$$

$$\lambda(a) \equiv \sqrt{\left(\mu v + c^2 a + \chi a^2 \right)^2 - 4 c^2 \chi a^3}. \tag{3.476b}$$

In the case of two distinct roots (3.477a), the four natural integrals for the first dependent variable are (3.477b):

$$\lambda \neq 0: \qquad \Phi_{\pm\pm}(x,t) = \int e^{at} \exp\left[\pm x\, b_\pm(a) \right] C_{\pm\pm}(a)\, da; \tag{3.477a, b}$$

in the case of double roots (3.478a) the corresponding values (3.478b) lead to the eight solutions (3.478c):

$$\lambda(a_{1-4}) = 0: \qquad \left[b(a_{1-4}) \right]^2 = \frac{\mu v + c^2 a_{1-4} + \chi (a_{1-4})^2}{2 c^2 \chi}, \tag{3.478a, b}$$

$$\Phi_{1-4}^\pm(x,t) = C_{1-4}^\pm \exp\left[a_{1-4}\, t \pm b(a_{1-4})\, x \right], \tag{3.478c}$$

since (3.476b) has four roots a_{1-4} in (3.478a). The second dependent variable follows from (3.471a) in the case of: (i) single roots (3.477a,b) leading to four solutions (3.479):

$$\Psi_{\pm\pm}(x,t) = \pm \frac{1}{\mu} \int \frac{e^{at}}{b_\pm(a)} \frac{\exp\left[\pm x\, b_\pm(a) \right]}{a^2 - c^2 \left[b_\pm(a) \right]^2} C_{\pm\pm}(a)\, da; \tag{3.479}$$

and (ii) double roots (3.478a–c) leading to eight solutions (3.480):

$$\Psi_{1-4}^\pm(x,t) = \pm \frac{C_{1-4}^\pm}{\mu\, b(a_{1-4})} \frac{\exp\left[a_{1-4}\, t \pm b(a_{1-4})\, x \right]}{a_{1-4}^2 - c^2 \left[b(a_{1-4}) \right]^2}. \tag{3.480}$$

In (3.477a, b; 3.479) [(3.478a–c; 3.480)], the parameter a is arbitrary (has a set of four fixed values) and since the simultaneous system (3.471a, b) is linear, the principle of superposition holds, and it is permissible to multiply by an arbitrary function of a followed by an integration in da (permissible only to multiply by an arbitrary constant). Thus, the sum (3.481a–c) [set (3.482a–d)] of (3.477a, b; 3.479) [(3.478a–c; 3.480)] specifies the general (eight special) integral(s), involving four arbitrary functions (one arbitrary constant each):

$$\lambda \neq 0: \qquad \{\Phi, \Psi(x,t)\} = \{\Phi_{++} + \Phi_{+-} + \Phi_{-+} + \Phi_{--}\,,\ \Psi_{++} + \Psi_{+-} + \Psi_{-+} + \Psi_{--}\}, \tag{3.481a–c}$$

$$\lambda = 0; \qquad \ell = 1,\ldots,4: \qquad \{\Phi, \Psi(x,t)\} = \{\Phi_\ell^+\,,\ \Phi_\ell^-\,,\ \Psi_\ell^+\,,\ \Psi_\ell^-\} \tag{3.482a–d}$$

corresponding to single (3.476a, b) [double (3.478a, b)] roots of the characteristic polynomial (3.475a, b) of the simultaneous system (3.471a, b) of wave and diffusion equations coupled through gradients. The characteristic polynomial can also be used to obtain a particular integral due to forcing by exponentials in non-resonant and resonant cases, both in general (Subsection 3.5.7) and in particular for the [Subsection 3.5.8 (3.5.9)] alternate harmonic system (coupled wave-diffusion system).

3.5.7 NON-RESONANT AND RESONANT FORCING BY EXPONENTIALS

Consider the system of simultaneous linear forced p.d.e. with constant coefficients (3.433a, b). A particular integral can be obtained most simply when the forcing terms are exponentials (3.483a):

$$B_j(x,y) = B_j \exp(a\ x + b\ y): \qquad \Phi_j(x,y) = C_j \exp(a\ x + b\ y), \qquad \text{(3.483a, b)}$$

then the particular integrals are also exponentials (3.436b) $\equiv$ (3.483b). Substituting (3.483a, b) into (3.433a, b) leads to (3.484):

$$0 = \exp(a\ x + b\ y)\left\{ \sum_{j=1}^{M} P_{ij}(a,b)\, C_j - B_j \right\} = 0; \qquad \text{(3.484)}$$

thus, the term in curly brackets vanishes, specifying a linear inhomogeneous system, whose solutions is (3.485a, b):

$$P_{N,M}(a,b) \neq 0: \qquad C_i = \left[P_{N,M}(a,b) \right]^{-1} \sum_{j=1}^{L} \bar{P}_{ij}(a,b)\, B_j, \qquad \text{(3.485a, b)}$$

where $\bar{P}_{ij}$ are the co-factors of P_{ij},

$$\sum_{j=1}^{L} P_{ij}(a,b)\, \bar{P}_{jk}(a,b) = P_{N,M}(a,b)\, \delta_{ij}; \qquad \text{(3.486)}$$

in (3.486) the identity matrix δ_{ij} appears and the characteristic polynomial (3.438d) is assumed to be non-zero (3.485a). The particular integral of (3.433a, b; 3.483a) is thus (3.483b; 3.485b), provided the characteristic polynomial (3.436d) does not vanish (3.485a). Thus, *the simultaneous system of linear p.d.e. with constant coefficients and exponential forcing term (3.433a, b; 3.483a)* $\equiv$ *(3.487a, b):*

$$i,j = 1,\ldots,L: \qquad \sum_{j=1}^{L} \left\{ P_{ij}\left(\frac{\partial}{\partial x}, \frac{\partial}{\partial y} \right) \right\} \Phi_j(x,t) = B_i \exp(a\ x + b\ y), \qquad \text{(3.487a, b)}$$

has particular integral (3.483b; 3.485b) $\equiv$ *(3.488b) in the non-resonant case when the characteristic polynomial (3.438d) does not vanish (3.485a)* $\equiv$ *(3.488a):*

$$P_{N,M}(a,b) \neq 0: \qquad \Phi_i(x,t) = \frac{\exp(a\ x + b\ y)}{P_{N,M}(a,b)} \times \sum_{j=1}^{L} \bar{P}_{i,j}(a,b)\, B_j. \qquad \text{(3.488a, b)}$$

*In the resonant case when the characteristic polynomial (3.438d) has root a (b)
of order r (s), and the lowest order non-zero derivative is (3.489a), a particular
integral is (3.489b):*

$$E(a,b) = \frac{\partial^{r+s}}{\partial a^r\, \partial b^s} \left[P_{N,M}(a,b) \right] \neq 0:$$

$$\Phi_i(x,t) = \left[E(a,b) \right]^{-1} \frac{\partial^{r+s}}{\partial a^r\, \partial b^s} \left\{ e^{ax+by} \sum_{j=1}^{L} \bar{P}_{ij}(a,b)\, C_j \right\}. \qquad (3.489a, b)$$

as shown before (Subsection 3.4.16). Next, the non-resonant and resonant forcing [Subsection 3.5.8 (3.5.9)] by exponentials of the alternate harmonic system
(Subsection 3.5.3) [coupled wave-diffusion system (Subsection 3.5.4)] is considered.

3.5.8 FORCING OF THE ALTERNATE HARMONIC SYSTEM

The alternate harmonic system (3.452a, b) of coupled linear first-order p.d.e. with
constant coefficients forced by exponentials (3.490a, b):

$$\frac{\partial \Phi}{\partial x} + \frac{\partial \Phi}{\partial y} + \frac{\partial \Psi}{\partial x} = - e^{ax+by}, \qquad 2\frac{\partial \Phi}{\partial y} + \frac{\partial \Psi}{\partial x} + \frac{\partial \Psi}{\partial y} = 2\, e^{ax+by},$$

$$(3.490a, b)$$

can be put in matrix form (3.491):

$$\begin{bmatrix} \dfrac{\partial}{\partial x} + \dfrac{\partial}{\partial y} & \dfrac{\partial}{\partial x} \\[2mm] 2\dfrac{\partial}{\partial y} & \dfrac{\partial}{\partial x} + \dfrac{\partial}{\partial y} \end{bmatrix} \begin{bmatrix} \Phi(x,y) \\ \Psi(x,y) \end{bmatrix} = e^{ax+by} \begin{bmatrix} -1 \\ 2 \end{bmatrix}. \qquad (3.491)$$

Seeking particular integrals in the same exponential form, (3.492a, b) leads to the
linear forced system (3.492c):

$$\Phi(x,y) = C_1\, e^{ax+by}:$$
$$\Psi(x,y) = C_2\, e^{ax+by}:$$

$$\begin{bmatrix} a+b & a \\ 2b & a+b \end{bmatrix} \begin{bmatrix} C_1 \\ C_2 \end{bmatrix} = \begin{bmatrix} -1 \\ 2 \end{bmatrix}. \qquad (3.492a\text{–}c)$$

The system (3.492c) is inverted (3.493c, d) provided that the characteristic polynomial does not vanish (3.493a, b):

$$P_{2,2}(a,b) = a^2 + b^2 \neq 0:$$

$$(a^2+b^2)\begin{bmatrix} C_1 \\ C_2 \end{bmatrix} = \begin{bmatrix} a+b & -a \\ -2b & a+b \end{bmatrix} \begin{bmatrix} -1 \\ 2 \end{bmatrix} = \begin{bmatrix} -3a-b \\ 2a+4b \end{bmatrix}. \qquad (3.493a\text{–}d)$$

Thus, a particular integral is (3.492a, b; 3.493d) ≡ (3.494b, c):

$$b \neq \pm i \, a: \qquad \left\{ \Phi(x,y), \Psi(x,y) \right\} = \frac{e^{ax + by}}{a^2 + b^2} \left\{ -3a - b, \ 2a + 4b \right\}, \qquad (3.494a\text{--}c)$$

in the non-resonant case (3.493a) ≡ (3.494a).

The resonant case (3.495a) corresponds to the forcing (3.490a, b) by the product of an exponential and circular cosine (3.495b–e):

$$b = \pm i \, a: \qquad \frac{\partial \Phi}{\partial x} + \frac{\partial \Phi}{\partial y} + \frac{\partial \Psi}{\partial x} = - \, Re\left\{ e^{ax \, \pm \, iay} \right\} = - \, e^{ax} \cos(ay),$$

$$2 \frac{\partial \Phi}{\partial y} + \frac{\partial \psi}{\partial x} + \frac{\partial \Psi}{\partial y} = 2 \, Re\left\{ e^{ax \, \pm \, iay} \right\} = 2 \, e^{ax} \cos(ay).$$

$$(3.495a\text{--}e)$$

The product of an exponential and a circular cosine is a solution (3.221a) of the unforced harmonic equation (3.212b) and, thus, could not be a solution of the forced harmonic equation, confirming that a singly resonant solution (3.489a, b) with (3.486a, b) is (3.496c–f):

$$E \equiv \frac{\partial}{\partial b} \left[P_{2,2}(a,b) \right] = \frac{\partial}{\partial b} \left(a^2 + b^2 \right) = 2 \, b:$$

$$\begin{bmatrix} \Phi(x,y) \\ \Psi(x,y) \end{bmatrix} = Re \left\{ \lim_{b \to \pm ia} \left[\frac{1}{2b} \frac{\partial}{\partial b} \left\{ \begin{bmatrix} -3a - b \\ 2a + 4b \end{bmatrix} e^{ax+by} \right\} \right] \right\}$$

$$= e^{ax} \, Re \left\{ \lim_{b \to \pm ia} \frac{e^{by}}{2b} \left\{ \begin{bmatrix} -1 \\ 4 \end{bmatrix} + y \begin{bmatrix} -3a - b \\ 2a + 4b \end{bmatrix} \right\} \right\}$$

$$= e^{ax} \, Re \left[\mp i \, \frac{e^{\pm iay}}{2a} \left\{ \begin{bmatrix} -1 \\ 4 \end{bmatrix} + a \, y \begin{bmatrix} -3 \mp i \\ 2 \pm 4 \, i \end{bmatrix} \right\} \right]$$

$$= \frac{e^{ax}}{2a} \, \sin(ay) \begin{bmatrix} -1 \\ 4 \end{bmatrix} + \frac{y}{2} \, e^{ax} \begin{bmatrix} -3\sin(ay) - \cos(ay) \\ 2\sin(ay) + 4\cos(ay) \end{bmatrix}.$$

$$(3.496a\text{--}f)$$

Thus, *the linear-coupled system of p.d.e. with constant coefficients (3.490a, b)* ≡ *(3.491) forced by exponentials has non-resonant (3.494a) solutions (3.494b, c); in the singly resonant case (3.495a), the resonant solutions are (3.496f), corresponding to the forced differential system (3.495c, e). The alternate harmonic (coupled wave-diffusion) system* [Subsection 3.5.3 (3.5.4)] *with exponential forcing has non-resonant and singly (singly and doubly) resonant solutions* [Subsection 3.5.8 (3.5.9)].

3.5.9　NON-RESONANT AND SINGLY AND DOUBLY RESONANT SOLUTIONS

In the case of the wave-diffusion system coupled by gradients (3.471a, b) with exponential forcing (3.497a, b):

$$\frac{\partial^2 \Phi}{\partial t^2} - c^2 \frac{\partial^2 \Phi}{\partial x^2} - \mu \frac{\partial \Psi}{\partial x} = B_1 \, e^{at+bx}, \tag{3.497a}$$

$$\frac{\partial \Psi}{\partial t} - \chi \frac{\partial^2 \Psi}{\partial x^2} - v \frac{\partial \Phi}{\partial y} = B_2 \, e^{at+bx}, \tag{3.497b}$$

the particular integrals (3.498a, b) satisfy (3.498c):

$$\Phi(x,t) = C_1 \, e^{at+bx}: \\ \Psi(x,t) = C_2 \, e^{at+bx}: \quad \begin{bmatrix} a^2 - c^2 \, b^2 & -\mu \, b \\ -v \, b & a - b^2 \, \chi \end{bmatrix} \begin{bmatrix} C_1 \\ C_2 \end{bmatrix} = \begin{bmatrix} B_1 \\ B_2 \end{bmatrix}; \tag{3.498a–c}$$

the system (3.498c) is inverted (3.499b), provided that the characteristic polynomial (3.475a) is non-zero (3.499a):

$$P_{3,4}(a,b) \neq 0: \quad P_{3,4}(a,b) \begin{bmatrix} C_1 \\ C_2 \end{bmatrix} = \begin{bmatrix} a - b^2 \, \chi & \mu \, b \\ v \, b & a^2 - c^2 \, b^2 \end{bmatrix} \begin{bmatrix} B_1 \\ B_2 \end{bmatrix}. \tag{3.499a, b}$$

Thus, in the non-resonant case (3.499a), a particular integral of (3.497a, b) is (3.500a, b):

$$\{\Phi(x,t), \ \Psi(x,t)\} = \frac{e^{at+bx}}{P_{3,4}(a,b)} \left\{ B_1 \left(a - b^2 \, \chi \right) + B_2 \, \mu \, b, \ B_1 \, v \, b + B_2 \left(a^2 - c^2 \, b^2 \right) \right\}. \tag{3.500a, b}$$

The four singly resonant cases occur (3.501a) for distinct roots (3.476a, b) with (3.489a) ≡ (3.501b, c):

$$\lambda \neq 0: \quad E_1(a,b) = \frac{\partial}{\partial b} \left[P_{3,4}(a,b) \right] = 4 \, c^2 \, \chi \, b^3 - 2 \left(\mu \, v + c^2 a + \chi \, a^2 \right) b, \tag{3.501a–c}$$

leading (3.489b) to the particular integrals (3.502a, b):

$$\begin{bmatrix} \Phi_{\pm\pm}(x,t) \\ \Psi_{\pm\pm}(x,t) \end{bmatrix} = \frac{e^{at}}{E_1\left(a, \ \pm b_{\pm}(a) \right)} \lim_{b \to \pm b_{\pm}(a)} \frac{\partial}{\partial b} \left\{ \begin{bmatrix} B_1 \left(a - b^2 \, \chi \right) + B_2 \, \mu \, b \\ B_1 \, v \, b + B_2 \left(a^2 - c^2 \, b^2 \right) \end{bmatrix} e^{bx} \right\}$$

$$= \frac{e^{at}}{E_1\left(a, \ \pm b_{\pm}(a) \right)} \exp\left[\pm x \, b_{\pm}(a) \right]$$

$$\times \left\{ \begin{bmatrix} \mp 2 \, \chi \, B_1 \, b_{\pm}(a) + B_2 \, \mu \\ \mp 2 \, c^2 \, B_2 \, b_{\pm}(a) + B_1 \, v \end{bmatrix} + x \begin{bmatrix} B_1 \left\{ a - \chi \left[b_{\pm}(a) \right]^2 \right\} \pm B_2 \, \mu \, b_{\pm}(a) \\ \pm B_1 \, v \, b_{\pm}(a) + B_2 \left\{ a^2 - c^2 \left[b_{\pm}(a) \right]^2 \right\} \end{bmatrix} \right\}. \tag{3.502a, b}$$

The two doubly resonant cases correspond (3.503a) to double roots (3.478a, b) implying (3.503b, c) from (3.489a):

$$\lambda(a_{1-4}) = 0: \qquad E_2(a,b) \equiv \frac{\partial^2}{\partial b^2} \left[P_{3,4}(a,b) \right] = 12 \ c^2 \ \chi \ b^2 - 2 \left(\mu \ v + c^2 \ a + \chi \ a^2 \right),$$

$$(3.503a-c)$$

leading (3.489b) to the particular integrals (3.504a, b):

$$\begin{bmatrix} \Phi_{1-4}^{\pm}(x,t) \\ \Psi_{1-4}^{\pm}(x,t) \end{bmatrix} = \frac{e^{at}}{E_2\left(a_{1-4} \ , \ \pm b(a_{1-4}) \right)} \lim_{b \to \pm b(a_{1-4})} \frac{\partial^2}{\partial b^2} \left\{ \begin{bmatrix} B_1\left(a_{1-4} - b^2 \ \chi\right) + B_2 \ \mu \ b \\ B_1 \ v \ b + B_2\left(a_{1-4}^2 - c^2 \ b^2\right) \end{bmatrix} e^{bx} \right\}$$

$$= \frac{e^{at}}{E_2\left(a_{1-4} \ , \ \pm b(a_{1-4}) \right)} \exp\left[\pm x \ b(a_{1-4}) \right]$$

$$\times \left\{ \begin{bmatrix} -2 \ \chi \ B_1 \\ -2 \ c^2 \ B_2 \end{bmatrix} + 2 \ x \begin{bmatrix} \mp 2 \ \chi \ B_1 \ b(a_{1-4}) + B_2 \ \mu \\ \mp 2 \ c^2 \ B_2 \ b(a_{1-4}) + B_1 \ v \end{bmatrix} \right.$$

$$\left. + x^2 \begin{bmatrix} B_1\left\{a_{1-4} - \chi \left[b(a_{1-4})\right]^2\right\} \pm B_2 \ \mu \ b_{\pm}(a_{1-4}) \\ \pm B_1 \ v \ b(a_{1-4}) + B_2\left\{a_{1-4}^2 - c^2 \left[b(a_{1-4})\right]^2\right\} \end{bmatrix} \right\}. \qquad (3.504a, b)$$

Thus, *the wave-diffusion system coupled by gradients and forced by exponentials (3.497a, b) has a non-resonant (3.499a) particular integral (3.500a, b) for the non-zero characteristic polynomial (3.475a). There are four singly resonant cases (3.502b; 3.501c) corresponding to single roots (3.501a–c; 3.476a, b), and eight doubly resonant cases (3.504b; 3.503c) corresponding to double roots (3.503a–c; 3.478a–c).* The method of the inverse characteristic polynomial (inverse matrix of polynomials) of partial derivatives applies [Subsections 3.4.31–3.4.46 (example 10.8)] to a single (simultaneous system) of linear forced p.d.e. with constant coefficients, in a straightforward way (with some additional steps and calculations).

3.6 LINEAR EQUATION WITH HOMOGENEOUS PARTIAL DERIVATIVES

The characteristic polynomial exists for a linear partial differential equation with constant (power) coefficients [Section 3.4 (3.6)]; in the latter case, each derivative of order n must be multiplied by a power with the same exponent (Section 3.6). The single (multiple) roots specify the natural integrals of the unforced equation (Subsection 3.6.1) that are powers (multiplied by logarithms), for example in the case of the analogue of the wave-diffusion equation (Subsection 3.6.2). The analogy between linear partial differential equations with constant (power) coefficients is most clear (Subsection 3.6.3) comparing ordinary (homogeneous) derivatives as they apply to: (i) exponentials (powers); and (ii) powers (logarithms). For example, the characteristic polynomial of the original wave-diffusion equation (Subsection 3.4.13) is: (i) distinct for the analogue wave-diffusion equation (Subsection 3.6.2); and (ii) identical for the homogeneous wave-diffusion equation (Subsection 3.6.4) that replaces ordinary by homogeneous derivatives (Subsection 3.6.3). Omitting dissipation leads to the analogy between the original (homogeneous) wave equations [Subsection 3.1.2 (3.6.5)] that have similarity solutions. Since the homogeneous wave equation is a partial differential equation of the second order, its solution becomes unique imposing two boundary or initial conditions (Subsection 3.6.6). The characteristic polynomial for a linear partial differential equation with homogeneous derivatives forced by powers (Subsection 3.6.7) specifies non-resonant and resonant solutions, taking as example again the analogue wave-diffusion equation with homogeneous derivatives (Subsection 3.6.8). The inverse polynomial of homogeneous partial derivatives can be used to obtain particular integrals for forcing (Subsection 3.6.9), for example by powers or logarithms, and this can be extended (Subsection 3.6.10) to products of powers by other functions like logarithms.

3.6.1 CHARACTERISTIC POLYNOMIAL AND POWER AND LOGARITHMIC SOLUTIONS

The method of solution of a linear partial differential equation with constant coefficients (3.194a), can be adapted to a case with non-uniform coefficients, namely, power coefficients with the same exponent as the order of derivation (3.505):

$$\sum_{n=0}^{N} \sum_{m=0}^{M} A_{nm} \, x^n \, y^m \, \frac{\partial^{n+m} \Phi}{\partial x^n \, \partial y^m} = B(x,y).$$
(3.505)

In the unforced case (3.506a), the solution can be sought as a product of powers (3.506b) leading to (3.506c, d):

$$B(x,y) = 0, \qquad \Phi(x,y) = C \, x^a \, y^b:$$

$$0 = \sum_{n=0}^{N} \sum_{m=0}^{M} A_{nm} \, x^n \, y^m \, \frac{\partial^{n+m}}{\partial x^n \, \partial y^m} \left(C \, x^a y^b \right)$$

$$= C \; x^a y^b \sum_{n=0}^{N} \sum_{m=0}^{M} A_{nm} \; a(a-1)...(a-n+1) \times b(b-1)...(b-m+1)$$

$$= \Phi(x,y) \; Q_{N,M}(a,b), \tag{3.506a–e}$$

where a non-trivial solution (3.507a) requires (3.506e) that the characteristic polynomial (3.507a, d) vanishes (3.507b):

$\Phi(x,y) \neq 0$:

$$0 = Q_{N,M}(a,b) \equiv \sum_{n=0}^{N} \sum_{m=0}^{M} A_{nm} \; a(a-1)...(a-n+1) \times b(b-1)...(b-m+1)$$

$$= \sum_{n=0}^{N} \sum_{m=0}^{M} A_{nm} \frac{a! \; b!}{(a-n)! \, (b-m)!}, \tag{3.507a–d}$$

and involves factorials (3.507c).

The single (3.508a) [(3.508b)] roots (3.507b) of the characteristic polynomial (3.507c, d):

$$Q_{N,M}(a,b) = Q_a \prod_{n=1}^{N} \left[a - a_n(b)\right] = Q_b \prod_{m=1}^{M} \left[b - b_m(a)\right] \tag{3.508a, b}$$

specify the particular integrals (3.509a) [(3.509b)]:

$$\Phi_n(x,y) = \int y^b \; x^{a_n(b)} \; C_n(b) \; db, \quad \Phi^m(x,y) = \int x^a \; y^{b_m(a)} \; C^m(a) \; da, \tag{3.509a, b}$$

in the general integral (3.510a) [(3.510b)]:

$$\Phi(x,y) = \sum_{n=1}^{N} \Phi_n(x,y) = \sum_{m=1}^{M} \Phi^m(x,y). \tag{3.510a, b}$$

In the case of multiple (3.511a, b) [(3.512a, b)] roots (3.507b) of the characteristic polynomial (3.507c, d):

$$\sum_{r=1}^{R} \alpha_r = N: \qquad Q_{N,M}(a,b) = Q_a \sum_{r=1}^{R} \left[a - a_r(b)\right]^{\alpha_r}, \tag{3.511a, b}$$

$$\sum_{s=1}^{S} \beta_s = M: \qquad Q_{N,M}(a,b) = Q_b \sum_{s=1}^{S} \left[b - b_s(a)\right]^{\beta_s}, \tag{3.512a, b}$$

the particular integrals (3.513a–c) [(3.514a–c)] involve parametric differentiation

$$p = 1, \ldots, \alpha_r: \qquad \Phi_{r,p}(x,y) = \lim_{a \to a_r(b)} \frac{\partial^p}{\partial a^p} \left[\Phi_r(a) \right]$$

$$= \int y^b \, x^{a_r(b)} \log^{p-1} x \, C_{r,p}(b) \, db, \qquad (3.513\text{a–c})$$

$$q = 1, \ldots, \beta_s: \qquad \Phi^{s,q}(x,y) = \lim_{b \to b_s(a)} \frac{\partial^q}{\partial b^q} \left[\Phi^s(b) \right]$$

$$= \int x^a \, y^{b_s(a)} \log^{q-1} y \, C^{s,q}(a) \, da. \qquad (3.514\text{a–c})$$

The general integral is (3.515a) [(3.515b)]:

$$\Phi(x,y) = \sum_{r=1}^{R} \sum_{p=1}^{\alpha_r} \Phi_{r,p}(x,y) = \sum_{s=1}^{S} \sum_{q=1}^{\beta_s} \Phi^{s,q}(x,y). \qquad (3.515\text{a, b})$$

Thus, *the general integral of the unforced (3.506a) linear partial differential equation with homogenous power coefficients (3.505) is (3.510a, b; 3.509a, b) [(3.515a, b; 3.513a–c; 3.514a–c)] in the case of single (3.508a, b) [multiple (3.511a, b; 3.512a, b)] roots (3.507b) of the characteristic polynomial (3.507c, d).* An example given next is the analogue of the wave-diffusion equation (Subsection 3.4.13) using homogeneous power coefficients (Subsection 3.6.2).

3.6.2 Analogue Wave-Diffusion Equation with Homogeneous Powers

The **analogue wave-diffusion equation** (3.257b) with homogeneous powers is (3.516):

$$x^2 \frac{\partial^2 \Phi}{\partial x^2} - \frac{t^2}{c^2} \frac{\partial^2 \Phi}{\partial t^2} - \frac{t}{\chi} \frac{\partial \Phi}{\partial t} = 0, \qquad (3.516)$$

where the constant coefficients $\{c, \chi\}$ are dimensionless. Seeking a solution as a product of powers (3.517a) leads to (3.517b):

$$\Phi(x,t) = C \, x^a \, t^b: \qquad 0 = C \, x^a \, t^b \left[a(a-1) - b(b-1)/c^2 - b/\chi \right]$$

$$\equiv \Phi(x,t) \, Q_{2,2}(a,b), \qquad (3.517\text{a–c})$$

specifying the characteristic polynomial (3.517c) $\equiv$ (3.518b):

$$0 = Q_{2,2}(a,b) = a^2 - a - b^2/c^2 - b(1/\chi - 1/c^2)$$

$$= \left[a - a_+(b) \right] \left[a - a_-(b) \right] = - c^{-2} \left[b - b_+(a) \right] \left[b - b_-(a) \right], \qquad (3.518\text{a–d})$$

whose roots (3.518a) with regard (3.518c) [(3.518d)] to $a(b)$ are (3.519a,b) [(3.519c,d)]:

$$2\, a_{\pm}(b) = 1 \pm \sqrt{1 + 4b^2 / c^2 + 4\, b\left(1 / \chi - 1 / c^2\right)}, \qquad (3.519a, b)$$

$$2\, b_{\pm}(a) = 1 - c^2 / \chi \pm \sqrt{\left(1 - c^2 / \chi\right)^2 + 4\, c^2\, a(a-1)}. \qquad (3.519c, d)$$

The double roots (3.520a–c) [(3.521a–c)] for a (b) occur (3.520d) [(3.521d)] for the values (3.520e, f) [(3.521e, f)] of b (a):

$$a_{+}(b^{\pm}) = a_{-}(b^{\pm}) \equiv a\,(b^{\pm}) = 1 / 2:$$

$$4\left(b^{\pm}\right)^2 + 4\, b^{\pm}\left(c^2 / \chi - 1\right) + c^2 = 0 \;\Rightarrow\; 2\, b^{\pm} = 1 - c^2 / \chi \pm \sqrt{\left(1 - c^2 / \chi\right)^2 - c^2}\,;$$

$$(3.520a–f)$$

$$b_{+}(a^{\pm}) = b_{-}(a^{\pm}) \equiv b(a^{\pm}) = \left(1 - c^2 / \chi\right) / 2:$$

$$4\, c^2\left(a^{\pm}\right)^2 - 4\, c^2\, a^{\pm} + \left(1 - c^2 / \chi\right)^2 = 0 \;\Rightarrow\; 2\, a^{\pm} = 1 \pm \sqrt{1 - \left(1 / c - c / \chi\right)^2}\,.$$

$$(3.521a–f)$$

Thus, *the general integral(s) of the unforced analogue wave-diffusion equation with homogeneous power coefficients (3.516) in the case of distinct (3.522a) [(3.523a)] roots (3.519a, b) [(3.519c, d)] is given by (3.522b, c) [(3.523b, c)] involving two arbitrary functions:*

$$b \neq b^{\pm} \qquad \Phi(x,t) = \Phi_{+}(x,t) + \Phi_{-}(x,t), \qquad \Phi_{\pm}(x,t) = \int x^{a_{\pm}(b)}\; t^b\; C_{\pm}(b)\; db,$$

$$(3.522a–c)$$

$$a \neq a^{\pm}: \qquad \Phi(x,t) = \Phi^{+}(x,t) + \Phi^{-}(x,t), \qquad \Phi^{\pm}(x,t) = \int x^{a}\; t^{b_{\pm}(a)}\; C^{\pm}(a)\; da;$$

$$(3.523a–c)$$

the exception to (3.522a–c) [(3.523a–c)] are four special integrals in the case of double (3.524a) [(3.525a)] roots (3.520a–f) [(3.521a–f)] that are given by (3.524b, c) [(3.525b, c)] involving one arbitrary constant for each:

$$b = b^{\pm}: \qquad \Phi_{\pm}(x,t) = C_{\pm}\, \sqrt{x}\; t^{b^{\pm}}, \qquad (3.524a–c)$$

$$a = a^{\pm}: \qquad \Phi^{\pm}(x,t) = C^{\pm}\, x^{a^{\pm}}\; t^{\left(1 - c^2 \chi\right)/2}. \qquad (3.525a–c)$$

Before proceeding to consider the forced linear partial differential equation with homogeneous power coefficients (Subsections 3.6.7–3.6.9), the characteristic polynomial is compared with the case of constant coefficients (Subsection 3.6.3), leading to some analogies between linear partial differential equations with ordinary and homogeneous derivatives (Subsections 3.6.4–3.6.6).

3.6.3   ORDINARY/HOMOGENEOUS DERIVATIVES OF EXPONENTIALS/POWERS/LOGARITHMS

The unforced linear partial differential equation with constant (3.194a) ≡ (3.526a) [homogeneous power (3.505; 3.506a) ≡ (3.527a)] coefficients has characteristic polynomials (3.200a) ≡ (3.526b) [(3.507c, d) ≡ (3.527b)]:

$$0 = \sum_{n=0}^{N} \sum_{m=0}^{M} A_{nm} \frac{\partial^{n+m} \Phi}{\partial x^n \, \partial y^m} : \qquad P_{N,M}(a,b) = \sum_{n=0}^{N} \sum_{m=0}^{M} A_{nm} \, a^n \, b^m, \qquad (3.526a, b)$$

$$0 = \sum_{n=0}^{N} \sum_{m=0}^{M} A_{nm} \, x^n \, y^m \frac{\partial^{n+m} \Phi}{\partial x^n \, \partial y^m} :$$

$$Q_{N,M}(a,b) = \sum_{n=0}^{N} \sum_{m=0}^{M} A_{nm} \, a(a-1)...(a-n+1) \, b(b-1)...(b-m+1). \qquad (3.527a, b)$$

*In the case of single roots (3.203a–c) [(3.508a, b)] of the characteristic polynomial, the particular integrals are (3.528c) [(3.529c)], leading by **ordinary (homogeneous) differentiation** (3.528a, b) [(3.529a, b)] to analogous multiplications (3.528d) [(3.529d)]:*

$$\partial_x \equiv \frac{\partial}{\partial x}, \qquad \partial_y \equiv \frac{\partial}{\partial y} : \qquad \Phi(x,y) = C \, e^{ax+by}, \qquad \partial_x^n \, \partial_y^m \, \Phi(x,y) = C \, a^n \, b^m \, e^{ax+by}$$
$$(3.528a\text{-}d)$$

$$\delta_x \equiv x \frac{\partial}{\partial x}, \qquad \delta_y \equiv y \frac{\partial}{\partial y} : \qquad \Psi(x,y) = C \, x^a \, y^b, \qquad \delta_x^n \, \delta_y^m \, \Psi(x,y) = C \, a^n \, b^m \, x^a \, y^b.$$
$$(3.529a\text{-}d)$$

In the case of multiple roots (3.206a–f) [(3.511a, b; 3.512a, b)] of the characteristic polynomial, power (3.530a) [logarithmic (3.531a)] terms appear that under ordinary (3.528a, b) [homogeneous (3.529a, b)] derivatives lead to analogous multiplications (3.530b) [(3.531b)]:

$$\partial_x^n \, \partial_y^m \left(C \, x^a \, y^b \right) = C \, x^{a-n} \, y^{b-m} \, a(a-1)...(a-n+1) \times b(b-1)...(b-m+1),$$
$$(3.530a, b)$$

$$\delta_x^n \, \delta_y^m \left(C \log^a x \, \log^b y \right) = C \log^{a-n} x \, \log^{b-m} y$$
$$\times a(a-1)...(a-n+1) \, b(b-1)...(b-m+1). \qquad (3.531a, b)$$

*The analogies extend to **ordinary (homogeneous) integration** of powers (3.532a) [logarithms (3.522b)]:*

$$\partial_x^{-n} \, \partial_y^{-m} \left(C \, x^a \, y^b \right) = \frac{C \, x^{a+n} \, y^{b+m}}{(a+1) \, ... \, (a+n) \, (b+1) \, ... \, (b+m)}, \qquad (3.532a)$$

$$\delta_x^{-n}\,\delta_y^{-m}\,\left(C\log^a x\,\log^b y\right)=\frac{C\,\log^{a+n} x\,\log^{b+m} y}{(a+1)\dots(a+n)\,(b+1)\dots(b+m)}. \qquad (3.532b)$$

In the case of the linear partial differential equation with constant (3.526a) [homogeneous power (3.527a)] coefficients, the differential operator is (3.533a) [(3.533b)], a characteristic polynomial (3.526b) [(3.527b)] of ordinary (3.528a, b) [homogeneous (3.529a, b)] derivatives:

$$\left\{P_{N,M}\left(\partial_x\,,\partial_y\right)\right\}\Phi(x,y)=B(x,y)=\left\{Q_{N,M}\left(\delta_x\,,\delta_y\right)\right\}\Psi(x,y) \qquad (3.533\text{a, b})$$

and in the integrals are made the substitutions of exponentials (powers) in (3.533c) and powers (logarithms) in (3.533d):

$$e^{ax+by}\;\rightarrow\;x^a\,y^b,\qquad x^a\,y^b\;\rightarrow\;\log^a x\,\log^b y. \qquad (3.533\text{c, d})$$

From (3.527a, b), bearing in mind (3.534a), it follows that homogeneous power coefficients are related to the homogeneous derivatives by (3.534b):

$$a\equiv x\,\frac{\partial}{\partial x}:\qquad x^n\,\frac{\partial^n}{\partial x^n}\,y^m\,\frac{\partial^m}{\partial y^m}=x\,\frac{\partial}{\partial x}\left(x\,\frac{\partial}{\partial x}-1\right)\dots\left(x\,\frac{\partial}{\partial x}-n+1\right)$$

$$\times\,y\,\frac{\partial}{\partial y}\left(y\,\frac{\partial}{\partial y}-1\right)\dots\left(y\,\frac{\partial}{\partial y}-m+1\right).$$
$$(3.534\text{a, b})$$

For example, the r.h.s. and l.h.s. of (3.534b) applied to powers both give (3.530a, b); the relation (3.534b) can also be proved directly (IV.1.304). Thus, *the forced linear partial differential equation with homogeneous power coefficients (3.535a) is equivalent (3.534b) to (3.535b) in terms of homogeneous derivatives:*

$$B(x,y)=\left\{\sum_{n=0}^{N}\sum_{m=0}^{M}A_{nm}\,x^n\,y^m\,\partial_x^n\,\partial_y^m\right\}\Phi(x,y)$$

$$=\left\{\sum_{n=0}^{N}\sum_{m=0}^{M}A_{nm}\,\delta_x\left(\delta_x-1\right)\dots\left(\delta_x-n+1\right)\delta_y\left(\delta_y-1\right)\dots\left(\delta_y-m+1\right)\right\}\Phi(x,y).$$
$$(3.535\text{a, b})$$

The analogies between ordinary and homogeneous derivatives may be exemplified by the original (homogeneous) wave-diffusion equation [Subsection 3.4.3 (3.6.4)].

3.6.4 ANALOGUE WAVE-DIFFUSION EQUATION WITH HOMOGENEOUS DERIVATIVES

The analogue wave-diffusion equation (3.516) has a characteristic polynomial (3.518b) distinct from the characteristic polynomial (3.258b) of the original wave-diffusion equation (3.257b). To have the same characteristic polynomial, the ordinary derivatives in the original wave-diffusion equation (3.257b) ≡ (3.536a) must be replaced by homogeneous derivatives (3.529a, b), leading to the **homogeneous wave-diffusion equation** (3.536b):

$$\left\{ \frac{\partial^2}{\partial x^2} - \frac{1}{c^2} \frac{\partial^2}{\partial t^2} - \frac{1}{\chi} \frac{\partial}{\partial t} \right\} \Phi(x,t) = 0, \tag{3.536a}$$

$$\left\{ \left(x \frac{\partial}{\partial x} \right)^2 - \frac{1}{c^2} \left(t \frac{\partial}{\partial t} \right)^2 - \frac{1}{\chi} t \frac{\partial}{\partial t} \right\} \Psi(x,t) = 0, \tag{3.536b}$$

that has additional terms (3.537) ≡ (3.536b)

$$x^2 \frac{\partial^2 \Psi}{\partial x^2} + x \frac{\partial \Psi}{\partial x} - \frac{t^2}{c^2} \frac{\partial^2 \Psi}{\partial t^2} - t \left(\frac{1}{c^2} + \frac{1}{\chi} \right) \frac{\partial \Psi}{\partial t} = 0, \tag{3.537}$$

relative to the analogue wave-diffusion equation (3.516). The original (3.536a) [homogeneous (3.536b) ≡ (3.537)] wave-diffusion equation has exponential (3.538a) [power (3.538b)] solutions:

$$\Phi(x,t) = e^{at+bx}, \qquad\qquad \Psi(x,t) = t^a \, x^b, \tag{3.538a, b}$$

leading to the same characteristic polynomial (3.539a, b) ≡ (3.258b)

$$P_{2,2}(a,b) = b^2 - \frac{a^2}{c^2} - \frac{a}{\chi} = Q_{2,2}(a,b). \tag{3.539a, b}$$

The roots for b (a) are (3.540a) ≡ (3.258c) [(3.540b) ≡ (3.260a)]:

$$b_\pm(a) = \pm \left| \frac{a}{\chi} + \frac{a^2}{c^2} \right|^2, \qquad a_\pm(b) = \frac{c^2}{2\chi} \left\{ -1 \pm \left| 1 + \frac{4 \, b^2 \, \chi^2}{c^2} \right|^{1/2} \right\}, \tag{3.540a, b}$$

and lead: (i) for a, to the particular integrals (3.541a) ≡ (3.259a, b) [(3.541b)]:

$$\Phi_\pm(x,t) = \int C_\pm(a) \exp\left[a \, t + x \, b_\pm(a) \right] da, \tag{3.541a}$$

$$\Psi_\pm(x,t) = \int C_\pm(a) \, t^a \, x^{b_\pm(a)} \, da; \tag{3.541b}$$

and (ii) for b, to the particular integrals (3.542a) ≡ (3.260b, c) [(3.542b)]:

$$\Phi^\pm(x,t) = \int C^\pm(b) \exp\left[b \, x + t \, a_\pm(b) \right] db, \tag{3.542a}$$

$$\Psi^\pm(x,t) = \int C^\pm(b) \, x^b \, t^{a_\pm(b)} \, db. \tag{3.542b}$$

The general integral of the original (3.536a) ≡ (3.257b) [homogeneous (3.536b) ≡ (3.537)] wave-diffusion equation is the sum (3.543a, b) ≡ (3.261a, b) [(3.544a, b)] of two particular integrals for the same parameter a in (3.543a; 3.544a) and b in (3.543b; 3.544b)]:

$$\Phi(x,t) = \Phi_+(x,t) + \Phi_-(x,t) = \Phi^+(x,t) + \Phi^-(x,t), \quad (3.543a, b)$$

$$\Psi(x,t) = \Psi_+(x,t) + \Psi_-(x,t) = \Psi^+(x,t) + \Psi^-(x,t). \quad (3.544a, b)$$

In the absence of dissipation, that is for infinite diffusivity $\chi = \infty$ the original (homogeneous) wave-diffusion equation simplifies to the original (homogeneous) wave equation, which has all ordinary (homogeneous) derivatives of the same order and, hence, similarity solutions specified by functions of a linear combination (Sections V.1.6–V1.9) [a product of powers (Subsection 3.6.5)] of the independent variables.

3.6.5   SIMILARITY SOLUTIONS FOR THE ORIGINAL (HOMOGENEOUS) WAVE EQUATION

In the non-dissipative case of infinite diffusivity (3.545a), the original wave-diffusion equation (3.536a) ≡ (3.257b) reduces to the original wave equation (3.545b) ≡ (3.11a):

$$\chi = 0: \qquad \frac{\partial^2 \Phi}{\partial x^2} - \frac{1}{c^2}\frac{\partial^2 \Phi}{\partial t^2} = 0, \qquad (3.545a, b)$$

$$0 = \left\{ \left(x\frac{\partial}{\partial x}\right)^2 - \frac{1}{c^2}\left(t\frac{\partial}{\partial t}\right)^2 \right\}\Psi(x,t), \qquad (3.545c)$$

leading to the **homogeneous wave equation** (3.545c) ≡ (3.546) by replacing ordinary with homogeneous (3.529a, b) derivatives:

$$0 = x^2 \frac{\partial^2 \Psi}{\partial x^2} + x\frac{\partial \Psi}{\partial x} - \frac{1}{c^2}\left(t^2 \frac{\partial^2 \Psi}{\partial t^2} + t\frac{\partial \Psi}{\partial t}\right). \qquad (3.546)$$

The original (3.545a) ≡ (3.11a) [homogeneous (3.545b) ≡ (3.546)] wave equation has exponential (3.538a) [power (3.538b)] solutions leading to the same characteristic polynomial (3.547a, b) ≡ (3.223b) with roots (3.547c) ≡ (3.223c):

$$P_{2,2}(a,b) = b^2 - \frac{a^2}{c^2} = Q_{2,2}(a,b): \qquad a = \pm b\, c, \qquad (3.547a–c)$$

*The roots for b lead to particular integrals that are **similarity solutions** specified by a function of a linear combination of the variables (3.548a, b) [a product of powers of the variables (3.549a, b)]:*

$$\Phi_\pm(x,t) = \int C_\pm(a)\exp\left[a\,(t \pm c\,x)\right] da = f^\pm\,(t \pm c\,x), \qquad (3.548a, b)$$

$$\Psi_{\pm}(x,t) = \int C_{\pm}(a)\, t^a\, x^{\pm a/c}\, da = g^{\pm}\left(x^{\pm 1/c}\, t\right); \qquad (3.549a, b)$$

likewise, the roots for a lead to natural integrals that are similarity functions of a linear combination (3.560a, b) [a product of powers (3.561a, b)] of the variables:

$$\Phi^{\pm}(x,t) = \int C^{\pm}(b)\exp\left[b\left(x \pm \frac{t}{c}\right)\right] db = f_{\pm}\left(x \pm \frac{t}{c}\right), \qquad (3.550a, b)$$

$$\Psi^{\pm}(x,t) = \int C^{\pm}(b)\, x^b\, t^{\pm bc}\, db = g_{\pm}\left(t^{\pm c}\, x\right). \qquad (3.551a, b)$$

The similarity solutions (3.548a, b) ≡ (3.550a, b) ≡ (3.226a, b) ≡ (3.227b, c) [(3.549a, b) ≡ (3.551a, b)] involve the two independent variables (x,t) in a single linear combination $x \pm c\,t$ (product of powers $t^{\pm c}\, x$), and this applies to any linear partial differential equation with ordinary (homogeneous) derivatives all of the same order (Subsection 3.7.7). Both the homogeneous (original) wave equations are of the second order on position and time, and have a unique solution imposing two initial or boundary conditions [Subsection(s) 3.6.6 (3.3.3–3.3.4)].

3.6.6 INITIAL CONDITIONS AND UNICITY OF SOLUTION OF THE HOMOGENEOUS WAVE EQUATION

The homogeneous wave equation (3.545c) ≡ (3.546) has general integral (3.544b) that is the sum (3.552) of the particular integrals (3.551b):

$$\Psi(x,t) = g_{+}\left(t^c\, x\right) + g_{-}\left(t^{-c}\, x\right). \qquad (3.552)$$

The solution is unique, specifying two initial conditions. For example: (i) the value at all positions (3.553a) at time unity; and (ii) the value of the first-order derivative with regard to time at all positions at time zero (3.553b–d):

$$\Psi(x,1) = g_{+}(x) + g_{-}(x) \qquad (3.553a)$$

$$\dot{\Psi}(x,1) \equiv \lim_{t \to 1} \frac{\partial \Psi(x,t)}{\partial t} = \lim_{t \to 1} c\, x \left[t^{c-1}\, g'_{+}\left(t^c\, x\right) - t^{-c-1}\, g'_{-}\left(t^{-c}\, x\right)\right]$$

$$= c\, x \left[g'_{+}(x) - g'_{-}(x)\right], \qquad (3.553b–d)$$

where the prime denotes a derivative with regard to the argument. The latter condition (3.553d) is integrated (3.554):

$$g_{+}(x) - g_{-}(x) = \frac{1}{c} \int^{x} \dot{\Psi}(\xi,1)\, \frac{d\xi}{\xi}. \qquad (3.554)$$

Thus, (3.553a; 3.554) can be solved for the two similarity functions:

$$2\, g_\pm(x) = \Psi(x,1) \pm \frac{1}{c} \int^x \frac{\dot\Psi(\xi,1)}{\xi} \, d\xi .$$

(3.555a, b)

Substitution of (3.555a, b) in (3.552) leads to *the general integral (3.556) of the homogeneous wave equation (3.545c) ≡ (3.546), satisfying the two initial conditions (3.553a–d):*

$$2\,\Psi(x,t) = \Psi(t^c\ x,1) + \Psi(t^{-c}\ x,1) + \frac{1}{c} \int^x \left[\dot\Psi(t^c\ \xi,1) - \dot\Psi(t^{-c}\ \xi,1) \right] \frac{d\xi}{\xi} .$$

(3.556)

It can be checked that (3.556): (i) satisfies the homogeneous wave equation (3.545c) ≡ (3.546) because it is of the form (3.552); (ii) for $t = 1$, agrees with the first boundary condition (3.553a); and (iii) the second boundary condition (3.553b) is also met, as follows from

$$2 \lim_{t \to 1} \frac{\partial \Psi(x,t)}{\partial t} = \lim_{t \to 1} \left\{ c\ x \left[t^{c-1}\ \Psi'(t^c\ x,1) - t^{-c-1}\ \Psi'(t^{-c}\ x) \right] \right.$$

$$\left. + \frac{1}{c\ x} \left[c\ x\ \dot\Psi(t^c\ x,1) + c\ x\ \dot\Psi(t^{-c}\ x,1) \right] \right\}$$

$$= 2\ \dot\Psi(x,1).$$

(3.557)

The general integral (3.556) applies to any choice of functions in the initial conditions (3.553a, b).

Choosing the initial conditions (3.558a, b):

$$\Psi(x,1) = e^x , \qquad \lim_{t \to 1} \frac{\partial \Psi(x,t)}{\partial t} \equiv \dot\Psi(x,1) = e^{-x}\ c\ x , \qquad (3.558a, b)$$

the similarity functions (3.555a, b) are given by:

$$2\, g_\pm(x) = e^x \pm \int^x e^{-\xi}\ d\xi = e^x \mp e^{-x} ,$$

(3.559a, b)

implying:

$$g_+(x) = \sinh x , \qquad\qquad g_-(x) = \cosh x ; \qquad (3.560a, b)$$

substituting (3.560a, b) in (3.552), it follows that *the general integral of the homogeneous wave equation (3.545c) ≡ (3.546) with initial conditions (3.558a, b) is:*

$$\Psi(x,t) = \sinh(t^c\ x) + \cosh(t^{-c}x).$$

(3.561)

It can be checked that: (i) both terms on the r.h.s. of (3.561) satisfy the homogeneous wave equation (3.545c) because of the identity (3.562a, b)

$$\left(x\frac{\partial}{\partial x}\right)^2\left[\sinh,\cosh\left(t^{\pm c}\ x\right)\right]=x\ t^{\pm c}\cosh,\sinh\left(t^{\pm c}\ x\right)+x^2\ t^{\pm 2c}\cosh,\sinh(t^{\pm c}\ x)$$

$$=\frac{1}{c^2}\left(t\frac{\partial}{\partial t}\right)^2\left[\sinh,\cosh\left(t^{\pm c}\ x\right)\right] \qquad (3.562\text{a, b})$$

(ii) the first initial condition (3.558a) is met by:

$$\Psi(x,1)=\sinh x+\cosh x=e^x; \qquad (3.563\text{a, b})$$

and (iii) the second boundary condition (3.558b) is met by:

$$\lim_{t\to 1}\frac{\partial\Psi(x,t)}{\partial t}=c\ x\ t^{c-1}\cosh\left(t^c\ x\right)-c\ x\ t^{-c-1}\sinh\left(t^{-c}\ x\right)$$

$$=c\ x\left(\cosh x-\sinh x\right)=e^{-x}\ c\ x. \qquad (3.564\text{a, b})$$

The linear partial differential equations with homogeneous power coefficients are considered both for unforced (forced) cases [Subsections 3.6.1–3.6.6 (3.6.7–3.6.10)]. The forcing is considered first by powers leading to non-resonant and resonant particular integrals (Subsection 3.6.7).

3.6.7   NON-RESONANT AND RESONANT FORCING BY POWERS

For the linear partial differential equation with homogeneous power coefficients (3.505) forced by a product of powers (3.565)

$$\sum_{n=0}^{N}\sum_{m=0}^{M}A_{nm}\ x^n\ y^m\ \frac{\partial^{n+m}\Phi}{\partial x^n\ \partial y^m}=B\ x^a\ y^b, \qquad (3.565)$$

a particular integral of similar form (3.506b) leading to (3.566a, b) is sought:

$$B\ x^a\ y^b=C\sum_{n=0}^{N}\sum_{m=0}^{M}A_{nm}\ x^a\ y^b\ a(a-1)...(a-n+1)$$

$$\times\ b(b-1)...(b-m+1)$$

$$=\Phi_*(x,y)\ Q_{N,M}(a,b). \qquad (3.566\text{a, b})$$

Excluding roots (3.567a) of the characteristic polynomial (3.507b, c) leads to the non-resonant particular integral (3.567b):

$$Q_{N,M}(a,b)\neq 0: \qquad\qquad \Phi_*(x,y)=\frac{B\ x^a\ y^b}{Q_{N,M}(a,b)}. \qquad (3.567\text{a, b})$$

In the resonant case when a (b) is a root of multiplicity r (s) of the characteristic polynomial with lowest order non-zero partial derivative (3.568a), the particular integral is (3.568b)

$$F(a,b) = \frac{\partial^{r+s}}{\partial a^r \, \partial b^s} \left[Q_{N,M}(a,b) \right]: \quad \Phi_*(x,y) = \frac{B \; x^a \; y^b}{F(a,b)} \, \log^r x \, \log^s y, \quad (3.568a, b)$$

where the numerator is (3.569):

$$\frac{\partial^{r+s}}{\partial a^r \, \partial b^s} \left(x^a \; y^b \right) = x^a \; y^b \log^r x \, \log^s y. \tag{3.569}$$

Thus, *the linear partial differential equation with homogeneous power coefficients (3.505) and power forcing (3.565) has particular integral (3.567b) [(3.568b)] in the non-resonant (3.567a) [resonant (3.568a)] case when the characteristic polynomial does not vanish (has roots of multiplicity respectively r, s in a, b).* As an example, the non-resonant and singly/doubly resonant particular integrals of the analogue wave-diffusion equation with homogeneous power coefficients (Subsection 3.6.2) forced by powers (Subsection 3.6.8) are considered.

3.6.8 Non-Resonant, Singly, and Doubly Resonant Solutions

Since the characteristic polynomial is of the second degree (3.518a–d) and has single (3.519a, b) and double (3.520a–c; 3.521a–c) roots, there are non-resonant and singly and doubly resonant solutions of the analogue wave-diffusion equation with homogeneous power coefficients (3.516) forced by powers (3.570):

$$x^2 \frac{\partial^2 \Phi}{\partial x^2} - \frac{t^2}{c^2} \frac{\partial^2 \Phi}{\partial t^2} - \frac{t}{\chi} \frac{\partial \Phi}{\partial t} = B \; x^a \; t^b. \tag{3.570}$$

The non-resonant (3.567b) particular integral (3.571c) assumes that the characteristic polynomial (3.518b–d) does not vanish (3.567a) $\equiv$ (3.571a, b):

$$a \neq a_\pm(b) \quad \text{and} \quad b \neq b_\pm(a): \quad \Phi(x,t) = \frac{B \; x^a \; t^b}{a(a-1) - b(b-1)/c^2 - b/\chi}. \tag{3.571a–c}$$

Bearing in mind the first-order derivatives (3.572a) [(3.572b)] of the characteristic polynomial (3.518b) with regard to a (b):

$$\frac{\partial}{\partial a} \left[Q_{2,2}(a,b) \right] = 2 \; a - 1, \quad \frac{\partial}{\partial b} \left[Q_{2,2}(a,b) \right] = -2 \; b/c^2 - 1/\chi + 1/c^2, \tag{3.572a, b}$$

the single (3.573a) [(3.574a)] roots lead to singly resonant (3.568b) particular integrals (3.573b) [(3.574b)]

$a = a_\pm(b)$:
$$\Phi_{*\pm}(x,t) = \frac{B\ x^{a_\pm(b)}\ t^b\ \log x}{2\ a_\pm(b) - 1},$$
(3.573a, b)

$b = b_\pm(a)$:
$$\Phi_*^\pm(x,t) = - c^2\ \frac{B\ x^a\ t^{b_\pm(a)}\ \log t}{2\ b_\pm(a) - 1 + c^2/\chi}.$$
(3.574a, b)

Bearing in mind the second-order derivatives (3.575a) [(3.575b)] of the characteristic polynomial (3.518b):

$$\frac{\partial^2}{\partial a^2}\left[Q_{2,2}(a,b)\right] = 2, \qquad \frac{\partial^2}{\partial b^2}\left[Q_{2,2}(a,b)\right] = - 2/c^2, \quad (3.575a, b)$$

the doubly resonant (3.568b) particular integrals (3.576b) [(3.577b)] correspond to (3.576a; 3.520a–f) [(3.577a; 3.521a–f)]:

$b = b^\pm$:
$$\Phi_{**\pm}(x,t) = \frac{B}{2}\ \sqrt{x}\ t^{b^\pm}\ \log^2 t,$$
(3.576a, b)

$a = a^\pm$:
$$\Phi_{**}^\pm(x,t) = - c^2\ \frac{B}{2}\ x^{a^\pm}\ t^{(1-c^2\chi)/2}\ \log^2 x.$$
(3.577a, b)

Thus, *the analogue wave-diffusion equation with homogeneous power coefficients (3.505) forced by powers (3.570) has particular: (i) integral (3.571c) in the non-resonant case (3.571a, b) excluding the roots (3.519a–d) of the characteristic polynomial (3.518b–d); (ii) integrals (3.573b) [(3.574b)] in the four singly resonant cases corresponding to distinct (3.573a; 3.519a, b) [(3.574a; 3.519c, d)] roots (3.519a) [(3.519b)] of the characteristic polynomial (3.518b–d); and (iii) integrals (3.576b) [(3.577b)] in the doubly resonant case (3.576a) [(3.577a)] of double roots (3.520a–f) [(3.521a–f)] of the characteristic polynomial (3.518b).* The method of the inverse characteristic polynomial of ordinary (3.528a, b) [homogeneous (3.529a, b)] derivatives can be used to obtain a particular integral of the forced linear partial differential equation with constant (homogeneous power) coefficients [Subsections 3.4.35–3.4.39 (3.6.9)].

3.6.9 INVERSE CHARACTERISTIC POLYNOMIAL OF HOMOGENEOUS PARTIAL DERIVATIVES

The forced linear partial differential equation with homogeneous power coefficients (3.505) ≡ (3.535a, b) ≡ (3.578a) involving the characteristic polynomial (3.518b) of homogeneous partial derivatives (3.529a, b) has a particular integral (3.578b):

$$B(x,y) = \left\{Q_{N,M}(\delta_x, \delta_y)\right\}\Phi(x,y): \qquad \Phi(x,y) = \left\{Q_{N,M}(\delta_x, \delta_y)\right\}^{-1} B(x,y),$$
(3.578a, b)

involving the **inverse characteristic polynomial of homogeneous partial deriva-tives,** *which can be interpreted as four ordinary derivatives (Subsections 3.4.35–3.4.39), and leads to finite expressions for forcing by powers of logarithms (3.531a, b; 3.532b).*

As an example, the analogue wave-diffusion equation with homogeneous power coefficients (3.516) forced by a linear product of logarithms (3.579) is considered:

$$b \log x \log t = \{x^2 \, \partial_x^2 - c^{-2} \, t^2 \, \partial_t^2 - \chi^{-1} \, t \, \partial_t\} \, \Phi(x,t) = \{Q_{2,2}(\delta_x, \delta_t)\} \, \Phi(x,t)$$

$$= \{\delta_x^2 - \delta_x - c^{-2} \, \delta_t^2 + c^{-2} \, \delta_t - \chi^{-1} \, \delta_t\} \, \Phi(x,t), \qquad (3.579a\text{–}c)$$

where (3.534b) is used to obtain (3.579c) in agreement (3.579b) with the characteristic polynomial (3.518b) of homogeneous partial derivatives (3.529a, b). The inverse of (3.579c) is

$$\Phi(x,t) = \left\{ \delta_x^2 - \delta_x - c^{-2} \, \delta_t^2 + (c^{-2} - \chi^{-1}) \, \delta_t \, \right\}^{-1} b \log x \log t, \qquad (3.580)$$

and the integration with regard to x (t) requires expansion to first (third) order in (3.581a–d) [(3.582a–d)]:

$$\Phi_1(x,t) = - \, b \, \delta_x^{-1} \, \{1 - \delta_x - (c^{-2} - \chi^{-1}) \, \delta_x^{-1} \, \delta_t + c^{-2} \, \delta_x^{-1} \, \delta_t^2\}^{-1} \log x \log t$$

$$= - \, b \, \delta_x^{-1} \, \{1 + \delta_x + (c^{-2} - \chi^{-1}) \, \delta_x^{-1} \, \delta_t\} \log x \log t$$

$$= - \, b \, \delta_x^{-1} \, \left\{ \log x \log t + \log t + \frac{c^{-2} - \chi^{-1}}{2} \log^2 x \right\}$$

$$= - \, b \, \log x \log t \left(1 + \frac{1}{2} \log x \right) - b \, \frac{\chi - c^2}{6 \, c^2 \, \chi} \log^3 x; \qquad (3.581a\text{–}d)$$

$$\Phi_2(x,t) = (c^{-2} - \chi^{-1})^{-1} \, \delta_t^{-1} \, \{1 - [c^{-2} / (c^{-2} - \chi^{-1})] \, \delta_t$$

$$- (c^2 - \chi^{-1})^{-1} \, (\delta_x - \delta_x^2) \, \delta_t^{-1}\} \, b \log x \log t$$

$$= \frac{b \, c^2 \, \chi}{\chi - c^2} \, \delta_t^{-1} \left\{ 1 + \frac{\chi}{\chi - c^2} \, \delta_t + \frac{c^2 \, \chi}{\chi - c^2} \, \delta_x \, \delta_t^{-1} \right\} \log x \log t$$

$$= \frac{b \, c^2 \, \chi}{\chi - c^2} \, \delta_t^{-1} \left\{ \log x \log t + \frac{\chi}{\chi - c^2} \, \log x + \frac{c^2 \chi / 2}{\chi - c^2} \, \log^2 t \right\}$$

$$= \frac{b \, c^2 \, \chi}{\chi - c^2} \left\{ \frac{1}{2} \, \log x \log^2 t + \frac{\chi}{\chi - c^2} \, \log x \log t + \frac{c^2 \chi / 6}{\chi - c^2} \, \log^3 t \right\}. \qquad (3.582a\text{–}d)$$

Thus, *the analogue wave-diffusion equation with homogeneous power coefficients and linear space-time forcing by logarithms (3.579a–c) has particular integrals (3.581d) and (3.582d), whose difference (3.583a, b) is a particular integral of the unforced equation (3.516):*

$$\Phi(x,t) = \Phi_1(x,t) - \Phi_2(x,t)$$

$$= -b \frac{\chi - c^2}{6 c^2 \chi} \log^3 x - b \log x \log t \left[1 + \left(\frac{c \chi}{\chi - c^2} \right)^2 \right]$$

$$- \frac{b}{2} \log^2 x \log t - \frac{b c^2 \chi / 2}{\chi - c^2} \log x \log^2 t - \frac{b}{6} \left(\frac{c^2 \chi}{\chi - c^2} \right)^2 \log^3 t.$$

$$(3.583a, b)$$

The inverse characteristic polynomial of homogeneous partial derivatives can also be applied to products of powers by other smooth functions like polynomials of logarithms (Subsection 3.6.10).

3.6.10 FORCING BY THE PRODUCT OF POWERS AND LOGARITHMS

The analogue of (3.389a, b) changing ordinary (3.528a, b) to homogeneous (3.529a, b) derivatives is: *the linear forced differential equation with homogeneous partial derivatives and constant coefficients (3.584b) forced by the product of powers by a smooth function (3.584a) has particular integrals (3.584c):*

$$\Psi \in \mathcal{D}^\infty (|R^2): \quad \left\{ \varrho_{N,M}(\delta_x, \delta_y) \right\} \Phi(x,y) = x^a \, y^b \, \Psi(x,y):$$

$$\Phi(x,y) = x^a \, y^b \left\{ \varrho_{N,M}(a + \delta_x, b + \delta_y) \right\}^{-1} \Psi(x,y), \quad (3.584a\text{--}c)$$

where the inverse characteristic polynomial of homogeneous partial derivatives appears with a translation. As an example, the analogue wave-diffusion equation with homogeneous power coefficients (3.516) forced by a product of powers and logarithms (3.585) is considered:

$$x^2 \frac{\partial^2 \Phi}{\partial x^2} - \frac{t^2}{c^2} \frac{\partial^2 \Phi}{\partial t^2} - \frac{t}{\chi} \frac{\partial \Phi}{\partial t} = b \, t^2 \, x^3 \, \log t \, \log x, \quad (3.585)$$

whose particular integral (3.584b) is (3.586a–d):

$$\Phi(x,t) = \left\{ x^2 \, \partial_x^2 - c^{-2} \, t^2 \, \partial_t^2 - \chi^{-1} \, t \, \partial_t \right\}^{-1} b \, t^2 \, x^3 \, \log t \, \log x$$

$$= \left\{ \delta_x^2 - \delta_x - c^{-2} \left(\delta_t^2 - \delta_t \right) - \chi^{-1} \, \delta_t \right\}^{-1} b \, t^2 \, x^3 \, \log t \, \log x$$

$$= b \, t^2 \, x^3 \left\{ (3 + \delta_x)^2 - (3 + \delta_x) - c^{-2} \, (2 + \delta_t)^2 \right.$$

$$\left. + (c^{-2} - \chi^{-1})(2 + \delta_t) \right\}^{-1} \log t \, \log x$$

$$= b \, t^2 \, x^3 \left\{ 6 - \frac{2}{c^2} - \frac{2}{\chi} + 5 \, \delta_x - \left(\frac{3}{c^2} + \frac{1}{\chi} \right) \delta_t \right.$$

$$\left. + \delta_x^2 - c^{-2} \, \delta_t^2 \right\}^{-1} \log t \, \log x. \quad (3.586a\text{--}d)$$

Introducing the constant (3.587a), the particular integral (3.586d) is evaluated by (3.587b, c):

$$\gamma = 6 - \frac{2}{c^2} - \frac{2}{\chi}:$$

$$\Phi(x,t) = \frac{b\,t^2\,x^3}{\gamma}\left\{1 - \frac{5}{\gamma}\,\delta_x + \frac{1}{\gamma}\left(\frac{3}{c^2} + \frac{1}{\chi}\right)\delta_t\right\}\log t\,\log x$$

$$= \frac{b\,t^2\,x^3}{\gamma}\left[\log x\,\log t - \frac{5}{\gamma}\log t + \frac{1}{\gamma}\left(\frac{3}{c^2} + \frac{1}{\chi}\right)\log x\right].$$

$$(3.587a\text{–}c)$$

Thus, *the analogue wave-diffusion equation with homogeneous power coefficients (3.516) forced by a product of powers and logarithms (3.585) has particular integrals (3.587c).* Similar methods apply to a single (simultaneous system of) linear equation(s) forced by homogeneous power coefficients [Section 3.6 (3.7)] using the characteristic (matrix of) polynomial(s) of homogeneous derivatives.

3.7 SIMULTANEOUS SYSTEM OF PARTIAL DIFFERENTIAL EQUATIONS WITH POWER COEFFICIENTS

The analogy of ordinary (homogeneous) derivatives applies to: (i) linear ordinary differential equations with constant (power) coefficients [IV.1.3–IV.1.5 (IV.1.6–IV.1.8)]; (ii) linear simultaneous ordinary differential equations with constant (power) coefficients [IV.7.4–IV.7.5 (IV.7.6–IV.7.7)]; (iii) linear partial differential equations with constant (power) coefficients [Section 3.4 (3.6)]; and (iv) linear simultaneous partial differential equations with constant (power) coefficients [Section 3.5 (3.7)]. Concerning linear simultaneous partial differential equations with power coefficients (Section 3.7), the analogy applies to: (i) the general integral of the unforced system of simultaneous equations (Subsections 3.7.1–3.7.7); and (ii) a particular integral of the forced system of simultaneous equations, such as: (i) resonant and non-resonant forcing by powers (Subsections 3.7.8–3.7.9); and (ii) use of the inverse partial differential operator (Subsection 3.7.10–3.7.11).

A simultaneous system of linear partial differential equations with homogeneous power coefficients can be re-written in terms of a matrix of polynomials of homogeneous derivatives (Subsection 3.7.1); the determinant specifies the characteristic polynomial whose single or multiple roots determine the natural integrals, whose linear superposition specifies the general integral of the unforced system (Subsection 3.7.2). An example is a coupled system of two first-order partial differential equations with homogeneous derivatives (Subsection 3.7.4) analogous to a wave equation (Subsection 3.7.3). The general integral of the coupled system becomes unique with two boundary conditions that can be applied: (i) to one function and its derivative (Subsection 3.7.5); and (ii) to the two functions (Subsection 3.7.6).

The original (homogeneous) wave equation is a particular case of linear partial differential equations with all derivatives of the same order, which for constant (power) coefficients, lead to similarity solutions (Sections 1.6–1.9) [analogue similarity solutions (Subsection 3.7.7)] that involve functions of a linear combination (product of powers) of the independent variables. The forced simultaneous system of linear partial differential equations with homogeneous power coefficients: (i) have non-resonant or resonant solutions when forced by powers (Subsection 3.7.8); and (ii) other cases of forcing, like logarithms, can be addressed by the method of the inverse matrix of polynomials of homogeneous partial derivatives (Subsection 3.7.10). Both methods (i) [(ii)] are exemplified by the coupled system of first-order equations corresponding to the homogeneous wave equation either unforced (Subsections 3.7.3–3.7.6) or with power (logarithmic) forcing [Subsection 3.7.9 (3.7.11)].

3.7.1 MATRIX OF POLYNOMIALS OF HOMOGENEOUS DERIVATIVES

A simultaneous system of linear partial differential equations (3.588a, b) with homogeneous power coefficients, with two independent variables x, y, L dependent variables Φ_j and forcing functions:

$$i, j = 1, \ldots, L: \qquad \sum_{j=1}^{L} \left\{ P_{ij} \left(\frac{\partial}{\partial x}, \frac{\partial}{\partial y} \right) \right\} \Phi_j (x, y) = B_i (x, y), \qquad (3.588a, b)$$

involves a matrix of polynomials of ordinary derivatives multiplied by powers with exponents equal to the order of derivation (3.589a):

$$P_{ij}\left(\frac{\partial}{\partial x}, \frac{\partial}{\partial y}\right) \equiv \sum_{n=0}^{N_{ij}} \sum_{m=0}^{M_{ij}} A_{n,m}^{i,j}\ x^n\ y^m\ \frac{\partial^{n+m}}{\partial x^n\ \partial y^m}$$

$$= \sum_{n=0}^{N_{ij}} \sum_{m=0}^{M_{ij}} A_{n,m}^{i,j}\ x\frac{\partial}{\partial x}\left(x\frac{\partial}{\partial x}-1\right)\dots\left(x\frac{\partial}{\partial x}-n+1\right)$$

$$y\frac{\partial}{\partial y}\left(y\frac{\partial}{\partial y}-1\right)\dots\left(y\frac{\partial}{\partial y}-m+1\right)$$

$$= Q_{ij}\left(x\frac{\partial}{\partial x}, y\frac{\partial}{\partial y}\right), \tag{3.589a-c}$$

and use of the relation (3.534b) leads (3.589b) to a polynomial (3.589c) of homogeneous derivatives (3.529a, b). Thus, *a simultaneous system of linear partial differential equations (3.588a, b) with homogeneous power coefficients (3.589a) is equivalent to the system (3.590a, b):*

$$i,j = 1,\dots,L: \qquad \sum_{j=1}^{L}\left\{Q_{ij}\left(x\frac{\partial}{\partial x}, y\frac{\partial}{\partial y}\right)\right\}\Phi_j(x,y) = B_i(x,y), \tag{3.590a, b}$$

involving polynomials (3.589b, c) of homogeneous derivatives (3.528a, b). The transformation from (3.588a, b; 3.589a) to (3.590a, b; 3.589b, c) is used to solve both the unforced (forced) systems [Subsections 3.7.2–3.7.7 (3.7.8–3.7.11)].

3.7.2   CHARACTERISTIC POLYNOMIAL OF HOMOGENEOUS DERIVATIVES

Considering the property (3.529a–d), the unforced (3.591a) system (3.590a, b) has solutions with the same powers and distinct coefficients (3.591a), leading, on substitution in (3.590b), to (3.591c–e):

$$B_i(x,y) = 0: \qquad \Phi_j(x,y) = C_j\ x^a\ y^b:$$

$$0 = \sum_{j=1}^{L}\left\{Q_{ij}\left(x\frac{\partial}{\partial x}, y\frac{\partial}{\partial y}\right)\right\}\Phi_j(x,y) = x^a\ y^b\sum_{j=1}^{L} Q_{ij}(a,b)\ C_j$$

$$= \sum_{j=1}^{L} Q_{ij}(a,b)\ \Phi_j(x,y). \tag{3.591a-e}$$

A non-trivial solution (3.592a) requires the determinant to be zero (3.592b), and this specifies the characteristic polynomial (3.592c):

$$\{\Phi_1(x,y),\dots,\Phi_L(x,y)\} \neq \{0,\dots,0\}: \quad 0 = Det\{Q_{ij}(a,b)\} \equiv Q_{N,M}(a,b). \tag{3.592a-c}$$

The single roots (3.508a, b) of the characteristic polynomial (3.592c) specify the natural integrals (3.509a, b), and the general integral is: (i) for the first independent variable (3.510a) [(3.510b)] the sum of the natural integrals (3.593b) [(3.594b)] with coefficient unity (3.593a) [(3.594a)]:

$$E_{1,n} = 1: \qquad \Phi_i(x,y) = \sum_{n=1}^{N} E_{i,n} \int y^b \, x^{a_n(b)} \, C_n(b) \, db, \qquad (3.593a, b)$$

$$E^{1,m} = 1: \qquad \Phi_i(x,y) = \sum_{m=1}^{M} E^{i,m} \int x^a \, y^{b_m(a)} \, C^m(a) \, da; \qquad (3.594a, b)$$

and (ii) for all other dependent variables, a sum of the same natural integrals with coefficients specified by compatibility of the system (3.591e).

In the case of multiple roots (3.511a, b) [(3.512a, b)] of the characteristic polynomial (3.592c), the natural integrals are replaced by (3.513a–c) [(3.514a–c)] in (3.515a) [(3.515b)] leading to (3.595a, b) [(3.596a, b)]:

$$E_{1,r,p} = 1: \qquad \Phi_i(x,y) = \sum_{r=1}^{R} \sum_{p=1}^{\alpha_r} E_{i,r,p} \int y^b \, x^{a_r(b)} \, \log^{p-1} x \, C_{r,p}(b) \, db, \qquad (3.595a, b)$$

$$E^{1,s,q} = 1: \qquad \Phi_i(x,y) = \sum_{s=1}^{S} \sum_{q=1}^{\beta_s} E^{i,s,q} \int x^a \, y^{b_m(a)} \, \log^{q-1} y \, C^{s,q}(a) \, da. \qquad (3.596a, b)$$

Thus, *the simultaneous linear unforced system of partial differential equations with homogeneous derivatives (3.591c) has general integral (3.593a, b; 3.594a, b) [(3.595a, b; 3.596a, b)] in the case of single (3.508a, b) [multiple (3.511a, b; 3.512a, b)] roots of the characteristic polynomial (3.592c), where: (i) for the first dependent variable $\Phi_1(x,y)$, the natural integrals (3.509a, b) [(3.513a–c; 3.514a–c)] are added with coefficient unity (3.593a; 3.594a) [(3.595a; 3.596a)]; (ii) all remaining dependent variables $\Phi_2(x.y),\ldots, \Phi_L(x.y)$ have coefficients determined by the compatibility of the unforced system (3.591c).* As an example, a coupled system of first-order partial differential equations with homogeneous derivatives that leads to a homogeneous analogue of the wave equation is considered next (Subsection 3.7.3).

3.7.3 System of Homogeneous Derivatives Analogous to the Wave Equation

In the Cartesian one-dimensional wave equation (3.222b) ≡ (3.597d), the wave speed can be suppressed (3.597d) via the change of independent variable (3.597a, b):

$$y \equiv c\,t, \qquad F(x,t) = \Phi(x,y): \qquad \frac{\partial^2 F}{\partial x^2} - \frac{1}{c^2} \frac{\partial^2 F}{\partial t^2} = 0 \quad \Leftrightarrow \quad \frac{\partial^2 \Phi}{\partial x^2} - \frac{\partial^2 \Phi}{\partial y^2} = 0.$$

$$(3.597a-d)$$

The **homogeneous wave equation** with homogeneous derivatives is (3.598a, b):

$$0 = \left\{ \left(x \frac{\partial}{\partial x} \right)^2 - \left(y \frac{\partial}{\partial y} \right)^2 \right\} \Phi(x.y) = x^2 \frac{\partial^2 \Phi}{\partial x^2} + x \frac{\partial \Phi}{\partial x} - y \frac{\partial \Phi}{\partial y} - y^2 \frac{\partial^2 \Phi}{\partial y^2},$$

(3.598a, b)

corresponding to (3.545c) ≡ (3.546) setting $c = 1$ and $t = y$ in agreement with (3.597a). The system of coupled first-order partial differential equations with homogeneous derivatives (3.599a, b):

$$x \frac{\partial \Phi}{\partial x} = y \frac{\partial \Psi}{\partial y}, \qquad\qquad x \frac{\partial \Psi}{\partial x} = y \frac{\partial \Phi}{\partial y}, \qquad (3.599\text{a, b})$$

leads by elimination (3.600a–c)

$$\left(x \frac{\partial}{\partial x} \right)^2 \Phi = x \frac{\partial}{\partial x} \left(y \frac{\partial \Psi}{\partial y} \right) = y \frac{\partial}{\partial y} \left(x \frac{\partial \Psi}{\partial x} \right) = \left(y \frac{\partial}{\partial y} \right)^2 \Phi, \quad (3.600\text{a–c})$$

to the homogeneous wave equation with homogeneous derivatives (3.600c) ≡ (3.598a, b).

The system (3.599a, b) can be put in matrix form (3.601c) and has solutions (3.591b) ≡ (3.601a, b) leading to (3.601d, e):

$$\Phi(x,y) = C_1 \ x^a \ y^b:$$
$$\Psi(x,y) = C_2 \ x^a \ y^b:$$

$$0 = \begin{bmatrix} x \dfrac{\partial}{\partial x} & -y \dfrac{\partial}{\partial y} \\[2mm] -y \dfrac{\partial}{\partial y} & x \dfrac{\partial}{\partial x} \end{bmatrix} \begin{bmatrix} \Phi(x,y) \\[1mm] \Psi(x,y) \end{bmatrix}$$

$$= x^a \ y^b \begin{bmatrix} a & -b \\ -b & a \end{bmatrix} \begin{bmatrix} C_1 \\ C_2 \end{bmatrix} = \begin{bmatrix} a & -b \\ -b & a \end{bmatrix} \begin{bmatrix} \Phi(x,y) \\ \Psi(x,y) \end{bmatrix}.$$

(3.601a–e)

A non-trivial solution (3.602a) requires that the determinant be zero (3.602b) specifying the characteristic polynomial (3.602c) whose roots are (3.602d):

$$\{\Phi(x,y), \Psi(x,y)\} \neq \{0,0\}: \qquad 0 = a^2 - b^2 = Q_{2,2}(a,b), \qquad b = \pm a. \qquad (3.602\text{a–d})$$

The natural integrals for single roots (3.602d) in (3.594b) are (3.603a–c):

$$\Phi_{\pm}(x,y) = \int x^a \ y^{\pm a} \ C_{\pm}(a) \ da = \int (x \ y^{\pm 1})^a \ C_{\pm}(a) \ da = f_{\pm}(x \ y^{\pm 1})$$

$$= g_+(xy), \ g_-\left(\frac{x}{y}\right), \tag{3.603a–e}$$

which is a function (3.603d) of xy and a function (3.603e) of x/y. The multiple roots of (3.602c, d) are $a = 0 = b$ and lead in (3.601a, b) to trivial solutions stating that both Φ and Ψ are constant. The two natural integrals can be considered (Subsection 3.7.4) as solutions of the homogeneous wave equation (3.598a, b) and of the coupled system (3.599a, b).

3.7.4 CORRESPONDING WAVE AND COUPLED SYSTEM OF HOMOGENEOUS DERIVATIVES

It can be checked that a function (3.603d) ≡ (3.605a) of xy leads to (3.604b–e) where a prime denotes a derivative with regard to the argument:

$$\Phi_+(x,y) = g_+(x\ y): \qquad x\,\frac{\partial \Phi_+}{\partial x} = x\ y\ g_+'\ (xy) = y\,\frac{\partial \Phi_+}{\partial y},$$

$$\left(x\,\frac{\partial}{\partial x}\right)^2 \Phi_+(x,y) = x\ y\ g_+'(xy) + x^2\ y^2\ g_+''(xy) = \left(y\,\frac{\partial}{\partial y}\right)^2 \Phi_+(x,y) \tag{3.604a–e}$$

and, thus, satisfies the homogeneous wave equation (3.598a). It can also be checked that a function (3.603e) ≡ (3.605a) of x/y leads to (3.605b–e)

$$\Phi_-(x,y) = g_-\left(\frac{x}{y}\right): \qquad x\,\frac{\partial \Phi_-}{\partial x} = \frac{x}{y}\ g_-'\left(\frac{x}{y}\right) = -\ y\,\frac{\partial \Phi_-}{\partial y},$$

$$\left(x\,\frac{\partial}{\partial x}\right)^2 \Phi_-(x,y) = \frac{x}{y}\ g_-'\left(\frac{x}{y}\right) + \frac{x^2}{y^2}\ g_-''\left(\frac{x}{y}\right) = \left(y\,\frac{\partial}{\partial y}\right)^2 \Phi_-(x,y), \tag{3.605a–e}$$

and, thus, also satisfies the homogeneous wave equation (3.598a). This leads to *the comparison between the **original (homogeneous) wave equation** with unit velocity* $c = 1$ *in (3.545b) [(3.545c)] with ordinary (3.597d) ≡ (3.606c) [homogeneous (3.598a) ≡ (3.607c)] derivatives whose general integral (3.606d) [(3.607d)] is the sum of two twice differentiable functions (3.606a, b) [(3.607a, b)]:*

$$f_{\pm} \in \mathcal{D}^2 \ (\text{I } R): \qquad \left\{\frac{\partial^2}{\partial x^2} - \frac{\partial^2}{\partial y^2}\right\}\Phi(x,y) = 0, \qquad \Phi(x,y) = f_+(x-y) + f_-(x+y), \tag{3.606a–d}$$

$$g_{\pm} \in \mathcal{D}^2 \ (\text{I } R): \qquad \left\{\left(x\,\frac{\partial}{\partial x}\right)^2 - \left(y\,\frac{\partial}{\partial y}\right)^2\right\}\Phi(x,y) = 0, \qquad \Phi(x,y) = g_+(xy) + g_-\left(\frac{x}{y}\right) \tag{3.607a–d}$$

of the sum and difference (product and ratio) of the two independent variables x, y.

The general integral (3.607d) specifies one dependent variable in the system (3.599a, b) and the other may be determined by (3.599a) ≡ (3.608a) leading to (3.608b):

$$\frac{\partial \Psi}{\partial y} = \frac{x}{y} \frac{\partial}{\partial x}\left[g_+(xy) + g_-\left(\frac{x}{y}\right)\right] = x\, g_+'(xy) + \frac{x}{y^2}\, g_-'\left(\frac{x}{y}\right): \qquad\qquad \text{(3.608a, b)}$$

$$\Psi(x,y) = g_+(xy) - g_-\left(\frac{x}{y}\right), \qquad\qquad \text{(3.608c)}$$

which may be integrated (3.608c). Thus, *the coupled system of unforced first-order linear partial differential equations with homogeneous derivatives (3.599a, b) has general integral (3.609c, d) involving two differentiable functions (3.609a, b):*

$$g_\pm \in \mathcal{D}\,(\mid R): \qquad \Phi(x,y) = g_+(xy) + g_-\left(\frac{x}{y}\right), \qquad \Psi(x,y) = g_+(xy) - g_-\left(\frac{x}{y}\right).$$

$$\text{(3.609a–d)}$$

It can also be checked that (3.599b) is also satisfied by (3.609a–d) since (3.609c) [(3.609d)] implies (3.610a, b) [(3.611a, b)], that coincide (3.611c) ≡ (3.599b):

$$y\frac{\partial \Phi}{\partial y} = y\frac{\partial}{\partial y}\left[g_+(xy) + g_-\left(\frac{x}{y}\right)\right] = x\,y\,g_+'(xy) - \frac{x}{y}\, g_-'\left(\frac{x}{y}\right), \qquad \text{(3.610a, b)}$$

$$x\frac{\partial \Psi}{\partial x} = x\frac{\partial}{\partial x}\left[g_+(xy) - g_-\left(\frac{x}{y}\right)\right] = x\,y\,g_+'(xy) - \frac{x}{y}\, g_-'\left(\frac{x}{y}\right) = y\frac{\partial \Phi}{\partial y}. \quad \text{(3.611a–c)}$$

The arbitrary differentiable functions (3.609a, b) appear in the solution (3.609c, d) of coupled linear differential equations with homogeneous derivatives (3.599a, b); since the system (3.599a, b) is of the second order (3.600a–c), its solution becomes unique, imposing two boundary conditions, for example: (i) on one function and its derivative (Subsection 3.7.5); and (ii) on two functions (Subsection 3.7.6).

3.7.5 BOUNDARY CONDITIONS ON ONE FUNCTION AND ITS DERIVATIVE

Choosing boundary conditions on one function (3.612a) and its derivative (3.612b–d), from (3.609c) follow:

$$\Phi(x,1) = g_+(x) + g_-(x), \qquad\qquad \text{(3.612a)}$$

$$\dot{\Phi}(x,1) \equiv \lim_{y \to 1}\frac{\partial \Phi(x,y)}{\partial y} = \lim_{y \to 1}\left[x\,g_+'(xy) - \frac{x}{y^2}\, g_-'\left(\frac{x}{y}\right)\right]$$

$$= x\left[g_+'(x) - g_-'(x)\right]. \qquad\qquad \text{(3.612b–d)}$$

The system (3.612a, d) coincides with (3.553a, d) with $c = 1$ and, thus, the solution (3.613a, b) is (3.555a, b) with $c = 1$:

$$2\, g_{\pm}(x) = \Phi(x,1) \pm \int^{x} \frac{\Phi(\xi,1)}{\xi}\, d\xi. \tag{3.613a, b}$$

Substitution of (3.613a, b) in (3.609c, d) leads to (3.614a, b):

$$2\, \Phi(x,y) = \Phi(xy,1) + \Phi\left(\frac{x}{y}, 1\right) + \int^{x}\left[\dot{\Phi}(\xi y,1) - \dot{\Phi}\left(\frac{\xi}{y}, 1\right)\right]\frac{d\xi}{\xi}, \tag{3.614a}$$

$$2\, \Psi(x,y) = \Phi(xy,1) - \Phi\left(\frac{x}{y}, 1\right) + \int^{x}\left[\dot{\Phi}(\xi y,1) + \dot{\Phi}\left(\frac{\xi}{y}, 1\right)\right]\frac{d\xi}{\xi}, \tag{3.614b}$$

where (3.614a) coincides with (3.556) with $c = 1$, and (3.614b) satisfies the compatibility relations (3.599a, b). Thus, *the coupled system of first-order partial differential equations with homogeneous derivatives (3.599a, b) with boundary conditions on one function (3.612a) and its derivative (3.612b–d) has the unique solution (3.614a, b).*

Choosing the particular boundary conditions (3.615a, b) similar to (3.558a, b) with $= 1$:

$$\Phi(x,1) = e^{x}, \qquad \dot{\Phi}(x,1) = \lim_{y \to 1} \frac{\partial \Phi(x,y)}{\partial y} = e^{-x}\, x, \tag{3.615a–c}$$

leads in (3.613a, b) to:

$$2\, g_{\pm}(x) = e^{x} \pm \int^{x} e^{-\xi}\, d\xi = e^{x} \mp e^{-x} = 2\sinh x,\ 2\cosh x. \tag{3.616a, b}$$

Substitution of (3.616a, b) in (3.609c, d) specifies the general integrals

$$\Phi(x,y) = \sinh(xy) + \cosh\left(\frac{x}{y}\right), \qquad \Psi(x,y) = \sinh(xy) - \cosh\left(\frac{x}{y}\right), \tag{3.617a, b}$$

where (3.617a) coincides with (3.561) with $c = 1$ and (3.617b) is compatible with the system (3.599a, b). Thus, *the coupled system of first-order partial differential equations with homogeneous derivatives (3.599a, b) with boundary conditions on one function (3.615a) and its derivative (3.615b, c) has solution (3.617a, b).* It can be checked that: (i–ii) the boundary conditions (3.615a) [(3.615b, c)] are met by (3.618a) [(3.618b–d)]:

$$\Phi(x,1) = \sinh x + \cosh x = e^{x}, \tag{3.618a}$$

$$\dot{\Phi}(x,1) = \lim_{y \to 1} \frac{\partial \Phi(x,y)}{\partial y} = \lim_{y \to 1} \left[x \cosh(xy) - \frac{x}{y^2} \sinh\left(\frac{x}{y}\right) \right]$$

$$= x \left(\cosh x - \sinh x \right) = x \ e^{-x}; \qquad (3.618\text{b–d})$$

and (iii–iv) the coupled partial differential equations (3.559a) [(3.559b)] are also satisfied by (3.619a) [(3.619b)]:

$$y \frac{\partial \Psi}{\partial y} = x \ y \cosh(xy) + \frac{x}{y} \sinh\left(\frac{x}{y}\right) = x \ \frac{\partial \Phi}{\partial x}, \qquad (3.619\text{a})$$

$$x \frac{\partial \Psi}{\partial x} = x \ y \cosh(xy) - \frac{x}{y} \sinh\left(\frac{x}{y}\right) = y \ \frac{\partial \Phi}{\partial y}. \qquad (3.619\text{b})$$

Instead of two boundary conditions (3.612a–d) on one function and its derivative (Subsection 3.7.5), the two boundary conditions can be applied to the two functions (Subsection 3.7.6).

3.7.6 BOUNDARY CONDITIONS APPLIED TO TWO FUNCTIONS

Choosing boundary conditions (3.620a, b) applied to the two functions (3.609c, d):

$$\Phi(x,1) = g_+(x) + g_-(x), \qquad \Psi(x,1) = g_+(x) - g_-(x), \qquad (3.620\text{a, b})$$

leads to:

$$2 \ g_\pm(x) = \Phi(x,1) \pm \Psi(x,1). \qquad (3.621\text{a, b})$$

Substituting (3.621a, b) in (3.609c, d), it follows that *the coupled system of linear first-order partial differential equations with homogeneous derivatives (3.599a, b) with boundary conditions (3.620a, b) has solutions:*

$$2 \ \Phi(x,y) = \Phi(xy,1) + \Phi\left(\frac{x}{y}, 1\right) + \Psi(xy,1) - \Psi\left(\frac{x}{y}, 1\right), \qquad (3.622\text{a})$$

$$2 \ \Psi(x,y) = \Phi(xy,1) - \Phi\left(\frac{x}{y}, 1\right) + \Psi(xy,1) + \Psi\left(\frac{x}{y}, 1\right). \qquad (3.622\text{b})$$

It can be checked that (3.622a, b): (i) satisfy (3.599a, b) because they are of the form (3.609c, d); and (ii) reduce to the boundary conditions (3.620a, b) for $= 1$.
Choosing the particular boundary conditions (3.623a, b):

$$\Phi(x,1) = e^x, \qquad \Psi(x,1) = -e^{-x}, \qquad (3.623\text{a, b})$$

the general integral (3.622a, b) becomes:

$$\Phi(x,y) = \frac{e^{xy} - e^{-xy}}{2} + \frac{e^{x/y} + e^{-x/y}}{2} = \sinh(xy) + \cosh\left(\frac{x}{y}\right), \qquad (3.624a, b)$$

$$\Psi(x,y) = \frac{e^{xy} - e^{-xy}}{2} - \frac{e^{x/y} + e^{-x/y}}{2} = \sinh(xy) - \cosh\left(\frac{x}{y}\right). \qquad (3.624c, d)$$

Thus, *the coupled system of linear first-order partial differential equations with homogeneous derivatives (3.599a, b) has the same solution (3.617a, b) ≡ (3.624b, d) with distinct boundary conditions: (i) on one function and its derivative (3.615a–c); and (ii) on the two functions (3.623a, b). This proves that the boundary conditions are equivalent* because: (a) the first boundary condition is the same (3.615a) ≡ (3.623a); and (b) the second boundary conditions (3.615b, c) and (3.623b) are equivalent on account of (3.599b) as follows from (3.625a–f):

$$\dot{\Phi}(x,1) \equiv \lim_{y \to 1} \frac{\partial \Phi}{\partial y} = \lim_{y \to 1} \frac{x}{y} \frac{\partial \Psi}{\partial x} = x \lim_{y \to 1} \frac{\partial \Psi}{\partial x} = x \frac{\partial}{\partial x}(- e^{-x}) = x\, e^{-x} = \Psi(x,1).$$

$$(3.625a–f)$$

The analogy between solutions (3.606a–d) [(3.607a–d)] of the original (homogeneous) unforced wave equation extends (Subsection 3.7.6) to general linear unforced partial differential equations with constant (homogeneous power) coefficients and all derivatives of the same order leading to similarity (analogue similarity) solutions.

3.7.7 Similarity Solutions and Homogeneous Analogues

The analogy applies equally to single equations [Sections 3.4 (3.6)] and simultaneous systems [Sections 3.5 (3.7)] and is stated more simply in the former case: *a linear partial differential equation with ordinary (homogeneous) derivatives and constant coefficients (3.626a) [(3.626b)] and all derivatives of the same order:*

$$\sum_{n=0}^{N} \sum_{m=0}^{N-n} A_{n,m} \frac{\partial^N \Phi}{\partial x^n\, \partial y^{N-n}} = 0 = \sum_{n=0}^{N} \sum_{m=0}^{N-n} A_{n,m} \left\{ \left(x \frac{\partial}{\partial x} \right)^n \left(y \frac{\partial}{\partial y} \right)^m \right\} \Psi(x,y),$$

$$(3.626a, b)$$

has (Table 3.9) the characteristic polynomial (3.627a, c) that, in the case of single roots (3.627b):

$$P_{N,N}(a,b) = \sum_{n=0}^{N} \sum_{m=0}^{N-n} A_{n,m}\, a^n\, b^{N-n} = A_{0,N} \prod_{n=1}^{N} (b - \alpha_n\, a) = Q_{N,N}(a,b),$$

$$(3.627a–c)$$

leads to the general integral (3.628a) [(3.629a)]:

$$\Phi(x,y) = \sum_{n=1}^{N} \int \exp\left[a\left(x - \alpha_n\, y\right)\right] C_n(a)\, da = \sum_{n=1}^{N} f_n\left(x - \alpha_n\, y\right), \quad (3.628\text{a, b})$$

$$\Psi(x,y) = \sum_{n=1}^{N} \int \left(x\, y^{\alpha_n}\right)^a C_n(a)\, da = \sum_{n=1}^{N} f_n\left(x\, y^{\alpha_n}\right), \quad (3.629\text{a, b})$$

*which is the sum of **similarity (homogeneous similarity) functions** (3.628b) [(3.629b)] that are N times differentiable (3.630a) and whose variable is a linear combination (3.630b) [a product of powers (3.630c)] of the independent variables* x, y:

$$f_n(\xi) \in \mathcal{D}^N(|\,R): \qquad\qquad \xi = x - \alpha_n\, y\,, \quad x\, y^{\alpha_n}. \qquad (3.630\text{a–c})$$

In the case of multiple roots of the characteristic polynomial (3.631a, b) [(3.631a, c)]:

$$\sum_{r=1}^{R} \beta_r = N: \qquad P_{N,N}(a,b) = A_{0,N} \prod_{r=1}^{R} (b - \alpha_r\, a)^{\beta_r} = Q_{N,N}(a,b) \qquad (3.631\text{a–c})$$

the similarity (homogeneous similarity) functions are multiplied by powers (3.632a) [logarithms (3.632b)]:

$$\Phi(x,y) = \sum_{r=1}^{R} \sum_{p=1}^{\beta_r} x^{p-1}\, f_{r,p}(x - \alpha_r\, y), \qquad (3.632\text{a})$$

$$\Psi(x,y) = \sum_{r=1}^{R} \sum_{p=1}^{\beta_r} \log^{p-1} x\, f_{r,p}(x\, y^{\alpha_r}). \qquad (3.632\text{b})$$

In the particular case (3.602a–d) of the homogeneous wave equation (3.598a, b), the values $\alpha_n = \pm\,1$ with $N = 2$ in (3.629b) lead to (3.603a–e). The unforced (forced) simultaneous system of linear partial differential equations with homogeneous power coefficients is considered in Subsections 3.7.1–3.7.7 (3.7.8–3.7.11).

3.7.8 FORCING OF SYSTEM OF HOMOGENEOUS DERIVATIVES BY POWERS

The simultaneous system of linear partial differential equations (3.590a, b) with forcing by powers (3.633a) has similar power solutions with distinct coefficients (3.633b):

$$B_i(x,y) = B_i\, x^a\, y^b, \qquad\qquad \Phi_j(x,y) = C_j\, x^a\, y^b, \qquad (3.633\text{a, b})$$

related by substitution of (3.633a, b) in (3.590a, b) leading to (3.634a–c):

$$B_i \ x^a \ y^b = \sum_{j=1}^{L} \left\{ Q_{ij} \left(x\frac{\partial}{\partial x} \ , \ y\frac{\partial}{\partial y} \right) \right\} C_j \ x^a \ y^b$$

$$= x^a \ y^b \sum_{j=1}^{L} Q_{ij}(a,b) \ C_j = \sum_{j=1}^{L} Q_{ij}(a,b) \ \Phi_j(x,y), \qquad (3.634a\text{–}c)$$

The system (3.634c) can be inverted (3.635c), provided that the determinant that coincides with the characteristic polynomial (3.635a) $\equiv$ (3.592c) is non-zero (3.635b):

$$Det \left\{ Q_{ij}(a,b) \right\} = Q_{N,M}(a,b) \neq 0: \qquad \Phi_i(x,y) = \frac{x^a \ y^b}{Q_{N;M}(a,b)} \sum_{j=1}^{L} \bar{Q}_{ij}(a,b) \ B_j,$$

$$(3.635a\text{–}c)$$

where $\bar{Q}_{ij}$ are (3.635d) the co-factors or minors of the matrix Q_{ij}

$$\sum_{j=1}^{L} Q_{ij}(a,b) \ \bar{Q}_{jk}(a,b) = \delta_{ij} \ Q_{N,M}(a,b), \qquad (3.635d)$$

in the non-resonant particular integral (3.635c). In the resonant case, when a (b) is a root of multiplicity r (s) of the characteristic polynomial, and the lowest order non-zero derivative is (3.636a, b), the particular integrals are (3.636c):

$$F(a,b) = \frac{\partial^{r+s}}{\partial a^r \ \partial b^s} \left[Q_{N,M}(a,b) \right] \neq 0: \qquad (3.636a, b)$$

$$\Phi_i(x,y) = \frac{1}{F(a,b)} \frac{\partial^{r+s}}{\partial a^r \ \partial b^s} \left\{ x^a \ y^b \sum_{j=1}^{L} \bar{Q}_{ij}(a,b) \ B_j \right\}. \qquad (3.636c)$$

Thus, *the simultaneous system of linear partial differential equations with homogeneous power coefficients forced by powers (3.590a, b; 3.633a) $\equiv$ (3.637a, b):*

$$i, j = 1, \ldots, L: \qquad \sum_{j=1}^{L} \left\{ Q_{ij} \left(x\frac{\partial}{\partial x} \ , \ y\frac{\partial}{\partial y} \right) \right\} \Phi_j(x,y) = B_i \ x^a \ y^b, \qquad (3.637a, b)$$

has particular integrals (3.635c) [(3.636c)] in the non-resonant (3.635a, b) [resonant (3.636a, b)] case when (a,b) are not roots [are roots of multiplicity r (s)] of the characteristic polynomial (3.592c). As an example, the forcing by powers of the system corresponding to the homogeneous wave equation (Subsections 3.7.3–3.7.6) is considered next (Subsection 3.7.9).

3.7.9 Non-Resonant and Resonant Forcing of the Homogeneous Wave System

The system of coupled linear first-order partial differential equations with homogeneous derivatives (3.599a, b) forced by powers (3.638a, b):

$$x \frac{\partial \Phi}{\partial x} - y \frac{\partial \Psi}{\partial y} = - B \, x^a \, y^b, \qquad x \frac{\partial \Psi}{\partial x} - y \frac{\partial \Phi}{\partial y} = 2 \, B \, x^a \, y^b, \qquad (3.638a, b)$$

can be written in matrix form (3.639)

$$\begin{bmatrix} x \dfrac{\partial}{\partial x} & - y \dfrac{\partial}{\partial y} \\[2mm] - y \dfrac{\partial}{\partial y} & x \dfrac{\partial}{\partial x} \end{bmatrix} \begin{bmatrix} \Phi(x,y) \\[2mm] \Psi(x,y) \end{bmatrix} = B \, x^a \, y^b \begin{bmatrix} -1 \\ 2 \end{bmatrix}. \qquad (3.639)$$

The system (3.638a, b) $\equiv$ (3.639) has particular integrals (3.640a, b) leading to (3.640c) by substitution in (3.638a, b):

$$\begin{aligned} \Phi(x,y) &= C_1 \, x^a \, y^b: \\ \Psi(x,y) &= C_2 \, x^a \, y^b: \end{aligned} \qquad \begin{bmatrix} a & -b \\ -b & a \end{bmatrix} \begin{bmatrix} C_1 \\ C_2 \end{bmatrix} = B \begin{bmatrix} -1 \\ 2 \end{bmatrix} \qquad (3.640a\text{–}c)$$

The system (3.640c) is inverted (3.641c, d), provided that the determinant that coincides with the characteristic polynomial (3.592d) $\equiv$ (3.641a) is non-zero (3.641b):

$$Q_{2,2}(a,b) = a^2 - b^2 \neq 0: \qquad (a^2 - b^2) \begin{bmatrix} C_1 \\ C_2 \end{bmatrix} = B \begin{bmatrix} a & b \\ b & a \end{bmatrix} \begin{bmatrix} -1 \\ 2 \end{bmatrix} = B \begin{bmatrix} 2b - a \\ 2a - b \end{bmatrix}.$$

$$(3.641a\text{–}d)$$

Substitution of (3.641d) in (3.640a, b) specifies the particular integrals (3.642b):

$$a \neq \pm \, b: \qquad \begin{bmatrix} \Phi(x,y) \\ \Psi(x,y) \end{bmatrix} = \frac{B \, x^a \, y^b}{a^2 - b^2} \begin{bmatrix} 2b - a \\ 2a - b \end{bmatrix}. \qquad (3.642a, b)$$

in the non-resonant case (3.642a).

The two resonant cases correspond to (3.643a), implying (3.643b, c) for the derivatives of the characteristic polynomial (3.592c):

$$a = \pm \, b: \qquad \frac{\partial}{\partial a} \big[Q_{2,2}(a,b) \big] = 2 \, a, \qquad \frac{\partial}{\partial b} \big[Q_{2,2}(a,b) \big] = - \, 2 \, b. \qquad (3.643a\text{–}c)$$

The particular integrals in the resonant case (3.643a, c) [(3.643a, b)] are (3.644a, b) [(3.645a, b)]:

$$
\begin{bmatrix} \Phi_1(x,y) \\ \Psi_1(x,y) \end{bmatrix} = -\lim_{b \to \pm a} \frac{B}{2b} \frac{\partial}{\partial b} \left\{ x^a \ y^b \begin{bmatrix} 2\,b-a \\ 2\,a-b \end{bmatrix} \right\}
\tag{3.644a}
$$

$$
= \mp \frac{B \ x^a \ y^{\pm a}}{2\,a} \left\{ \begin{bmatrix} 2 \\ -1 \end{bmatrix} + \log y \begin{bmatrix} -a \pm 2\,a \\ 2\,a \mp a \end{bmatrix} \right\},
\tag{3.644b}
$$

$$
\begin{bmatrix} \Phi_2(x,y) \\ \Psi_2(x,y) \end{bmatrix} = \lim_{a \to \pm b} \frac{B}{2\,a} \frac{\partial}{\partial a} \left\{ x^a y^b \begin{bmatrix} 2b-a \\ 2a-b \end{bmatrix} \right\}
\tag{3.645a}
$$

$$
= \pm \frac{B}{2b} x^{\pm b} \ y^b \left\{ \begin{bmatrix} -1 \\ 2 \end{bmatrix} + \log x \begin{bmatrix} 2b \mp b \\ -b \pm 2b \end{bmatrix} \right\},
\tag{3.645b}
$$

corresponding to the forcings (3.646a) [(3.646b)]:

$$
\begin{bmatrix} x \dfrac{\partial}{\partial x} & -\,y \dfrac{\partial}{\partial y} \\ -\,y \dfrac{\partial}{\partial y} & x \dfrac{\partial}{\partial x} \end{bmatrix} \begin{bmatrix} \Phi(x,y) \\ \Psi(x,y) \end{bmatrix} = \begin{bmatrix} -B \\ 2\,B \end{bmatrix} x^a \ y^{\pm a}, \quad y^b \ x^{\pm b}.
\tag{3.646a, b}
$$

Thus, *the coupled linear system of first-order partial differential equations with homogeneous derivatives forced by powers (3.638a, b) ≡ (3.639) has a non-resonant (3.642a) particular integral (3.642b). The two pairs of singly resonant cases (3.644a, b) [(3.645a, b)] are particular integrals of (3.646a) [(3.646b)].* The forcing by powers can be extended to other functions, such as logarithms (Subsection 3.7.10), using the method of the inverse matrix of polynomials of homogeneous partial derivatives.

3.7.10　Inverse Matrix of Polynomials of Homogeneous Partial Derivatives

The simultaneous linear system of partial differential equations with homogeneous derivatives (3.590a, b) with forcing (3.647):

$$
\Sigma_{j=1}^{L} \left\{ Q_{ij}(\delta_x, \delta_y) \right\} \Phi_j(x,y) = \Psi_i(x,y)
\tag{3.647}
$$

has particular integral (3.648):

$$
\Phi_i(x,y) = \left\{ Q_{N,M}(\delta_x, \delta_y) \right\}^{-1} \Sigma_{j=1}^{L} \left\{ \bar{Q}_{ij}(\delta_x, \delta_y) \right\} \Psi(x,y),
\tag{3.648}
$$

involving the matrix of co-factors (3.635d) and the inverse of the characteristic polynomial (3.592c) of homogeneous derivatives. The inverse characteristic polynomial of homogeneous partial derivatives is interpreted as before (Subsections 3.4.21–3.4.29).

As an example, the coupled system of linear first-order partial differential equations with homogeneous derivatives (3.599a, b) forced by the product of logarithms (3.649a, b) is considered:

$$x \frac{\partial \Phi}{\partial x} - y \frac{\partial \Psi}{\partial y} = - b \log x \log y, \qquad x \frac{\partial \Psi}{\partial x} - y \frac{\partial \Phi}{\partial y} = 2 b \log x \log y. \qquad \text{(3.649a, b)}$$

The system (3.649a, b) can be written in matrix form (3.650):

$$\begin{bmatrix} \delta_x & - \delta_y \\ - \delta_y & \delta_x \end{bmatrix} \begin{bmatrix} \Phi(x,y) \\ \Psi(x,y) \end{bmatrix} = b \log x \log y \begin{bmatrix} -1 \\ 2 \end{bmatrix} \qquad \text{(3.650)}$$

that can be inverted (3.651a–c):

$$\begin{bmatrix} \Phi(x,y) \\ \Psi(x,y) \end{bmatrix} = \left(\delta_x^2 - \delta_y^2\right)^{-1} \begin{bmatrix} \delta_x & \delta_y \\ \delta_y & \delta_x \end{bmatrix} \begin{bmatrix} -1 \\ 2 \end{bmatrix} b \log x \log y \qquad \text{(3.651a)}$$

$$= b \left(\delta_x^2 - \delta_y^2\right)^{-1} \begin{bmatrix} - \delta_x + 2\, \delta_y \\ - \delta_y + 2\, \delta_x \end{bmatrix} \log x \log y \qquad \text{(3.651b)}$$

$$= b \left(\delta_x^2 - \delta_y^2\right)^{-1} \begin{bmatrix} - \log y + 2 \log x \\ - \log x + 2 \log y \end{bmatrix}. \qquad \text{(3.651c)}$$

From (3.651a–c) follow (Subsection 3.7.11) two pairs of solutions of the forced system (3.649a, b) plus one pair of solutions of the unforced system (3.599a, b).

3.7.11 ONE (TWO) PAIR(S) OF SOLUTIONS OF THE UNFORCED (FORCED) SYSTEM

Since the second-order homogeneous derivatives of logarithms are zero, the system (3.651c) can be interpreted (3.652a, b) [(3.653a, b)] for $x(y)$:

$$\begin{bmatrix} \Phi_1(x,y) \\ \Psi_1(x,y) \end{bmatrix} = b\, \delta_x^{-2} \begin{bmatrix} - \log y + 2 \log x \\ - \log x + 2 \log y \end{bmatrix} = b \log^2 x \begin{bmatrix} \dfrac{\log x}{3} - \dfrac{\log y}{2} \\ - \dfrac{\log x}{6} + \log y \end{bmatrix},$$

$$\text{(3.652a, b)}$$

$$\begin{bmatrix} \Phi_2(x,y) \\ \Psi_2(x,y) \end{bmatrix} = -\, b\, \delta_y^{-2} \begin{bmatrix} \log y - 2\log x \\ -\log x + 2\log y \end{bmatrix} = -\, b\, \log^2 y \begin{bmatrix} \dfrac{\log y}{6} - \log x \\[2mm] -\dfrac{\log x}{2} + \dfrac{\log y}{3} \end{bmatrix}.$$

$$(3.653\text{a, b})$$

Thus, *the coupled system of linear first-order partial differential equations with homogeneous derivatives and bilinear logarithmic forcing (3.649a, b) $\equiv$ (3.650) have two distinct particular integrals (3.652a,b) and (3.653a,b). Their difference (3.654a–d):*

$$\Phi(x,y) = \Phi_1(x,y) - \Phi_2(x,y) = \frac{b}{6}\left(2\log^3 x + \log^3 y\right) - \frac{b}{2}\log x \log y\left(2\log y + \log x\right),$$

$$(3.654\text{a, b})$$

$$\Psi(x,y) = \Psi_1(x,y) - \Psi_2(x,y) = \frac{b}{6}\left(2\log^3 y - \log^3 x\right) + \frac{b}{2}\log x \log y\left(2\log x - \log y\right),$$

$$(3.654\text{c, d})$$

are solutions of the unforced system (3.599a, b). The method of characteristic polynomials (Table 3.7) applies to single (simultaneous systems of) linear differential equations with constant coefficients [Sections 3.4 (3.5)] and homogeneous power coefficients [Sections 3.6 (3.7)], and also to linear finite difference equations with constant coefficients [Sections 3.8 (3.9)].

TABLE 3.9
Similarity Solutions and Analogues

Coefficients of Differential Equation[a]	Constant (3.626a)	Homogeneous Powers (3.626b)
Characteristic polynomial	(3.627a)	(3.627c)
Single roots	(3.627b)	(3.627b)
General integral	(3.628a, b; 3.630a, b)	(3.629a, b; 3.630a, c)
Multiple roots	(3.631a, b)	(3.631a, c)
General integral	(3.632a)	(3.632b)

[a] Linear partial differential equation with all derivatives of the same order.

Note: Table 3.9 offers a comparison of linear partial differential equations with all derivatives of the same order and constant (homogeneous power) coefficients, that is with ordinary (homogenous) derivatives, in the unforced case leading to natural integrals that are similarity functions of a linear combination (product of powers) of the independent variables.

3.8 PARTIAL FINITE DIFFERENCE EQUATIONS

An ordinary (partial) finite difference equation is a relation among the values of a function at a set of points in one (several) dimensions [section IV.1.9 (V.3.8)]. It is linear if it involves no powers or products of the function and it has constant coefficients if they do not depend on the point. An unforced linear partial finite difference equation corresponds to a characteristic polynomial of finite differences in each direction (subsection 3.8.1), whose single or multiple roots determine the general solution (subsection 3.8.2); the characteristic polynomial can also be used to obtain particular solutions for forcing with positive integer exponents, including non-resonant (resonant) cases if the bases of the powers are not (are) roots of the characteristic polynomial (subsection 3.8.6). As an example of a linear partial finite difference equation of the second order with constant coefficients, an analogue of the wave-diffusion equation (subsection 3.4.13) is taken, including: (i) the general solution in the unforced case for single or double roots (subsection 3.8.3); and (ii) particular non-resonant and singly resonant solutions in the case of forcing by powers (subsection 3.8.7). The uniqueness of solution specified by three initial values is considered for the analogue finite difference wave-diffusion (wave) equation [subsection 3.8.4 (3.8.5)].

3.8.1 CHARACTERISTIC POLYNOMIAL OF PARTIAL FINITE DIFFERENCES

Considering a rectangular network of points with spacings $h\ (k)$ in the $x\ (y)$ -direction (3.655a) [(3.655b)] the values of a function of two variables at the grid points (Figure 3.4) specify a matrix (3.655c, d):

$$x_n = x + n\ h, \qquad y_m = y + m\ k \qquad \Phi(x + n\ h\ ,\ y + m\ k) = \Phi(x_n\ ,y_m) \equiv \Phi_{n,m}.$$
$$(3.655a\text{–}d)$$

A **partial finite difference equation** of order $N\ (M)$ in $x\ (y)$ is a relation (3.656) among the values of the function in a rectangle $N\ h \times M\ k$:

$$0 = F\left(\Phi_{n,m},\ \Phi_{n+1,m},\ \Phi_{n,m+1},\ \Phi_{n+1,\ m+1}\ ,\cdots,\Phi_{n+N,m},\ \cdots,\ \Phi_{n,m+M},\cdots,\Phi_{n+N,m+M}\right).$$
$$(3.656)$$

The partial finite difference equations is **linear** iff it does not involve products or powers of the function:

$$B_{p,q} = \Sigma_{n=0}^{N} \sum_{m=0}^{M} A_{n,m}^{p,q}\ \Phi_{p+n,q+m}, \qquad (3.657)$$

where the coefficients may depend on the point; in the case of **constant coefficients** (3.658):

$$B_{p,q} = \sum_{n=0}^{N} \sum_{m=0}^{M} A_{n,m}\ \Phi_{p+n,q+m}, \qquad (3.658)$$

the coefficients do not depend on the point.

A **partial forward finite difference** moves one step forward in the x (y) -direction (3.659a–c) [(3.660a–c)]:

$$\Delta_x \Phi_{n.m} = \Delta_x \ \Phi(x+n\ h\ ,\ y+m\ k) = \Phi(x+n\ h+h\ ,\ y+m\ k) = \Phi_{n+1,m},$$
$$\text{(3.659a–c)}$$

$$\Delta_y \Phi_{n.m} = \Delta_y \ \Phi(x+n\ h\ ,\ y+m\ k) = \Phi(x+n\ h\ ,\ y+m\ k+k) = \Phi_{n,m+1}.$$
$$\text{(3.660a–c)}$$

Repeated application of (3.659c; 3.660c) leads to (3.661a–c):

$$\Delta_x^n \ \Delta_y^m \ \Phi_{p,q} = \Delta_x^n \ \Delta_y^m \ \Phi(x+p\ h\ ,\ y+q\ k) \qquad \text{(3.661a)}$$

$$= \Phi(x+p\ h+n\ h\ ,\ y+q\ k+m\ k) = \Phi_{p+n,q+m}. \qquad \text{(3.661b, c)}$$

Substitution of (3.661a–c) in (3.658) shows that *a linear partial finite difference equation with constant coefficients (3.658)* $\equiv$ *(3.662a):*

$$B_{p,q} = \sum_{n=0}^{N} \sum_{m=0}^{M} A_{n,m} \ \Delta_x^n \ \Delta_y^m \ \Phi_{p,q} = \left\{ P_{N,M}\left(\Delta_x\ ,\ \Delta_y\right) \right\} \Phi_{p,q}, \qquad \text{(3.662a, b)}$$

is equivalent (3.662a) $\equiv$ *(3.662b) to a* **characteristic polynomial of partial finite differences** *(3.663):*

$$P_{N,M}\left(\Delta_x\ ,\ \Delta_y\right) \equiv \sum_{n=0}^{N} \sum_{m=0}^{M} A_{n,m} \ \Delta_x^n \ \Delta_y^m, \qquad \text{(3.663)}$$

applied to the matrix. The characteristic polynomial of partial finite differences can be used to obtain the general (a particular) solution in the unforced (forced) case [subsections 3.8.2–3.8.5 (3.8.6–3.8.7)].

3.8.2 GENERAL SOLUTION OF THE UNFORCED EQUATION

Considering the unforced equation (3.664a), the solution may be sought as a product of powers with non-negative integer exponents n (m) in (3.664b):

$$B_{p,q} = 0: \qquad\qquad \Phi_{n.m} = C\ a^n\ b^m; \qquad\qquad \text{(3.664a, b)}$$

with basis a (b) that satisfy the substitution of (3.664b) in (3.658; 3.664a) $\equiv$ (3.665a) leading to (3.665b) $\equiv$ (3.665c):

$$0 = \sum_{n=0}^{N} \sum_{m=0}^{M} A_{n,m} \, \Phi_{p+n,q+m} = C \, a^p \, b^q \sum_{n=0}^{N} \sum_{m=0}^{M} A_{n,m} \, a^n \, b^m$$

$$= C \, a^p \, b^q \, P_{N,M}(a,b), \tag{3.665a–c}$$

where the characteristic polynomial (3.666b) must vanish (3.666a):

$$0 = P_{N,M}(a,b) = \sum_{n=0}^{N} \sum_{m=0}^{M} A_{n,m} \, a^n \, b^m, \tag{3.666a, b}$$

as for a linear partial differential equation with constant coefficients (3.666a, b) $\equiv$ (3.200a, b). The single roots of the characteristic polynomial (3.667a) [(3.667b)]:

$$P_{N,M}(a,b) = a_0 \prod_{n=1}^{N} \left[a - a_n(b) \right] = b_0 \prod_{m=1}^{M} \left[b - b_m(a) \right] \tag{3.667a, b}$$

lead to the **natural sequences** (3.668a) [(3.668b)]:

$$\Phi_{n,m}^1 = \int C_1(b) b^m \left[a_n(b) \right]^n \, db, \tag{3.668a}$$

$$\Phi_{n,m}^2 = \int C_2(a) \, a^n \left[b_m(a) \right]^m \, da, \tag{3.668b}$$

whose sum (3.669a) [(3.669b)] specifies the general solution:

$$\Phi_{n,m} = \sum_{n=1}^{N} \Phi_{n,m}^1 = \sum_{m=1}^{M} \Phi_{n,m}^2. \tag{3.669a, b}$$

In the case of multiple roots (3.670a, b) [(3.671a, b)]:

$$\sum_{r=1}^{R} \alpha_r = N: \qquad P_{N,M}(a,b) = a_0 \prod_{r=1}^{R} \left[a - a_r(b) \right]^{\alpha_r}, \tag{3.670a, b}$$

$$\sum_{s=1}^{S} \beta_s = M: \qquad P_{N,M}(a,b) = b_0 \prod_{s=1}^{S} \left[b - b_s(a) \right]^{\beta_s}, \tag{3.671a, b}$$

the natural sequences are similar to (3.668a) [(3.668b)] multiplied by (3.672a) [(3.672b)] for multiple roots

$$\frac{\partial^p(a^n)}{\partial a^p} = a^{n-p}\, n\,(n-1)...(n-p+1), \tag{3.672a}$$

$$\frac{\partial^q(b^m)}{\partial b^q} = b^{m-q}\, m\,(m-1)...(m-q+1), \tag{3.672b}$$

up to the order of multiplicity (3.673a, b) [(3.674a, b)]:

$$p = 1,...,\alpha_r: \qquad \Phi_{n,m}^{1,r,p} = n^{p-1} \int C_{1,p}(b)\, b^m\, [a_r(b)]^n\, db, \tag{3.673a, b}$$

$$q = 1,...,\beta_s: \qquad \Phi_{n,m}^{2,s,q} = m^{q-1} \int C_{2,q}(a)\, a^n\, [b_s(a)]^m\, da, \tag{3.674a, b}$$

and the general solution is the sum (3.675a) [(3.675b)]:

$$\Phi_{n,m} = \sum_{r=1}^{R} \sum_{p=1}^{\alpha_r} \Phi_{n,m}^{1,r,p} = \sum_{s=1}^{S} \sum_{q=1}^{\beta_s} \Phi_{n,m}^{2,s,q}. \tag{3.675a, b}$$

In the passage from (3.672a) [(3.672b)] to (3.673a, b) [(3.674a, b)], two changes were made: (i) multiplication by a constant a^p (b^q) that remains a solution of the unforced equation; and (ii) consideration of only the highest powers n^p (m^q) since the lower powers $n^{p-1},...$ $(m^{q-1},...)$ already appear in preceding solutions. Thus, *the general solution of a linear unforced partial finite difference equation with constant coefficients (3.658; 3.664a) $\equiv$ (3.665a) is the sum (3.669a, b) [(3.675a, b)] of natural sequences (3.668a, b) [(3.673a, b; 3.674a, b)] in the case of single (3.667a, b) [multiple (3.670a, b; 3.671a, b)] roots of the characteristic polynomial (3.666a, b).* As an example, a finite difference analogue of the wave-diffusion equation is considered next (subsection 3.8.3).

3.8.3 FINITE DIFFERENCE ANALOGUE OF THE WAVE-DIFFUSION EQUATION

The wave-diffusion equation (3.257a) has for a finite difference analogue (3.676):

$$0 = \Phi_{n,m+2} - \frac{1}{c^2}\, \Phi_{n+2,m} - \frac{1}{\chi}\, \Phi_{n+1,m} \tag{3.676}$$

where c (χ) is the speed of propagation (diffusivity). The **analogue finite difference wave-diffusion equation** (3.676) is a linear unforced partial finite difference equation of the second order with constant coefficients that has solutions (3.677a) that, substituted in (3.676), lead to (3.677b) $\equiv$ (3.677c):

$$\Phi_{n,m} = C\, a^n\, b^m: \qquad 0 = C\, a^n\, b^m \left(b^2 - \frac{a^2}{c^2} - \frac{a}{\chi}\right) = \Phi_{n,m}\, P_{2,2}(a,b), \tag{3.677a–c}$$

involving the characteristic polynomial (3.678a) ≡ (3.258a):

$$P_{2,2}(a,b) = b^2 - \frac{a^2}{c^2} - \frac{a}{\chi} = \left[b - b_+(a)\right]\left[b - b_-(a)\right] = - c^{-2}\left[a - a_+(b)\right]\left[a - a_-(b)\right],$$

$$(3.678\text{a–c})$$

with roots (3.678b) [(3.678c)] given by (3.679a, b) ≡ (3.258c) [(3.679c, d) ≡ (3.260a)]:

$$b_\pm(a) = \pm\sqrt{\frac{a^2}{c^2} + \frac{a}{\chi}}, \qquad a_\pm(b) = -\frac{c^2}{2\,\chi} \pm \frac{c^2}{2}\sqrt{\frac{1}{\chi^2} + 4\,\frac{b^2}{c^2}}. \qquad (3.679\text{a–d})$$

In the case of distinct roots (3.680a) [(3.681a)], the natural sequences (3.680b, c) [(3.681b, c)]:

$$a \neq -\frac{c^2}{\chi}: \qquad \Phi_{n,m}^\pm = \int a^n \left[b_\pm(a)\right]^m C_\pm(a)\,da, \qquad (3.680\text{a–c})$$

$$b^2 \neq -\frac{c^2}{4\,\chi^2}: \qquad \pm\Phi_{n,m} = \int b^m \left[a_\pm(b)\right]^n C^\pm(b)\,db, \qquad (3.681\text{a–c})$$

lead to the general solution (3.682a) [(3.682b)]:

$$\Phi_{n,m} = \Phi_{n,m}^+ + \Phi_{n,m}^- = {}^+\Phi_{n,m} + {}^-\Phi_{n,m}. \qquad (3.682\text{a, b})$$

In the case (3.679a, b) the double root is $a = 0 = b$, leading to a trivial constant solution in (3.677a). The solutions are not trivial in the case (3.679c, d) of double non-zero roots (3.683b–d), that exist only for (3.683a):

$$b^\pm = \pm\,i\,\frac{c}{2\,\chi}: \qquad a_+(b^\pm) = a_-(b^\pm) \equiv a(b^\pm) = -\frac{c^2}{2\,\chi}; \qquad (3.683\text{a–d})$$

in this case there exist two special integrals (3.684a–c)

$$\pm\Phi_{n,m} = F_\pm\left[a(b^\pm)\right]^n (b^\pm)^m = F_\pm\left(-\frac{c^2}{2\,\chi}\right)^n \left(\pm\,i\,\frac{c}{2\,\chi}\right)^m$$

$$= F_\pm(-)^n (\pm\,i)^m\, c^{2n+m} (2\,\chi)^{-n-m}. \qquad (3.684\text{a–c})$$

each involving one arbitrary constant. Thus, *the analogue finite difference wave-diffusion equation is a linear partial finite difference equation with constant coefficients that in the unforced case (3.676) has a general solution (3.680a–c; 3.682a) [(3.681a–c; 3.682b)] for distinct roots (3.679a, b) [(3.679c, d)] of the characteristic polynomial (3.678a–c); in the case of double roots (3.683a–d), there are two special solutions (3.684a–c) each involving one arbitrary constant.* The preceding solutions may be made unique by a choice of initial values (subsection 3.8.4).

3.8.4 STARTING CONDITIONS AND UNICITY OF SOLUTION

The general solution (3.682a) [(3.682b)] is the sum of natural sequences (3.680b, c) [(3.681b, c)] where the integrations may be suppressed and the functions (C_+, C_-) replaced by constants, leading to (3.685a) [(3.685b)]:

$$\Phi_{n,m} = a^n \left\{ C_+ \left[b_+(a) \right]^m + C_- \left[b_-(a) \right]^m \right\} \tag{3.685a}$$

$$= b^m \left\{ C^+ \left[a_+(b) \right]^m + C^- \left[a_-(b) \right]^m \right\} \tag{3.685b}$$

that involve three arbitrary elements: (i) the base a (b) that appears in (3.679a, b) [(3.679c, d)]; and (ii) the constant coefficients $C_\pm$ $(C^\pm)$. The solutions become unique determining the unknowns from three independent and compatible starting conditions, as shown next for (3.685a; 3.679a, b) $\equiv$ (3.686):

$$\Phi_{n,m} = a^n \left(\frac{a^2}{c^2} + \frac{a}{\chi} \right)^{m/2} \left[C_+ + (-)^n C_- \right]. \tag{3.686}$$

The values of the function (3.686) at all points of a rectangular grid (Figure 3.4) can be determined from the values at three points, for example the lower left corner (3.687a–c):

$$\Phi_{0,0} = C_+ + C_-, \quad \Phi_{1,0} = a \left(C_+ - C_- \right), \quad \Phi_{0,1} = \left| \frac{a^2}{c^2} + \frac{a}{\chi} \right|^{1/2} (C_+ + C_-). \tag{3.687a–c}$$

The ratio of (3.687a, c) specifies the value of a in (3.688a) $\equiv$ (3.688b, c):

$$\frac{a^2}{c^2} + \frac{a}{\chi} = \left(\frac{\Phi_{0,1}}{\Phi_{0,0}} \right)^2 : \quad a = -\frac{c^2}{2\chi} \left\{ 1 \pm \left| 1 - \left(\frac{2\chi \, \Phi_{0,1}}{c \, \Phi_{0,0}} \right)^2 \right|^{1/2} \right\}. \tag{3.688a–c}$$

From (3.687a, b) follow the values of $C_\pm$:

$$2 C_\pm = \Phi_{0,0} \pm \frac{\Phi_{1,0}}{a}, \tag{3.689}$$

that substituted in (3.686) together with (3.688a) yield:

$$\Phi_{n,m} = \frac{a^n}{2} \left(\frac{a^2}{c^2} + \frac{a}{\chi} \right)^{m/2} \left[\Phi_{0,0} + \frac{\Phi_{1,0}}{a} + (-)^n \left(\Phi_{0,0} - \frac{\Phi_{1,0}}{a} \right) \right]$$

$$= \left(\frac{\Phi_{0,1}}{\Phi_{0,0}} \right)^m \left\{ \frac{a^n}{2} \Phi_{0,0} \left[1 + (-)^n \right] + \frac{a^{n-1}}{2} \Phi_{1,0} \left[1 - (-)^n \right] \right\}. \tag{3.690a, b}$$

Setting (3.691a) [(3.691b)] the sequence (3.690a, b) may be split into even (3.691c) and odd (3.691d) sub-sequences:

$$n = \{2\ r\ ; 2\ r+1\}: \qquad \{\Phi_{2r,m}\ ; \Phi_{2r+1,m}\} = a^{2r} \left(\frac{\Phi_{0,1}}{\Phi_{0,0}}\right)^m \{\Phi_{0,0}\ ; \Phi_{1,0}\}. \qquad \text{(3.691a–d)}$$

Thus *the analogue finite difference wave-diffusion equation (3.676) has general solutions (3.685a) [(3.685b)] where* $(a, C_{\pm})[(b, C^{\pm})]$ *are arbitrary constants determined by three starting values. For example, the solution (3.685a) is specified at all points of a rectangular grid (Figure 3.4) by the three values (3.687a–c) at the lower left corner, by: (i) the sequence (3.690a, b); and (ii) the equivalent to the set of even (3.691a, c) and odd (3.691b, d) sub-sequences. In both cases, a can take two values (3.688b, c).* The sequences simplify in the case of the analogue finite difference wave equation (subsection 3.8.5).

3.8.5 ANALOGUE FINITE DIFFERENCE WAVE EQUATION

Excluding dissipation in the analogue finite difference wave-diffusion equation (3.676), by setting the diffusivity equal to infinity (3.692a) leads to the **analogue finite difference wave equation** (3.692b):

$$\chi = \infty: \qquad\qquad \Phi_{n+2,m} = c^2\ \Phi_{n,m+2}. \qquad \text{(3.692a, b)}$$

In this case, the general solution (3.686) becomes:

$$\Phi_{n,m} = c^{-m}\ a^{n+m}\ \left[C_+ + (-)^n\ C_-\right]. \qquad \text{(3.693)}$$

The values of the function at all points of a rectangular grid (Figure 3.4) can be determined from the values at three points, for example the lower left corner:

$$\Phi_{0,0} = C_+ + C_-, \qquad \Phi_{1,0} = a\ (C_+ - C_-), \qquad \Phi_{0,1} = \frac{a}{c}\ (C_+ + C_-), \qquad \text{(3.694a–c)}$$

in agreement with (3.687a–c) with $\chi = \infty$. From the ratio of (3.694a, c) follows (3.695a) the value of a:

$$a = c\ \frac{\Phi_{0,1}}{\Phi_{0,0}}, \qquad 2\ C_{\pm} = \Phi_{0,0} \pm \frac{\Phi_{1,0}}{a} = \Phi_{0,0}\left(1 \pm \frac{1}{c}\frac{\Phi_{1,0}}{\Phi_{0,1}}\right), \qquad \text{(3.695a–c)}$$

and (3.694a, b) lead to (3.695b) that becomes (3.695c) on substitution of (3.695a).
Substitution of (3.695a, c) in (3.693) leads to the sequence solution:

$$\Phi_{n,m} = \frac{c^{-m}}{2}\left(c\ \frac{\Phi_{0,1}}{\Phi_{0,0}}\right)^{n+m}\left[\Phi_{0,0} + \frac{1}{c}\frac{\Phi_{1,0}}{\Phi_{0,1}} + (-)^n\left(\Phi_{0,0} - \frac{1}{c}\frac{\Phi_{1,0}}{\Phi_{0,1}}\right)\right]$$

$$= \frac{c^n}{2}\left(\Phi_{0,1}\right)^{n+m}\left(\Phi_{0,0}\right)^{1-n-m} \times \left\{1 + (-)^n + \frac{1}{c}\frac{\Phi_{1,0}}{\Phi_{0,1}}\left[1 - (-)^n\right]\right\}. \qquad \text{(3.696a, b)}$$

Setting (3.697a) [(3.697b)] the sequence (3.696a, b) splits into an even (3.697c) and an odd (3.697d) sub-sequence:

$$n = \{2\,r\,;2\,r+1\}: \quad \left\{\Phi_{2r,m}\,;\Phi_{2r+1,m}\right\} = c^{2r}\left(\frac{\Phi_{0,1}}{\Phi_{0,0}}\right)^{2r+m}\left\{\Phi_{0,0}\,;\Phi_{1,0}\right\}. \quad (3.697\text{a--d})$$

It can be checked that (3.697c, d) corresponds to (3.691c, d) with (3.695a). The sequences (3.697c, d) converge as $r \to \infty$ if (3.698a) holds; for example, the values (3.698b--d) lead to the sub-sequences (3.698e, f):

$$\Phi_{0,1} < \Phi_{0,0}; \qquad \left\{\Phi_{0,0}\,;\Phi_{0,1}\,;\Phi_{1,0}\right\} = \{2\,;1\,;3\}: \qquad (3.698\text{a--d})$$

$$\left\{\Phi_{2r,m}\,;\Phi_{2r+1,m}\right\} = c^{2r}\,2^{-2r-m}\,\{2\,,3\}. \qquad (3.698\text{e, f})$$

Thus, *the analogue finite difference unforced wave equation (3.692b) has for solution the sequence (3.693) where a and $C_\pm$ are arbitrary constants. The solution is specified at all points of a rectangular grid (Figure 3.4) by the values at three points, for example the lower left corner (3.694a–c) leading to the sequence (3.696a, b), which splits into even (3.697a, c) and odd (3.697b, d) sub-sequences. The sub-sequences (3.697c, d) converge for large order $r \to \infty$ if (3.698a) holds, and for the particular values (3.698b–d) the sub-sequences simplify to (3.698e, f). The method of the characteristic polynomial of partial finite differences applies both to unforced (forced) equations [subsections 3.8.1–3.8.5 (3.8.6–3.8.7)].*

3.8.6 RESONANT/NON-RESONANT FORCING BY INTEGRAL POWERS

Considering the linear partial finite difference equation with constant coefficients (3.658) ≡ (3.699b) forced by powers with non-negative integer exponents (3.699a):

$$B\,a^p\,b^q = B_{p,q} = \sum_{n=0}^{N}\sum_{m=0}^{M} A_{n,m}\,\Phi_{p+n,q+m}, \qquad (3.699\text{a, b})$$

the solution is sought in the form (3.664b) leading to (3.700a) ≡ (3.700b):

$$B = C\sum_{n=0}^{N}\sum_{m=0}^{M} A_{n,m}\,a^n\,b^m = C\,P_{N,M}(a,b), \qquad (3.700\text{a, b})$$

involving the characteristic polynomial (3.666b). If the characteristic polynomial is not zero (3.701a), it can be divided in (3.700b), and the substitution in (3.666b) specifies the particular solution (3.701b):

$$P_{N,M}(a,b) \neq 0: \qquad \Phi_{n,m} = \frac{B\,a^n\,b^m}{P_{N,M}(a,b)}. \qquad (3.701\text{a, b})$$

If a (b) is a root of multiplicity r (s) of the characteristic polynomial, the lowest-order non-zero derivative is (3.702a, b) and the particular solution is (3.702c):

$$E(a,b) \equiv \frac{\partial^{r+s}}{\partial a^r \, \partial b^s} \left[P_{N,M}(a,b) \right] \neq 0: \qquad \Phi_{n,m} = \frac{B}{E(a,b)} \frac{\partial^{r+s}}{\partial a^r \, \partial b^s} \left(a^n \, b^m \right).$$

$$(3.702\text{a--c})$$

In (3.702b) appear (3.703a, b):

$$\frac{\partial^r}{\partial a^r}(a^n) = a^{n-r} \; n \; (n-1)...(n-r+1), \tag{3.703a}$$

$$\frac{\partial^s}{\partial b^s}(b^m) = b^{m-s} \; m \; (m-1)...(m-s+1). \tag{3.703b}$$

In (3.703a) [(3.703b)], the second factor can be replaced by n^r (m^s) because all lower powers (3.673a, b) [(3.674a, b)] are particular integrals of the unforced equation and correspond to zeros of the characteristic polynomial. Thus, in (3.702a, b) can be made the substitutions (3.704a, b)

$$\frac{\partial^r}{\partial a^r}(a^n) \;\; \rightarrow \;\; a^{n-r} \; n^r, \qquad \frac{\partial^s}{\partial b^s}(b^m) \;\; \rightarrow \;\; b^{m-s} \; m^s \tag{3.704a, b}$$

and the general solution becomes (3.705)

$$\Phi_{n,m} = \frac{B}{E(a,b)} \; n^r \; m^s \; a^{n-r} \; b^{m-s}. \tag{3.705}$$

Thus, *the linear partial finite difference equation with constant coefficients (3.658) forced (3.699a) by powers (3.699b) has particular solution: (i) given by (3.701b) in the non-resonant case when the bases (a,b) are not (3.701a) roots of the characteristic polynomial (3.666b); and (ii) given by (3.705) when the base a (b) is a root (3.702a, b) of multiplicity r (s) of the characteristic polynomial.* In the case of the analogue finite difference wave-diffusion equation (subsection 3.8.3), the characteristic polynomial is of the second degree and, hence, has single (double) roots that lead to non-resonant (singly and doubly resonant) particular solutions when forced by powers with positive integer exponents (subsection 3.8.7).

3.8.7 NON-RESONANT AND SINGLY/DOUBLY RESONANT SOLUTIONS

Consider the analogue finite difference wave-diffusion equation (3.676) forced by powers (3.706):

$$\Phi_{n,m+2} - \frac{1}{c^2} \Phi_{n+2,m} - \frac{1}{\chi} \Phi_{n+1,m} = B \; a^n \; b^m. \tag{3.706}$$

The solution is sought in the form (3.677a) leading to (3.707a) by substitution in (3.706):

$$B = \left(b^2 - \frac{a^2}{c^2} - \frac{a}{\chi} \right) C = P_{2,2}(a,b) \; C, \tag{3.707a, b}$$

where the characteristic polynomial (3.678a) appears in (3.707b). In the case of distinct roots (3.679a, b) ≡ (3.708a) and (3.679c, d) ≡ (3.708b) the particular solution of (3.706) is obtained solving (3.707a) for C and substituting in (3.677a) leading to the non-resonant (3.701a, b) solution (3.708c):

$$a\ \chi + c^2 \neq 0 \neq c^2 + 4\ b^2\ \chi^2: \qquad \Phi_{*n,m} = \frac{B\ a^n\ b^m}{b^2 - a^2 / c^2 - a / \chi}. \qquad (3.708\text{a–c})$$

In the singly resonant case (3.705; 3.701a, b) of distinct roots for $a\ (b)$: (i) the denominator in (3.708c) is replaced (3.678a) by (3.709a) [(3.709b)]:

$$\frac{\partial}{\partial b}\big[P_{2,2}(a,b)\big] = 2\ b, \qquad \frac{\partial}{\partial a}\big[P_{2,2}(a,b)\big] = -\frac{2\ a}{c^2} - \frac{1}{\chi}; \qquad (3.709\text{a, b})$$

and (ii) in the numerator appears (3.705) the factor $m\ (n)$, leading to (3.710a, b) [(3.711a, b)]:

$$b = b_\pm(a): \qquad \Phi^\pm_{**n,m} = \frac{B}{2}\ m\ a^n\ \big[b_\pm(a)\big]^{m-2}, \qquad (3.710\text{a, b})$$

$$a = a_\pm(b): \qquad {}^\pm\Phi_{**n,m} = -\frac{B\ n}{2a / c^2 + 1 / \chi}\ b^m\ \big[a_\pm(b)\big]^{n-1}. \qquad (3.711\text{a, b})$$

In the case of double root for $a\ (b)$: (i) the denominator becomes (3.712a) [(3.712b)]

$$\frac{\partial^2}{\partial b^2}\big[P_{2,2}(a,b)\big] = 2, \qquad \frac{\partial^2}{\partial a^2}\big[P_{2,2}(a,b)\big] = -\frac{2}{c^2}; \qquad (3.712\text{a, b})$$

(ii) in the numerator appears (3.705) the factor $m^2\ (n^2)$; (iii) the zero double root (3.713a, b) gives a trivial zero solution, that is the r.h.s. of (3.706) is zero, so the simplest particular solution is zero (3.713c)

$$a = a_\pm(b) = 0: \qquad \Phi^\pm_{**n,m} = 0; \qquad (3.713\text{a–c})$$

$$b = b_\pm(a) = \pm \frac{i\ c}{2\ \chi}, \qquad a(b^\pm) = -\frac{c^2}{2\ \chi}:$$

$$\pm\Phi_{***n,m} = -\frac{B\ n^2}{2 / c^2}\big[a(b^\pm)\big]^{n-2}\ (b^\pm)^m = -\frac{B\ c^2\ n^2}{2}\left(-\frac{c^2}{2\ \chi}\right)^{n-2}\left(\pm \frac{i\ c}{2\ \chi}\right)^m$$

$$= (-)^{n-1}\ (\pm\ i)^m\ 2^{1-n-m}\ B\ n^2\ c^{2n+m-2}\ \chi^{2-n-m}; \qquad (3.714\text{a–e})$$

and (iv) the non-zero double root (3.683a–d) ≡ (3.714a, b) leads to the doubly resonant solution (3.714c–e).

Thus, *the analogue finite difference wave-diffusion equation forced by powers (3.706) has particular solutions: (i) given by (3.708c) in the non-resonant case*

(3.708a, b) ≡ (3.680a; 3.681a) when the characteristic polynomial (3.678a–c) is not zero; (ii) given by (3.710a, b) [(3.711a, b)] in the singly resonant case for distinct roots (3.679a, b) [(3.679c, d)] of the characteristic polynomial (3.678b, c) [(3.678b, d)]; and (iii) zero (3.713c) [non-zero (3.714a–e)] doubly resonant for zero (3.713a, b) [non-zero (3.683a–c)] double roots of the characteristic polynomial. As an example of checking a solution, it is shown that the doubly resonant solution (3.714e) satisfies (3.706) ≡ (3.715) with (3.714a, b):

$$\Phi_{n,m+2} - \frac{1}{c^2}\,\Phi_{n+2,m} - \frac{1}{\chi}\,\Phi_{n+1,m} = B\left(-\frac{c^2}{2\,\chi}\right)^n\left(\pm\frac{i\,c}{2\,\chi}\right)^m. \qquad (3.715)$$

Substitution of (3.714e) in the l.h.s. of (3.715) leads to (3.716):

$$\Phi_{n,m+2} - \frac{1}{c^2}\,\Phi_{n+2,m} - \frac{1}{\chi}\,\Phi_{n+1,m} = -\frac{B\,c^2}{2}\left(-\frac{c^2}{2\,\chi}\right)^{n-2}\left(\pm\frac{i\,c}{2\,\chi}\right)^m G, \qquad (3.716)$$

where the factor G in (3.716) is given by (3.717a):

$$G = \left(\pm\frac{i\,c}{2\,\chi}\right)^2 n^2 - \frac{1}{c^2}\left(-\frac{c^2}{2\,\chi}\right)^2 (n+2)^2 - \frac{1}{\chi}\left(-\frac{c^2}{2\,\chi}\right)(n+1)^2 = -\frac{c^2}{2\,\chi^2}, \qquad (3.717a, b)$$

that simplifies to (3.717b) because the factors of n^2 and n vanish. Substitution of (3.717b) in (3.716) leads to (3.715), confirming that (3.714e) is its solution that coincides with the r.h.s. of (3.706) for (a, b) given by (3.714a, b). The method of the characteristic polynomial (matrix of polynomials) of partial finite differences applies to the solution of single (simultaneous systems) of linear partial finite difference equations with constant coefficients [section 3.8 (3.9)].

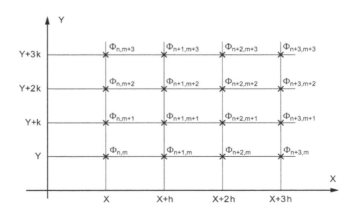

FIGURE 3.4 A function of two variables, either two spatial coordinates or space-time, can be discretised on a grid, for example rectangular, to approximate a partial differential equation by a multiple finite difference equation.

3.9 SIMULTANEOUS SYSTEM OF PARTIAL FINITE DIFFERENCE EQUATIONS

A simultaneous system of linear partial finite difference equations with constant coefficients consists of a matrix of polynomials of partial finite differences applied to a set of discretised functions equated to a set of discretised forcing functions (Subsection 3.9.1), and the determinant of the polynomials specifies the characteristic polynomial of the system. In the unforced case, the single or multiple roots of the characteristic polynomial determine the natural sequences, which specify the general solution by superposition, with coefficients ensuring the compatibility of the system (Subsection 3.9.2). The particular solution forced by powers is also determined using the characteristic polynomial, leading to non-resonant and resonant cases (Subsection 3.9.6). As an example, a system of two coupled finite difference equations with constant coefficients, which is equivalent to the analogue finite difference wave equation (Subsection 3.8.5), is considered, for which the following are obtained: (i) the general solution in the unforced case as a superposition of natural sequences for single or double roots of the characteristic polynomial (Subsection 3.9.3); (ii/iii) the unique solution imposing three starting conditions all on the same function (Subsection 3.9.4) or two on one function and one on the other (Subsection 3.9.5); and (iv) a particular solution with forcing by powers in non-resonant and singly resonant cases (Subsection 3.9.7). This completes the six classes of differential and finite difference equations and simultaneous systems that have a characteristic polynomial used to obtain both general (particular) solutions in the unforced (forced) cases (Subsection 3.9.8), as summarised in Tables 3.10 to 3.15.

3.9.1 CHARACTERISTIC POLYNOMIAL FOR SIMULTANEOUS PARTIAL FINITE DIFFERENCES

Consider a set (3.718a) of functions of two variables with values (3.718b) at a rectangular grid of points (Figure 3.4), leading to (3.718c) using partial finite differences (3.659a–c; 3.660a–c; 3.661a–c):

$$j = 1,\ldots,L: \qquad \Phi^j_{n,m} = \Phi^j\left(x + n\ h\ ,\ y + m\ k\right) = \Delta^n_x\ \Delta^m_y\ \Phi^j_{n,m}\ ; \qquad (3.718\text{a–c})$$

a simultaneous system of partial finite difference equations is a set (3.719a) of relations among the values of the functions at various grid points (3.719b):

$$i = 1,\ldots,L: \qquad F_i\left(\Phi^j_{n,m}\ ,\ \Phi^j_{n+1,m}\ ,\ldots,\ \Phi^j_{n+N,m}\ ,\ldots,\ \Phi^j_{n+N,m+M}\right). \qquad (3.719\text{a, b})$$

The system is linear if there are no products or powers of the functions and, thus, consists of a matrix of polynomials of partial finite differences applied to the discretised functions and equated to a set of discretised forcing functions:

$$i,j = 1,\ldots,L: \qquad \sum_{j=1}^{L} P_{ij}\left(\Delta_x\ ,\ \Delta_y\right)\Phi^j_{p,q} = B^i_{p,q}\ , \qquad (3.720\text{a, b})$$

167

with polynomials possibly of different degrees:

$$P_{ij}\left(\Delta_x, \Delta_y\right) = \sum_{n=0}^{N_{ij}} \sum_{m=0}^{M_{ij}} A_{n,m}^{p,q} \, \Delta_x^n \, \Delta_y^m. \tag{3.721}$$

If the coefficients depend (3.721) [do not depend (3.722)] on the point:

$$P_{ij}\left(\Delta_x, \Delta_y\right) = \sum_{n=0}^{N_{ij}} \sum_{m=0}^{M_{ij}} A_{n,m} \, \Delta_x^n \, \Delta_y^m, \tag{3.722}$$

the simultaneous system has variable (3.721) [constant (3.722)] coefficients. Thus, *a simultaneous system of linear partial finite difference equations (3.720a, b) has constant coefficients if the matrix of polynomials of partial finite differences has constant coefficients (3.722)*. In this case, the unforced system has natural solutions that are sequences of powers with non-negative integer exponents (Subsection 3.9.2).

3.9.2 Natural Sequences of Powers with Integer Exponents

Considering the unforced (3.723a; 3.720a, b) simultaneous system with (3.722) constant coefficients (3.723b):

$$B_{p,q}^j = 0: \qquad\qquad 0 = \sum_{j=1}^{L} \left\{ P_{ij}\left(\Delta_x, \Delta_y\right)\right\} \Phi_{p,q}^j, \tag{3.723a, b}$$

the particular solutions are **natural sequences** of powers with non-negative integer exponents (3.724a) leading to (3.724b, c) by substitution in (3.723a):

$$\Phi_{n,m}^j = C_j \, a^n \, b^m: \qquad 0 = a^p \, b^q \sum_{j=1}^{L} P_{ij}(a,b) \, C_j = \sum_{j=1}^{L} P_{ij}(a,b) \, \Phi_{n,m}^j. \tag{3.724a–c}$$

A non-trivial solution (3.725a) implies that the determinant of the matrix, which specifies the characteristic polynomial (3.725c), must be zero (3.725b):

$$\left\{ \Phi_{n,m}^1, \dots, C_{n,m}^L \right\} \neq \{0, \dots, 0\}: \qquad 0 = Det\left\{ P_{ij}(a,b)\right\} = P_{N,M}(a,b). \tag{3.725a–c}$$

Thus, *a linear unforced simultaneous system of partial finite difference equations (3.723b) with constant coefficients (3.722) has characteristic polynomial (3.725c) that must vanish. The single (3.667a) [(3.667b)] roots of the characteristic polynomial correspond to the natural sequences (3.668a) [(3.668b)] whose sum (3.726b) [(3.727b)] specifies the first function (3.726a) [(3.727a)]:*

$$E_{n,m}^{1,1} = 1: \qquad \Phi_{n,m}^j = \sum_{n=1}^{N} E_{n,m}^{1,j} \, \Phi_{n,m}^1, \qquad (3.726a, b)$$

$$E_{n,m}^{2,1} = 1: \qquad \Phi_{n,m}^j = \sum_{m=1}^{M} E_{n,m}^{2,j} \, \Phi_{n,m}^2, \qquad (3.727a, b)$$

whereas all the other functions are linear combinations with coefficients $E_{n,m}^{1,j}\left(E_{n,m}^{2,j}\right)$ determined by compatibility of the system. The multiple roots of the characteristic polynomial (3.670a, b) [(3.671a, b)] lead to the natural sequences (3.673a, b) [(3.674a, b)] whose sum (3.728b) [(3.729b)] specifies the first function (3.728a) [(3.729a)]:

$$E_{n,m}^{1,1,p} = 1: \qquad \Phi_{n,m}^{\gamma} = \sum_{r=1}^{R} \sum_{p=1}^{\alpha_r} E_{n,m}^{1,j,r,p} \, \Phi_{n,m}^{1,r,p}, \qquad (3.728a, b)$$

$$E_{n,m}^{2,1,q} = 1: \qquad \Phi_{n,m}^{\gamma} = \sum_{s=1}^{S} \sum_{q=1}^{\beta_s} E_{n,m}^{2,j,s,q} \, \Phi_{n,m}^{2,s,q}, \qquad (3.729a, b)$$

with all other functions linear combinations with coefficients $E_{n,m}^{1,r,p}\left(E_{n,m}^{2,s,q}\right)$ determined by compatibility of the system. As an example, a coupled system of partial finite difference equations analogue to the wave equation (Subsection 3.9.3) is considered.

3.9.3 FINITE DIFFERENCE SYSTEM ANALOGOUS TO THE WAVE EQUATION

The coupled system of first-order linear unforced partial finite difference equations with constant coefficients (3.730a, b):

$$\Phi_{n+1,m} = c \, \Psi_{n,m+1}, \qquad \Psi_{n+1,m} = c \, \Phi_{n,m+1} \qquad (3.730a, b)$$

can be written in matrix form (3.731a):

$$\begin{bmatrix} \Delta_x & -c\,\Delta_y \\ -c\,\Delta_y & \Delta_x \end{bmatrix} \begin{bmatrix} \Phi_{n,m} \\ \Psi_{n,m} \end{bmatrix} = 0: \qquad 0 = \Delta_x^2 - c^2 \, \Delta_y^2 = P_{2,2}\left(\Delta_x, \Delta_y\right),$$

$$(3.731a\text{--}c)$$

showing that the characteristic polynomial (3.731c) specified by the determinant (3.731b) resembles the one-dimensional Cartesian wave equation (3.223b). The same result (3.732a–d) can be obtained from (3.730a, b):

$$\Phi_{n+2,m} = c\ \Psi_{n+1,m+1} = c^2\ \Phi_{n,m+2}, \tag{3.732a, b}$$

$$\Psi_{n,m+2} = \frac{1}{c}\ \Phi_{n+1,m+1} = \frac{1}{c^2}\ \Psi_{n+2,m}, \tag{3.732c, d}$$

leading to (3.733a, b):

$$\left(\Delta_x^2 - c^2\ \Delta_y^2\right)\left\{\Phi_{n,m}\ ;\Psi_{n,m}\right\} = 0, \tag{3.733a, b}$$

in agreement with (3.731b, c). The characteristic polynomial (3.731c) ≡ (3.734a) has roots (3.734b) corresponding (3.724a) to the natural sequences (3.734c, d):

$$P_{2,2}(a,b) = a^2 - b^2\ c^2, \qquad a = \pm\ c\ b: \qquad \Phi_{n,m}^{\pm} = b^m\ (\pm\ c\ b)^n = (\pm\ c)^n\ b^{n+m}. \tag{3.734a–d}$$

Substituting (3.734d) ≡ (3.735a) in (3.730b) leads to (3.735b), implying (3.735c, d):

$$\Psi_{n+1,m}^{\pm} = c\ \Phi_{n,m+1}^{\pm} = (\pm)^n\ c^{n+1}\ b^{n+m+1}: \qquad \Psi_{n,m}^{\pm} = (\pm)^{n-1}\ c^n\ b^{n+m} = \pm\ \Phi_{n,m}^{\pm}. \tag{3.735a–d}$$

It can be checked that (3.730a) ≡ (3.736a) is also satisfied by (3.735d) ≡ (3.736c):

$$\Phi_{n+1,m}^{\pm} = c\ \Psi_{n,m+1}^{\pm} = (\pm)^{n-1}\ c^{n+1}\ b^{n+m+1}: \qquad \Phi_{n,m}^{\pm} = (\pm)^n\ c^n\ b^{n+m}, \tag{3.736a–c}$$

because (3.736c) ≡ (3.734d).

Thus, *the coupled linear system of unforced first-order partial finite difference equations with constant coefficients (3.730a, b), comparable to a Cartesian one-dimensional wave equation (3.733a, b), has general solutions (3.737a, b) [(3.737c, d)] that are a linear combination of (3.734d) [(3.735d)] with constant coefficients $C_{\pm}$:*

$$\Phi_{n,m} = C_+\ \Phi_{n,m}^+ + C_-\ \Phi_{n,m}^- = b^{n+m}\ c^n\left[C_+ + (-)^n\ C_-\right], \tag{3.737a, b}$$

$$\Psi_{n,m} = C_+\ \Psi_{n,m}^+ + C_-\ \Psi_{n,m}^- = b^{n+m}\ c^n\left[C_+ + (-)^{n+1}\ C_-\right] \tag{3.737c, d}$$

where: (i) the sequence (3.737a, b) is equivalent with (3.734b) to (3.693), which is a solution of the analogue finite difference wave equation (3.692b) ≡ (3.732b) that follows from the coupled system (3.730a, b); and (ii) the sequence (3.737c, d) follows from (3.733a, b) by compatibility (3.735a–d; 3.736a–c) with the same coupled system (3.730a, b). The arbitrary constants b, $C_{\pm}$ are determined by three independent and compatible starting conditions, which can be imposed as follows: (i) all three starting conditions on one function (Subsection 3.9.4); and (ii) two starting conditions on one function and the third starting condition on the other function (Subsection 3.9.5).

3.9.4   THREE STARTING CONDITIONS ON ONE FUNCTION

The values of the two functions (3.737a, b; 3.737c, d) at all points of a rectangular grid (Figure 3.4) can be determined from three values. Choosing the values of the first function (3.737a, b) at the lower left corner:

$$\Phi_{0,0} = C_+ + C_-, \quad \Phi_{0,1} = b\left(C_+ + C_-\right), \quad \Phi_{1,0} = b\,c\left(C_+ - C_-\right), \quad (3.738\text{a–c})$$

leads from the ratio (3.738a, b) to (3.739a):

$$b = \frac{\Phi_{0,1}}{\Phi_{0,0}}, \qquad 2\,C_{\pm} = \Phi_{0,0} \pm \frac{\Phi_{1,0}}{b\,c} = \Phi_{0,0}\left(1 \pm \frac{1}{c}\frac{\Phi_{1,0}}{\Phi_{0,1}}\right), \qquad (3.739\text{a–c})$$

and from (3.738a, c) to (3.739b), which becomes (3.739c) on substitution of (3.739a). Substitution of (3.739a, c) in (3.737b; 3.737d) specifies *the general solution of the coupled system of first-order linear unforced partial finite difference equations with constant coefficients (3.730a, b) for the two functions at all points of a rectangular grid from the values of the first function at the lower left corner:*

$$\Phi_{n,m} = \frac{c^n}{2}\left(\Phi_{0,1}\right)^{n+m} \Phi_{0,0}^{1-n-m}\left\{1+(-)^n + \frac{1}{c}\frac{\Phi_{1,0}}{\Phi_{0,1}}\left[1-(-)^n\right]\right\}, \qquad (3.740)$$

$$\Psi_{n,m} = \frac{c^n}{2}\left(\Phi_{0,1}\right)^{n+m} \left(\Phi_{0,0}\right)^{1-n-m}\left\{1+(-)^{n+1} + \frac{1}{c}\frac{\Phi_{1,0}}{\Phi_{0,1}}\left[1-(-)^{n+1}\right]\right\}. \qquad (3.741)$$

Each sequence (3.740) [(3.741)] splits into two subsequences (3.742a, b) [(3.743a, b)] one of even (3.742c) [(3.743c)] and one of odd (3.742d) [(3.743d)] order:

$$n = \{2\,r,\ 2\,r+1\}: \qquad \left\{\Phi_{2r,m}\,;\Phi_{2r+1,m}\right\} = c^{2r}\left(\frac{\Phi_{0,1}}{\Phi_{0,0}}\right)^{2r+m}\left\{\Phi_{0,0}\,;\Phi_{1,0}\right\},$$

$$(3.742\text{a–d})$$

$$n = \{2\,r,\ 2\,r+1\}: \qquad \left\{\Psi_{2r,m}\,;\Psi_{2r-1,m}\right\} = c^{2r-1}\left(\frac{\Phi_{0,1}}{\Phi_{0,0}}\right)^{2r+m-1}\left\{\Phi_{1,0}\,;\Phi_{0,0}\right\}$$

$$= \left\{\frac{1}{c}\,\Phi_{2r+1,m-1}\,;\,c\,\Phi_{2r-2,m+1}\right\}. \qquad (3.743\text{a–f})$$

It can be checked that (3.742c, d; 3.743c, d) satisfy (3.743e, f) in agreement with (3.730a, b). The sequences (3.742c, d; 3.743c, d) converge for large order $r,m \to \infty$ if the condition (3.744a) is met, and choosing the initial values

(3.744b–d) ≡ (3.698b–d), the subsequences (3.742c, d) [(3.743c, d)] become (3.744e, f) [(3.744g, h)]:

$$\Phi_{0,1} < \Phi_{0,0}; \qquad \left\{\Phi_{0,0} \; ; \Phi_{0,1} \; ; \Phi_{1,0}\right\} = \left\{ 2\,,1\,,3 \right\}: \qquad (3.744a–d)$$

$$\left\{\Phi_{2r,m} \; ; \Phi_{2r+1,m}\right\} = c^{2r} \; 2^{-2r-m} \left\{ 2\,,3 \right\}, \qquad (3.744e, f)$$

$$\left\{\Psi_{2r,m} \; ; \Psi_{2r-1,m}\right\} = c^{2r-1} \; 2^{-2r-m+1} \left\{ 3\,,2 \right\}. \qquad (3.744g, h)$$

where (3.744e, f) ≡ (3.698e, f) coincide and (3.744g, h) agrees with (3.743e, f). Instead of taking three starting conditions for one function (3.738a–c), an alternative is to take two starting conditions for one function and the third starting condition for the other function (Subsection 3.9.5).

3.9.5 ONE (TWO) STARTING CONDITION(S) FOR ONE (ANOTHER) FUNCTION

For the same general solution (3.737b; 3.737d) of the coupled system of first-order linear unforced partial finite difference equations with constant coefficients (3.730a, b), the two starting conditions for the first function are retained (3.738a, b) ≡ (3.745a, b) and the third starting condition is applied to the second function (3.745c):

$$\Phi_{0,0} = C_+ + C_-, \qquad \Phi_{0,1} = b\left(C_+ + C_-\right), \qquad \Psi_{0,0} = C_+ - C_-. \quad (3.745a–c)$$

The ratio of (3.745a, b) specifies b in (3.746a) and from (3.745a, c) follows (3.746b–c):

$$b = \frac{\Phi_{0,1}}{\Phi_{0,0}}, \qquad 2\,C_\pm = \Phi_{0,0} \pm \Psi_{0,0}. \qquad (3.746a–c)$$

Substitution of (3.746a–c) in (3.737b; 3.737d) specifies *the general solution of the coupled system of first-order linear unforced partial finite difference equations with constant coefficients (3.730a, b), for the two functions at all points of a rectangular grid (Figure 3.4), from two (one) values of the first (3.745a, b) [second (3.745c)] function:*

$$\Phi_{n,m} = \frac{c^n}{2}\left(\frac{\Phi_{0,1}}{\Phi_{0,0}}\right)^{n+m} \times \left\{\Phi_{0,0}\left[1+(-)^n\right] + \Psi_{0,0}\left[1-(-)^n\right]\right\}, \qquad (3.747)$$

$$\Psi_{n,m} = \frac{c^n}{2}\left(\frac{\Phi_{0,1}}{\Phi_{0,0}}\right)^{n+m} \times \left\{\Phi_{0,0}\left[1+(-)^{n+1}\right] + \Psi_{0,0}\left[1-(-)^{n+1}\right]\right\}. \qquad (3.748)$$

Each sequence (3.747) [(3.748)] splits into two subsequences (3.749a, b) [(3.750a, b)], one of even (3.749c) [(3.750c)] and one of odd (3.749d) [(3.750d)] order:

$$n = \{2\,r\,,\,2\,r+1\}: \qquad \left\{\Phi_{2r,m}\;;\Phi_{2r-1,m}\right\} = c^{2r} \left(\frac{\Phi_{0,1}}{\Phi_{0,0}}\right)^{2r+m} \Phi_{0,0}\left\{1\;;\;\frac{1}{c}\,\frac{\Psi_{0,0}}{\Phi_{0,1}}\right\},$$

$$(3.749\text{a--d})$$

$$n = \{2\,r\,,\,2\,r+1\}: \qquad \left\{\Psi_{2r,m}\;;\Psi_{2r+1,m}\right\} = c^{2r} \left(\frac{\Phi_{0,1}}{\Phi_{0,0}}\right)^{2r+m} \left\{\;\Psi_{0,0}\;;\;c\,\Phi_{0,1}\right\}$$

$$\equiv \left\{\frac{1}{c}\,\Phi_{2r+1,m-1}\;;\;c\,\Phi_{2r,m+1}\right\}. \qquad (3.750\text{a--f})$$

It can be checked that (3.749c, d; 3.750c, d) satisfy (3.750e, f) in agreement with (3.730a, b). The subsequences (3.749c, d; 3.750c, d) converge for large order m , 2r → ∞ if the condition (3.751a) is met. Choosing the starting values (3.751b–d), the subsequences (3.749c, d) [(3.750c, d)] simplify to (3.751e, f) [(3.751g, h)]:

$$\Phi_{0,1} < \Phi_{0,0}; \qquad \left\{\Phi_{0,0}\;;\Phi_{0,1}\;;\Psi_{0,0}\right\} = \{\,2\;;1\;;3\,\}: \qquad (3.751\text{a--d})$$

$$\left\{\Phi_{2r,m}\;;\Phi_{2r+1,m}\right\} = c^{2r}\,2^{-2r-m}\,\{\,2\;;6\,/\,c\} \qquad (3.751\text{e, f})$$

$$\left\{\Psi_{2r,m}\;;\Psi_{2r+1,m}\right\} = c^{2r}\,2^{-2r-m}\,\{\,3\;;c\,\}. \qquad (3.751\text{g, h})$$

The method of the matrix of polynomials of partial finite differences applies both to unforced (forced) equations [Subsections 3.9.2–3.9.5 (3.9.6–3.9.7)].

3.9.6 SIMULTANEOUS PARTIAL FINITE DIFFERENCE EQUATIONS FORCED BY POWERS

The forcing by powers with non-negative integral exponents (3.752a) of the simultaneous linear system of partial finite difference equations (3.720b) ≡ (3.752b) with constant coefficients (3.722):

$$B^i_{p,q} = B_i\,a^p\,b^q = \sum_{j=1}^{L}\left\{P_{ij}\left(\Delta_x\,,\,\Delta_y\right)\right\}\Phi^j_{p,q}, \qquad (3.752\text{a, b})$$

leads to particular solutions (3.724a) satisfying (3.753) obtained by substitution in (3.752b):

$$B_i\,a^n\,b^m = \sum_{j=1}^{L} P_{ij}(a,b)\,C_j\,a^n\,b^m. \qquad (3.753)$$

If the (a,b) are not roots of the characteristic polynomial (3.754a), the inversion of (3.753) and substitution in (3.724a) lead to (3.754b, c):

$$P_{N,M}(a,b) \neq 0: \qquad \Phi^i_{n,m} = a^n\, b^m \sum_{j=1}^{L} \{P_{ij}(a,b)\}^{-1} B_j \qquad (3.754a,\ b)$$

$$= \frac{a^n\, b^m}{P_{N,M}(a,b)} \sum_{j=1}^{L} \overline{P}_{ij}(a,b)\, B_j, \qquad (3.754c)$$

where $\overline{P}_{ij}$ are the algebraic complements or minors (3.486).

If a (b) is a root of multiplicity r (s) of the characteristic polynomial (3.725c), the lowest order non-zero derivative is (3.755a, b) and the particular solution (3.754c) is replaced by (3.755c):

$$E(a,b) \equiv \frac{\partial^{r+s}}{\partial a^r\, \partial b^s} \big[P_{N,M}(a,b) \big] \neq 0:$$

$$\Phi^i_{n,m}(a,b) = \frac{1}{E(a,b)} \frac{\partial^{r+s}}{\partial a^r\, \partial b^s} \left\{ a^n\, b^m \sum_{j=1}^{L} \overline{P}_{ij}(a,b)\, B_j \right\}. \qquad (3.755a\text{–}c)$$

Thus, *a simultaneous system of linear partial finite difference equations (3.720a, b) with constant coefficients (3.722) forced by powers with non-negative integral exponents (3.752a, b) has particular non-resonant (3.754b, c) [resonant (3.755c)] solutions when the bases are not (3.752a) [are (3.755a, b)] roots of the characteristic polynomial (3.725c).* An example of forcing by powers (Subsection 3.9.7) is reconsidered for the coupled first-order system analogous to the wave equation (Subsection 3.9.3).

3.9.7   NON-RESONANT AND SINGLY RESONANT SOLUTIONS

Consider the coupled linear system of partial finite difference equations with constant coefficients (3.730a, b) that is analogous to the Cartesian one-dimensional wave equation (3.731a–c) in the case (3.752a, b) of forcing by powers with non-negative integer exponents (3.756a, b):

$$\Phi_{n+1,m} - c\, \Psi_{n,m+1} = -\, a^n\, b^m, \qquad \Psi_{n+1,m} - c\, \Phi_{n,m+1} = 2\, a^n\, b^m, \qquad (3.756a,\ b)$$

which can be written in matrix form (3.757) using partial finite differences (3.659a–c; 3.660a–c):

$$\begin{bmatrix} \Delta_x & -c\,\Delta_y \\ -c\,\Delta_y & \Delta_x \end{bmatrix} \begin{bmatrix} \Phi_{n,m} \\ \Psi_{n,m} \end{bmatrix} = a^n\, b^m \begin{bmatrix} -1 \\ 2 \end{bmatrix}. \qquad (3.757)$$

The particular solutions may be sought in the form (3.758a, b), leading to (3.758c) by substitution in (3.757):

$$\Phi_{n,m} = C_1 \, a^n \, b^m:$$
$$\Psi_{n,m} = C_2 \, a^n \, b^m:$$
$$\begin{bmatrix} a & -c\,b \\ -c\,b & a \end{bmatrix} \begin{bmatrix} C_1 \\ C_2 \end{bmatrix} = \begin{bmatrix} -1 \\ 2 \end{bmatrix}.$$

(3.758a–c)

The system (3.758c) may be inverted (3.759a, b):

$$\left(a^2 - c^2 \, b^2\right) \begin{bmatrix} C_1 \\ C_2 \end{bmatrix} = \begin{bmatrix} a\,c\,b \\ c\,b\,a \end{bmatrix} \begin{bmatrix} -1 \\ 2 \end{bmatrix} = \begin{bmatrix} 2\,c\,b - a \\ 2\,a - c\,b \end{bmatrix}.$$

(3.759a, b)

Substitution of (3.759b) in (3.758a, b) specifies the particular solution (3.760b) of the system (3.756a, b) in the non-resonant case (3.760a):

$$a \neq \pm \, c \, b: \qquad \begin{bmatrix} \Phi_{n,m} \\ \Psi_{n,m} \end{bmatrix} = \frac{a^n \, b^m}{a^2 - c^2 \, b^2} \begin{bmatrix} 2\,c\,b - a \\ 2\,a - c\,b \end{bmatrix}.$$

(3.760a, b)

In the resonant case, (3.761a) appears in the denominator (3.761b–d), leading to the particular integrals (3.761e, f):

$$a = \pm \, c \, b: \qquad E(a,b) = \frac{\partial}{\partial a} \left[P_{2,2}(a,b) \right] = \frac{\partial}{\partial a} \left(a^2 - c^2 \, b^2 \right) = 2 \, a:$$

$$\begin{bmatrix} \Phi^{\pm}_{n,m} \\ \Psi^{\pm}_{n,m} \end{bmatrix} = \lim_{a \to \pm \, cb} \frac{b^m}{2 \, a} \frac{\partial}{\partial a} \left\{ a^n \begin{bmatrix} 2\,c\,b - a \\ 2\,a - c\,b \end{bmatrix} \right\}$$

$$= \pm \frac{b^{m-1}}{2 \, c} \lim_{a \to \pm \, cb} \left\{ a^n \begin{bmatrix} -1 \\ 2 \end{bmatrix} + n \, a^{n-1} \begin{bmatrix} 2\,c\,b - a \\ 2\,a - c\,b \end{bmatrix} \right\},$$

(3.761a–f)

implying the pairs of singly resonant solutions (3.762a, b) and (3.762c, d):

$$\left\{ \Phi^{+}_{n,m} \, , \, \Psi^{+}_{n,m} \right\} = \frac{1}{2} \, c^{n-1} \, b^{n+m-1} \, \{n-1 \, , \, n+2\},$$

(3.762a, b)

$$\left\{ \Phi^{-}_{n,m} \, , \, \Psi^{-}_{n,m} \right\} = \frac{(-)^n}{2} \, c^{n-1} \, b^{n+m-1} \, \{3 \, n+1 \, , \, -3 \, n - 2\}.$$

(3.762c, d)

Thus, *the coupled system of linear first-order partial finite difference equations forced by powers with non-negative integral exponents (3.756a, b)* ≡ *(3.757) has particular non-resonant solutions (3.760a, b). The particular singly resonant solutions (3.762a, b) [(3.762c, d)] correspond to the upper (lower) sign in (3.763a–d):*

$$\Phi_{n+1,m}^{\pm} - c\,\Psi_{n,m+1}^{\pm} = -\,b^m\,(\pm\,b\,c)^n = -\,(\pm\,c)^n\,b^{n+m}, \qquad \text{(3.763a, b)}$$

$$\Psi_{n+1,m}^{\pm} - c\,\Phi_{n,m+1}^{\pm} = 2\,b^m\,(\pm\,b\,c)^n = 2\,(\pm\,c)^n\,b^{n+m}. \qquad \text{(3.763c, d)}$$

It may be checked that (3.762c, d) satisfies (3.756a) [(3.756b)] as follows from (3.764a–d) [(3.765a–d)]:

$$\Phi_{n+1,m}^{-} - c\,\Psi_{n,m+1}^{-} = \frac{(-)^n}{2}\,c^{n-1}\,b^{n+m}\left[-(3\,n+4)\,c + (3\,n+2)\,c\right]$$

$$= -\,(-)^n\,c^n\,b^{n+m} = -\,b^m\,(-\,b\,c)^n = -\,a^n\,b^m,$$

$$\text{(3.764a–e)}$$

$$\Psi_{n+1,m}^{-} - c\,\Phi_{n,m+1}^{-} = \frac{(-)^n}{2}\,c^{n-1}\,b^{n+m}\left[(3\,n+5)\,c - (3\,n+1)\,c\right]$$

$$= 2\,(-)^n\,c^n\,b^{n+m} = 2\,b^m\,(-\,c\,b)^n = 2\,a^n\,b^m,$$

$$\text{(3.765a–e)}$$

since (3.764e) ≡ (3.756a) [(3.765e) ≡ (3.756b)]. It can be checked similarly that (3.760a, b) satisfy (3.756a, b) and (3.762a–d) satisfy (3.756a, b; 3.761a) ≡ (3.763a–d). This concludes the theory and examples of six classes of partial differential and finite difference equations and simultaneous systems that have characteristic polynomials (Subsection 3.9.8).

3.9.8 Six Classes of Partial Differential Equations/Partial Finite Difference Equations and Simultaneous Systems with Characteristic Polynomials

The method of the characteristic polynomial is the simplest of the five methods of solution of linear partial differential equations (Section 3.3 and Table 3.2 in Section 3.4), and applies in all cases with constant coefficients, including some of the most frequent equations of mathematical physics, such as: (i) elliptic, hyperbolic, and parabolic (Table 3.3 in Section 3.4); and (ii) combinations and generalizations (Table 3.4 in Section 3.4) of (i). The method of characteristic polynomials applies to linear partial differential equations with constant coefficients specifying: (i) the general integral in the unforced case (Table 3.5 in Section 3.4); (ii) non-resonant and resonant particular solutions for exponential forcing (Table 3.5 in Section 3.4); (iii) particular integrals for other types of forcing (Table 3.6 in Section 3.4); and (iv) factorization and inverse characteristic polynomials when all derivatives are of the same order (Table 3.7 in Section 3.4). The method of characteristic polynomials is presented in greater detail for single linear partial differential equations, and more briefly for the five other cases involving simultaneous systems, homogeneous power coefficients, and/or multiple finite differences

(Table 3.8 in Section 3.4), because there are analogies, such as similarity solutions (Table 3.9 in Section 3.7).

The analogies among the six cases of linear partial differential (multiple finite difference) equations with constant (homogeneous) power coefficients (Table 3.10) include: (i) general solutions in the unforced case (Table 3.11); (ii) non-resonant and singly or multiple resonant solutions for exponential or power forcing (Table 3.12); (iii) the use of the inverse characteristic polynomials to obtain particular solutions for other types of forcing (Table 3.13); (iv) the examples of analogues (Table 3.14) among single equations (simultaneous systems), partial differential (multiple finite difference) equations, and constant (homogeneous power) coefficients; and (v) the examples (Table 3.15) of boundary and initial (starting) conditions to single equations (E) or simultaneous systems (S) of linear partial differential [finite difference (F)] equations with constant (A) or homogeneous power (B) (constant) coefficients. The most important equations of mathematical physics used as examples (Diagram 3.1 in Section 3.4) are combined as a particular case (Table 3.16) of a generalised equation of mathematical physics (note 3.1) and its solutions (Table 3.17) are considered (Diagram 3.2 in Section 3.4) with continuous (discretised) derivatives as partial differential (finite difference) equations [notes 3.1–3.4 (3.5–3.12)].

TABLE 3.10

Partial Differential and Finite Difference Single Equation (Simultaneous Systems)

Type	Differential (A)	Differential (B)	Finite Difference (A)
Section	3.4 (3.5)	3.6 (3.7)	3.8 (3.9)
Unforced	3.4.1–3.4.3 (3.5.1)	3.6.1, 3.6.3 (3.7.1)	3.8.1 (3.9.1)
Single/multiple roots	3.4.4–3.4.5 (3.5.1)	3.6.1 (3.7.2)	3.8.2 (3.9.2)
Examples	3.4.6–3.4.15 (3.5.2–3.5.6)	3.6.2, 3.6.4–3.6.6 (3.7.3–3.7.7)	3.8.3–3.8.5 (3.9.3–3.9.5)
Forcing	Exponential	Power	Power
Non-resonant/resonant	3.4.16 (3.5.7)	3.6.7 (3.7.8)	3.8.6 (3.9.6)
Examples	3.4.17–3.4.30 (3.5.8–3.5.9)	3.6.8 (3.7.9)	3.8.7 (3.9.7)
Inverse polynomial	3.4.31–3.4.32 (E10.8.1)	3.6.9 (3.7.10)	
Examples	3.4.33–3.4.39 (E10.8.2–E10.8.4)	3.6.9 (3.7.11)	
Shifted inverse polynomial (C)	3.4.40, 3.4.43 (–)	3.6.10 (–)	
Examples	3.4.40–3.4.41, 3.4.44 (–)	– (–)	

(A) Linear with constant coefficients (and ordinary derivatives).

(B) Linear with homogeneous power coefficients (or constant coefficients and homogeneous derivatives).

(C) Differential equation with all derivatives of the same order.

Note: Comparison of the method of characteristic polynomials applied to the six classes in Table 3.8 in Section 3.4, namely single (simultaneous systems) of linear partial differential (finite difference) equations with constant (homogeneous power) coefficients.

TABLE 3.11
Unforced Linear Partial Differential Equations and Simultaneous Systems

Case	Single Equation	Simultaneous System
Differential (A)	(3.194a)	(3.433a, b; 3.434) ≡ (3.435)
Characteristic polynomial	(3.200a, b)	(3.438c)
Single roots	(3.203a–c)	(3.203a–c)
Natural integrals	(3.204a, b)	(3.204a, b)
General integral	(3.205a, b)	(3.439a–d)
Multiple roots	(3.206a–f)	(3.206a–f)
Natural integrals	(3.207; 3.208)	(3.440a, b; 3.441a, b)
General integral	(3.209a, b)	(3.442a, b; 3.443a, b)
Differential (B)	(3.505) ≡ (3.535a, b)	(3.588a, b; 3.589a) ≡ (3.590a, b; 3.589b, c)
Characteristic polynomial	(3.507b, c)	(3.592b, c)
Single roots	(3.508a, b)	(3.508a, b)
Natural integrals	(3.509a, b)	(3.509a, b)
General integral	(3.510a, b)	(3.593a, b; 3.594a, b)
Multiple roots	(3.511a, b; 3.512a, b)	(3.511a, b; 3.512a, b)
Natural integrals	(3.513a–c; 3.514a–c)	(3.513a–c; 3.514a–c)
General integral	(3.515a, b)	(3.595a, b; 3.596a, b)
Finite difference (C)	(3.657; 3.655a–d)	(3.722; 3.718a–c)
Characteristic polynomial	(3.663)	(3.725c)
Single roots	(3.667a, b)	(3.667a, b)
Natural sequences	(3.668a, b)	(3.668a, b)
General solution	(3.669a, b)	(3.726a, b; 3.727a, b)
Multiple roots	(3.670a, b; 3.671a, b)	(3.670a, b; 3.671a, b)
Natural sequences	(3.673a, b; 3.674a, b)	(3.673a, b; 3.674a, b)
General solution	(3.675a, b)	(3.728a, b; 3.729a, b)

(A) Linear with constant coefficients (and ordinary derivatives).
(B) Linear with homogeneous power coefficients (or constant coefficients and homogeneous derivatives).
(C) Linear with finite differences and constant coefficients.
Note: General and special integrals consisting of linear combinations of natural integrals in the solution of unforced equations of the six classes in Table 3.8 in Section 3.4.

TABLE 3.12
Non-resonant/Resonant Forced Solutions

Case	Single Equation	Simultaneous System
Differential (A)	(3.279a, b)	(3.487a, b)
Forcing	Exponential	Exponential
Characteristic polynomial	(3.200a)	(3.438d)
Non-resonant solution	(3.280a, b)	(3.488a, b)
Resonant solution	(3.282a, b)	(3.489a, b)

(Continued)

TABLE 3.12 (*Continued*)
Non-resonant/Resonant Forced Solutions

Case	Single Equation	Simultaneous System
Differential (B)	(3.565)	(3.637a, b)
Forcing	Power	Power
Characteristic polynomial	(3.507b, c)	(3.592c; 3.589a–c)
Non-resonant solution	(3.567a, b)	(3.635a–c)
Resonant solution	(3.568a, b)	(3.636a–c)
Finite difference (C)	(3.699a, b)	(3.752b)
Forcing	Power	Power
Characteristic polynomial	(3.666b)	(3.725c)
Non-resonant solution	(3.701a, b)	(3.754a–c)
Resonant solution	(3.705; 3.702a, b)	(3.755a–c)

[A] Linear with constant coefficients.
[B] Linear with homogeneous power coefficients (or constant coefficients and homogeneous derivatives).
[C] Finite difference with constant coefficients

Note: As in Table 3.10, for the non-resonant and resonant solutions of the six classes of equations and simultaneous systems in Table 3.8 in Section 3.4, due to forcing by exponentials (powers) in cases with constant (homogeneous power) coefficients.

TABLE 3.13
Method of Inverse Characteristic Polynomial

Differential Equation	Inverse Polynomial	Shifted Inverse Polynomial
Single (A)	(3.363a–d; 3.364a–d) (3.365a–c; 3.366a, b; 3.367d; 3.368a, b) (3.369a–d; 3.371) (3.372a–c; 3.373a, b) (3.365a–c; 3.366a; 3.407a–c) (3.365a–c; 3.369a–d; 3.426a–e)	(3.389a, b) (3.390a–c) (3.391a–c) (3.392a–c)
System (A)	(10.153a, b; 10.154; 10.155)	–
Single (B)	(3.578a, b)	(3.584a–c)
System (B)	(3.647; 3.648)	–

[A] Linear with constant coefficients (and ordinary derivatives).
[B] Linear with homogeneous power coefficients (or constant coefficients and homogeneous derivatives).

Note: The method of the inverse characteristic polynomial of partial derivatives or finite differences is an alternative to those in Table 3.12 for other types of forcing, such as powers (powers or logarithms) for constant (homogeneous power) coefficients.

TABLE 3.14

Examples of Differential and Difference Equations

| | | | Forced | |
| | | | Non-resonant/Resonant | Inverse |
Equations or System	Type	Forced		Polynomial
Cauchy-Riemann system	SA	(3.447b, c) (3.449a; 3.450)	–	–
Alternate harmonic system	SA	(3.452a, b; 3.453) (3.449a; 3.458c)	(3.490a, b) (3.494a–c) (3.496a–f)	(10.150a, b) (10.163a, b)
Coupled wave diffusion	SA	(3.471a, b; 3.472) (3.477a, b; 3.479; 3.481a–c) (3.478a–c; 3.480; 3.482a–d)	(3.497a, b) (3.500a, b) (3.504a, b; 3.503a–c; 3.478a–c; 3.476b)	(10.151a, b) (10.177a–c) (10.176a, b)
Analogue wave diffusion	EB	(3.516; 3.518a–d) (3.522a–c; 3.519a, b) (3.523a–c; 3.519c, d) (3.524a–c; 3.520a–f) (3.525a–c; 3.521a–f)	(3.570) (3.573a, b) (3.574a, b) (3.576a, b) (3.577a, b)	(3.579a–c; 3.581a–d; 3.582a–d) (3.585; 3.587a–c)
Analogue homogeneous wave	SB	(3.599a, b; 3.609a–d)	(3.638a, b; 3.642a,b) (3.646a, b; 3.644b; 3.645b)	(3.649a, b; 3.652a, b; 3.653a, b)
Analogue wave diffusion	EF	(3.676) (3.680a–c; 3.682a, b) (3.683a–d; 3.684a–c)	(3.706) (3.708a–c) (3.710a, b; 3.711a, b) (3.713a, b; 3.714a–e)	–
Analogue	SF	(3.730a, b) (3.735d; 3.736c) (3.737a,b; 3.737c,d)	(3.756a, b) (3.760a, b) (3.763a–d) (3.762a–d)	–

[E:] single partial differential equation

[S:] simultaneous system of partial differential equations

[A:] linear with constant coefficients (and ordinary derivatives)

[B:] linear with homogeneous power coefficients (or constant coefficients and homogeneous derivatives)

[F:] multiple or partial finite difference with constant coefficients

Note: Examples of the six classes of equations in Table 3.8 in Section 3.4 for some of the equations in Tables 3.3 and 3.4 in Section 3.4, considering "analogous" equations across several of the classes.

TABLE 3.15

Application of Boundary (B)/Initial (I)/Starting (S) Conditions

Type	Name	Equation	B/I/S Condition	Solution
EA	Wave equation	(3.123a)	/ (3.123b, c) /	(3.113b; 3.122b)
			/ (3.124a–c) /	(3.124d–h)
			/ (3.125a–c) /	(3.125d–i)
		(3.129a)	(3.129b, c) //	(3.113b; 3.128c)
			(3.130a–c) //	(3.130d–h)
			(3.131a–c) //	(3.131d–f)
EA	Diffusion equation	(3.134a)	/ (3.134b) /	(3.134c)
			/ (3.135a) /	(3.135b)
		(3.138a)	(3.142a–d) //	(3.145)
			(3.146a, b) //	(3.146c)
SA	Cauchy-Riemann system	(3.447b, c)	(3.460a, b) //	(3.463a, b)
			(3.464a, b) //	(3.465a–d)
EB	Homogeneous wave equation	(3.545c) ≡ (3.546)	/ (3.553a–d) /	(3.556)
			/ (3.558a, b) /	(3.561)
SB	Homogeneous wave-coupled system	(3.599a, b)	(3.612a–d) //	(3.614a, b)
			(3.615a–c) //	(3.617a, b)
			(3.620a, b) //	(3.622a, b)
			(3.623a, b) //	(3.624b, d)
EF	Finite difference wave equation	(3.692b)	// (3.694a–c)	(3.696b) ≡ (3.697a–d)
			// (3.698b–d)	(3.698e, f)
SF	Finite difference-coupled wave system	(3.730a, b)	// (3.738a–c)	(3.740; 3.741) ≡ (3.742a–f; 3.743a–f)
			// (3.744b–d)	(3.744e–h)
			// (3.745a–c)	(3.747; 3.748) ≡ (3.749a–f; 3.750a–f)
			// (3.751b–d)	(3.751e–h)

[E]: single partial differential equation
[S]: simultaneous system of partial differential equations
[A]: linear with constant coefficients (and ordinary derivatives)
[B]: linear with homogeneous power coefficients (or constant coefficients and homogeneous derivatives)
[F]: multiple or partial finite difference.

Note: Application of boundary and initial (starting) conditions to partial differential (finite difference) equations and simultaneous systems.

NOTES: GENERALISED EQUATION OF MATHEMATICAL PHYSICS AND ENGINEERING (GEMPE)

The Generalised Equation of Mathematical Physics and Engineering (GEMPE) is a combined generalisation of the Laplace, wave, diffusion, telegraph, Helmholtz, Schrödinger, Klein-Gordon, bar, and beam vibration equations in (note 3.1) dimensional and dimensionless forms. Analytical solutions are obtained in the unforced case (note 3.2), for non-resonant and resonant exponential forcing (note 3.3) and for forcing by powers (note 3.4). The GEMPE is discretised by finite differences (note 3.5), solved in the unforced case (note 3.6) and for forcing by powers (note 3.7). The numerical solution of the discretised GEMPE raises convergence issues (note 3.9), considered first for the space-time causality condition in the wave equation (note 3.8) and then extended to the discretised GEMPE, leading to a further condition on the upper bound on temporal step size due to damping (note 3.10). The unicity of solution of the GEMPE in its original continuous (notes 3.1–3.4) [associated discretised (notes 3.5–3.12)] form requires the use of boundary and initial (also starting) conditions (note 3.11), that may depend on the roots of the characteristic polynomial (note 3.12).

Note 3.1 – Propagation Speed, Damping, Resilience, Stiffness and Tension

The Cartesian one-dimensional equation of mathematical physics and engineering (GEMPE) is:

$$\rho \frac{\partial^2 \Phi}{\partial t^2} - T \frac{\partial^2 \Phi}{\partial x^2} + E\,I\,\frac{\partial^4 \Phi}{\partial x^4} + \vartheta\,\frac{\partial \Phi}{\partial t} + k\,\Phi = F(x,t), \tag{3.766}$$

consisting of the following terms as regards (Figure 3.5) physical interpretation: (i) the first is the inertia force equal to the mass density per unit length ρ multiplied by the acceleration, that is the second order time derivative of the displacement; (ii) if $\Phi = \zeta$ is the transverse displacement of an elastic string the second term is (Chapter III.2) the vertical force due to a tangential traction T; (iii) if $\Phi = \zeta$ is the transverse displacement of an elastic bar, then the third term is the shear stress where E denotes the Young's modulus of the material and I is the moment of inertia of the cross-section; (iv) the fourth term is the kinematic friction force equal to the friction coefficient ϑ multiplied by the velocity, that is the time derivative of the displacement; (v) the fifth term is the restoring force due to an elastic support equal to the displacement multiplied by the resilience k of the springs; (vi) the forcing term is an external force per unit length F, specifying a loading, for example the weight $F = -\rho g$, where g is the acceleration of gravity. Equation (3.766) is linear, neglecting large displacements, slopes and velocities, and assumes a homogeneous (steady) medium for which all physical properties are independent of position (time) and thus all coefficients are constant.

Dividing (3.766) by the mass density, ρ, leads to the standard form of the one-dimensional GEMPE (3.767):

$$\frac{\partial^2 \Phi}{\partial t^2} - c^2 \frac{\partial^2 \Phi}{\partial x^2} + q^2 \frac{\partial^4 \Phi}{\partial x^4} + 2\,v\,\frac{\partial \Phi}{\partial t} + \omega_0^2\,\Phi = B(x,t), \qquad (3.767)$$

where (i) the elastic propagation speed (3.768a) has the dimensions (3.769a) of length per unit time; (ii) the bending stiffness parameter (3.768b) has the dimensions (3.769b) of a diffusivity, that is square of length per unit time; (iii/iv) the damping (3.768c) has the dimensions (3.769c) $\equiv$ (3.769d) of inverse time as the natural frequency (3.768d); (v) the forcing term is the external force per unit mass (3.768e) and has the dimensions of acceleration, that is length divided by the square of time (3.769e):

$$c \equiv \sqrt{\frac{T}{\rho}}, \qquad q \equiv \sqrt{\frac{E\,I}{\rho}}, \qquad \lambda \equiv \frac{\mu}{2\,\rho}, \qquad \omega_0 = \sqrt{\frac{k}{\rho}}, \qquad B(x,t) \equiv \frac{F(x,t)}{\rho},$$
$$(3.768\text{a--e})$$

$$[c] = L\,T^{-1}, \qquad [q] = L^2\,T^{-1}, \qquad [v] = T^{-1} = [\omega_0], \qquad [B] = L\,T^{-2}.$$
$$(3.769\text{a--e})$$

The generalised equation of mathematical physics and engineering (3.767) $\equiv$ (3.766) includes (Table 3.16) several cases, such as: (a) the harmonic oscillator for the first and fifth terms; (b) the damped harmonic oscillator for the first, fourth, and fifth terms; (c) the wave equation for the first and second terms; (d) the diffusion equation for the second and fourth terms; (e) the wave-diffusion equation for the first, second, and fourth terms; (f) the vibrations of an elastic bar for the first and third terms; (g) the vibrations of an elastic beam for the first, second, and third terms. The harmonic or Laplace equation corresponds to the first two terms with compression $T = -\,\rho$, also replacing time t by a spatial coordinate y. The standard form of the GEMPE is solved next in unforced (forced) cases [note(s) 3.2 (3.3–3.4)].

Note 3.2 – Natural and General Solutions in the Unforced Case

In the unforced case (3.770a), the solution of the generalised equation of mathematical physics and engineering in standard form (3.767) is sought as an exponential in space-time (3.770b):

$$B(x,t) = 0: \qquad\qquad \Phi(x,t) = C \exp(a\,t + b\,x), \qquad (3.770\text{a, b})$$

whose coefficients (a,b) are roots (3.771a) of the characteristic polynomial that is (3.771b) quadratic (3.771c) [biquadratic (3.771d)] in $a\,(b)$:

$$0 = P_{2,4}(a,b) = a^2 + 2\,v\,a + \omega_0^2 - c^2\,b^2 + q^2\,b^4$$

$$= \left[a - a_+(b)\right]\left[a - a_-(b)\right] = \left[b - b_+(a)\right]\left[b + b_-(a)\right]\left[b - b_-(a)\right]\left[b + b_-(a)\right].$$
$$(3.771\text{a--d})$$

The roots (3.772a, b) for a:

$$a_{\pm}(b) = -v \pm \zeta(b), \qquad [\zeta(b)]^2 = v^2 - \omega_0^2 + c^2 \, b^2 - q^2 \, b^4. \qquad (3.772a, b)$$

if distinct (3.773a) lead to two natural integrals (3.773b, c) where $C_{\pm}(a)$ are arbitrary functions, whose sum is the general integral (3.773d):

$$\zeta(b) \neq 0: \qquad \Phi_{\pm}(x,t) = \int e^{bx} \exp\left[t \, a_{\pm}(b)\right] C_{\pm}(b) \, db, \qquad (3.773a\text{–}c)$$

$$\Phi(x,t) = \Phi_{+}(x,t) + \Phi_{-}(x,t). \qquad (3.773d)$$

In the case (3.774a) of a double root (3.774b) corresponds to (3.774c) the four special integrals (3.774d–g) where $G_{\pm\pm}$ are arbitrary constants:

$$\zeta(\pm b^{\pm}) = 0, \qquad a_{\pm}(\pm b^{\pm}) = -v, \qquad 2 \, q^2 \, (b^{\pm})^2 = c^2 \pm \sqrt{c^4 + 4 \, q^2 \, (v^2 - \omega_0^2)}:$$

$$\Phi_{\pm\pm}(x,t) = e^{-\lambda t} \left\{ G_{++} \, e^{b^+ x}, \; G_{+-} \, e^{-b^+ x}, \; G_{-+} \, e^{b^- x}, \; G_{--} \, e^{-b^- x} \right\}. \qquad (3.774a\text{–}g)$$

The roots (3.775a, b) of (3.771a, b, d) for b:

$$2 \, q^2 \, [b_{\pm}(a)]^2 = c^2 \pm \xi(a), \qquad [\xi(a)]^2 = c^4 - 4 \, q^2 \left(\omega_0^2 + 2 \, v \, a + a^2\right), \qquad (3.775a, b)$$

if distinct (3.776a) lead to four natural integrals (3.776b–e) where $C^{\pm\pm}(a)$ are arbitrary functions, whose sum (3.776f) is the general integral:

$$\xi(a) \neq 0: \qquad \Phi^{\pm\pm}(x,t) = \int e^{at} \exp\left[\pm x \, b_{\pm}(a)\right] C^{\pm\pm}(a) \, da,$$

$$\Phi(x,t) = \Phi^{++}(x,t) + \Phi^{+-}(x,t) + \Phi^{-+}(x,t) + \Phi^{--}(x,t). \qquad (3.776a\text{–}f)$$

In the case (3.777a) of double roots (3.777b) corresponds to (3.777c) the four special integrals (3.777d, e) where $G^{\pm\pm}$ are arbitrary constants:

$$\xi(a^{\pm}) = 0, \qquad [b_{\pm}(a^{\pm})]^2 = \frac{c^2}{2 \, q^2}, \qquad a^{\pm} = -v \pm \sqrt{v^2 + \omega_0^2 - c^2/4 \, q^2}:$$

$$^{+}\Phi^{\pm}(x,t) = \exp\left(\frac{c \, x}{q \, \sqrt{2}}\right) \left\{ G^{++} e^{a^+ t}, \; G^{+-} e^{-a^+ t} \right\}$$

$$^{-}\Phi^{\pm}(x,t) = \exp\left(-\frac{c \, x}{q \, \sqrt{2}}\right) \left\{ G^{-+} e^{a^- t}, \; G^{--} e^{-a^- t} \right\}. \qquad (3.777a\text{–}g)$$

Since $b(a)$ in (3.772a, b) [(3.775a, b)] is arbitrary and equation (3.767) is linear, the principle of superposition holds: (i) in (3.773b, c) [(3.776b–e)] there is multiplication by an arbitrary function of $b(a)$ followed by integration in $db(da)$; (ii) thus (3.773d) [(3.776f)] is the general integral of (3.767) because it involves two (four) arbitrary functions and the partial differential equation (3.767) is of order two (four) in $t(x)$. In contrast in (3.774a–d) [(3.777a–d)] both a and b have fixed values, and thus the solutions (3.774d–g) [(3.777d–g)] are special integrals of (3.767) involving four arbitrary constants but no arbitrary functions.

The singular solutions, for example (3.777a–g) can be obtained as particular cases of the corresponding natural integrals (3.776a–f) using (3.778a) as arbitrary function (3.778b) a Dirac delta or unit impulse (Chapter III.3) in (3.777g)

$$b(a^\pm) = \pm \frac{c}{q\sqrt{2}}; \qquad C^{\pm\pm}(a) = G^{\pm\pm}\,\delta(a-a^\pm):$$

$$\Phi^{\pm\pm}(x,t) = \int e^{at}\exp\left[\pm x\,b(a)\right]G^{\pm\pm}\,\delta(a-a^\pm)\,da$$

$$= G^{\pm\pm}\,e^{a^\pm t}\exp\left(\pm \frac{c\,x}{q\sqrt{2}}\right) = {}^\pm\Phi^\pm(x,t), \qquad (3.778a\text{–}e)$$

that coincides with (3.777d–g) $\equiv$ (3.778d, e). Thus, *the unforced (3.770a) one-dimensional generalised equation of mathematical physics and engineering (3.767) has characteristic polynomial (3.771a–d) leading to the general integral (3.773a–d) [(3.776a–f)] for distinct roots (3.772a, b) [(3.775a, b)], and special integrals (3.774a–g) [(3.777a–g)] for double roots.* Next is considered non-resonant and resonant forcing by exponentials in space-time (note 3.3).

Note 3.3 – Non-Resonant and Singly/Doubly Resonant Solutions

The one-dimensional generalised equation of mathematical physics and engineering (3.767) forced by exponentials in space-time (3.779):

$$\frac{\partial^2\Phi}{\partial t^2} - c^2\,\frac{\partial^2\Phi}{\partial x^2} + q^2\,\frac{\partial^4\Phi}{\partial x^4} + 2\,v\,\frac{\partial\Phi}{\partial t} + \omega_0^2\,\Phi = B\,e^{at+bx}, \qquad (3.779)$$

has particular solution (3.780b) provided that the characteristic polynomial (3.771b) is not zero (3.780a):

$$P_{2,4}(a,b) \neq 0: \qquad \Phi(x,t) = \frac{B\,e^{at+bx}}{a^2 + 2\,v\,a + \omega_0^2 - c^2\,b^2 + q^2\,b^4}. \qquad (3.780a, b)$$

The distinct roots (3.772a, b) lead (3.781a, b) to two (3.781c) particular integrals (3.781d) singly resonant in time:

$$E_1(a,b) = \frac{\partial}{\partial a}[P_{2,4}(a,b)] = 2\,(v+a), \qquad a = a_\pm(b): \tag{3.781a–c}$$

$$\Phi_\pm(x,t) = \frac{B\,t}{2}\,\frac{\exp[\,b\,x + t\,a_\pm(b)\,]}{v + a_\pm(b)}. \tag{3.781d}$$

The distinct roots (3.775a, b) lead (3.782a, b) to four (3.782c) particular integrals (3.782d) singly resonant in space:

$$E_2(a,b) = \frac{\partial}{\partial b}[P_{2,4}(a,b)] = 4\,q^2\,b^3 - 2\,c^2\,b, \qquad b = \pm\,b_\pm(a): \tag{3.782a–c}$$

$$\Phi_{\pm\pm}(x,t) = \pm\,\frac{B\,x}{2\,b_\pm(a)}\,\frac{\exp[\,a\,t \pm x\,b_\pm(a)\,]}{2\,q^2\,[b_\pm(a)]^2 - c^2}. \tag{3.782d}$$

The double root (3.774a–c) leads (3.783a, b) to (3.783c, d) four particular integrals (3.783e) doubly resonant in time:

$$E_3(a,b) = \frac{\partial^2}{\partial a^2}[P_{2,4}(a,b)] = 2, \qquad b = \pm\,b^\pm, \qquad a_\pm = -\,v: \tag{3.783a–d}$$

$$\Phi^{\pm\pm}(x,t) = \frac{B\,t^2}{2}\,e^{-vt}\,\exp(\pm\,x\,b^\pm). \tag{3.783e}$$

The symmetric double roots (3.777c) lead (3.784a, b) to (3.784c–e) four particular integrals (3.784f) doubly resonant in space:

$$E_4(a,b) = \frac{\partial^2}{\partial b^2}[P_{2,4}(a,b)] = 12\,q^2\,b^2 - 2\,c^2, \qquad a = a^\pm, \qquad b_\pm(a^\pm) = \pm\,\frac{c}{q\,\sqrt{2}}:$$

$$E_4[a^\pm, b_\pm(a^\pm)] = 4\,c^2: \qquad {}^{\pm\pm}\Phi(x,t) = \frac{B\,x^2}{4\,c^2}\,\exp\!\left(a^\pm\,t \pm \frac{c\,x}{q\,\sqrt{2}}\right). \tag{3.784a–f}$$

Thus, *the one-dimensional generalised equation of mathematical physics and engineering (3.767) forced by exponentials in space-time (3.779) has characteristic polynomial (3.771a–d) leading to: (i) one particular non-resonant solution (3.780a, b); (ii) two (four) singly resonant solutions in time (3.781a–d) [space (3.782a–d)]; and (iii) four doubly resonant solutions in time (3.783a–e) [space (3.784a–f)].* Next is considered forcing by powers in space-time (note 3.4).

Note 3.4 – Forcing by Powers in Space-Time

The one-dimensional generalised equation of mathematical physics and engineering (3.767) forced (3.785) linearly in space-time (3.785):

$$\frac{\partial^2 \Phi}{\partial t^2} - c^2 \frac{\partial^2 \Phi}{\partial x^2} + q^2 \frac{\partial^4 \Phi}{\partial x^4} + 2 \, v \, \frac{\partial \Phi}{\partial t} + \omega_0^2 \, \Phi = B \, t \, x, \tag{3.785}$$

has particular integral (3.786):

$$\Phi(x,t) = \left(\omega_0^2 + 2 \, v \, \partial_t + \partial_t^2 - c^2 \, \partial_x^2 + q^2 \, \partial_x^4 \right)^{-1} B \, t \, x. \tag{3.786}$$

Since the forcing term is linear in space-time, it is sufficient to expand to order ∂_t in the particular integral (3.787a–d)

$$\Phi_1(x,t) = \left\{ \omega_0^2 + 2 \, v \, \partial_t \right\}^{-1} B \, t \, x = \frac{B \, x}{\omega_0^2} \left\{ 1 + \frac{2 \, v}{\omega_0^2} \, \partial_t \right\}^{-1} t$$

$$= \frac{B \, x}{\omega_0^2} \left\{ 1 - \frac{2 \, v}{\omega_0^2} \, \partial_t \right\} t = \frac{B \, x}{\omega_0^2} \left(t - \frac{2 \, v}{\omega_0^2} \right). \tag{3.787a–d}$$

If the forcing (3.788) was linear (quadratic) in time (space):

$$\frac{\partial^2 \Phi}{\partial t^2} - c^2 \frac{\partial^2 \Phi}{\partial x^2} + q^2 \frac{\partial^4 \Phi}{\partial x^4} + 2 \, v \, \frac{\partial \Phi}{\partial t} + \omega_0^2 \, \Phi = B \, t \, x^2, \tag{3.788}$$

it is necessary to expand to first (second) order in ∂_t (∂_x) in (3.789a–e):

$$\Phi_2(x,t) = \left\{ \omega_0^2 + 2 \, v \, \partial_t + \partial_t^2 - c^2 \, \partial_x^2 + q^2 \, \partial_x^4 \right\}^{-1} B \, t \, x^2$$

$$= \left\{ \omega_0^2 + 2 \, v \, \partial_t - c^2 \, \partial_x^2 \right\}^{-1} B \, t \, x^2$$

$$= \frac{B}{\omega_0^2} \left\{ 1 + \frac{2 \, v}{\omega_0^2} \, \partial_t - \frac{c^2}{\omega_0^2} \, \partial_x^2 \right\}^{-1} t \, x^2$$

$$= \frac{B}{\omega_0^2} \left\{ 1 - \frac{2 \, v}{\omega_0^2} \, \partial_t + \frac{c^2}{\omega_0^2} \, \partial_x^2 \right\} t \, x^2$$

$$= \frac{B}{\omega_0^2} \left(t \, x^2 - \frac{2 \, v \, x^2}{\omega_0^2} + \frac{2 \, c^2 \, t}{\omega_0^2} \right); \tag{3.789a–e}$$

Thus, *the one-dimensional generalised equation of mathematical physics and engineering (3.767) forced linearly in time and linearly (3.785) [quadratically*

(3.788)] in space has particular integral (3.787d) [(3.789e)]. A partial differential equation (notes 3.1–3.4) can be approximated by a partial finite difference equation (notes 3.6–3.7) by discretising the partial derivatives as partial finite differences (note 3.5).

Note 3.5 – Discretisation of Partial Derivatives of All Orders

A function of time and space (3.790a–c) may be discretised on a rectangular grid (Figure 3.6):

$$t = T + n\,h, \qquad x = X + m\,k: \qquad \Phi_{n,m} = \Phi(T + n\,h,\ X + m\,k), \qquad (3.790\text{a–c})$$

similar to the Cartesian grid (Figure 3.4) with the changes of variable $(x, y) \rightarrow (t, x)$ from (3.655a–d) to (3.730a–c). The first-order partial derivatives with regard to time (3.791a) [space (3.792a)] can be approximated by finite differences (3.791b) [(3.792b)]:

$$\frac{\partial \Phi}{\partial t} \equiv \lim_{h \to 0} \frac{\Phi(t+h,\ x) - \Phi(t,x)}{h} \sim h^{-1}\left(\Phi_{n+1,m} - \Phi_{n,m}\right), \quad (3.791\text{a, b})$$

$$\frac{\partial \Phi}{\partial x} \equiv \lim_{k \to 0} \frac{\Phi(t,\ x+k) - \Phi(t,x)}{k} \sim k^{-1}\left(\Phi_{n,m+1} - \Phi_{n,m}\right). \quad (3.792\text{a, b})$$

The discretisation extends to the second-order derivatives with regard to: (i) time (3.793a–c):

$$\frac{\partial^2 \Phi}{\partial t^2} = \lim_{h \to 0} h^{-1}\left[\partial_t\ \Phi(t+h, x) - \partial_t\ \Phi(t,x)\right]$$

$$= \lim_{h \to 0} h^{-2}\left\{\left[\Phi(t+2\,h, x) - \Phi(t+h, x)\right] - \left[\Phi(t+h, x) - \Phi(t,x)\right]\right\}$$

$$\sim h^{-2}\left(\Phi_{n+2,m} - 2\,\Phi_{n+1,m} + \Phi_{n,m}\right); \quad (3.793\text{a–c})$$

(ii) likewise for space (3.794):

$$\frac{\partial^2 \Phi}{\partial x^2} \sim k^{-2}\left(\Phi_{n,m+2} - 2\,\Phi_{n,m+1} + \Phi_{n,m}\right); \quad (3.794)$$

(iii) and for space-time (3.795a–c):

$$\frac{\partial^2 \Phi}{\partial x\,\partial t} = \lim_{k \to 0} k^{-1}\left[\Phi_t(t, x+k) - \Phi_t(t,x)\right]$$

$$= \lim_{h,k \to 0} (k\,h)^{-1}\left\{\left[\Phi(t+h, x+k) - \Phi(t, x+k)\right] - \left[\Phi(t+h, x) - \Phi(t,x)\right]\right\}$$

$$\sim (k\,h)^{-1}\left(\Phi_{n+1,m+1} - \Phi_{n,m+1} - \Phi_{n+1,m} + \Phi_{n,m}\right). \quad (3.795\text{a–c})$$

The discretisation can be extended to derivatives of any order as a binomial expansion, for example (IV.4.513) ≡ (3.796a, b):

$$\frac{\partial^m \Phi}{\partial x^m} = \lim_{k \to 0} k^{-m} \sum_{s=0}^{m} (-)^s \binom{m}{s} \Phi(t, x+s\,k) \sim k^{-m} \sum_{s=0}^{m} \frac{m!\,(-)^s}{s!\,(m-s)!} \Phi_{n,m+s},$$

$$(3.796a, b)$$

that becomes (3.797) in the case of the fourth-order spatial derivative:

$$k^4 \frac{\partial^4 \Phi}{\partial x^4} = \Phi_{n,m+4} - 4\,\Phi_{n,m+3} + 6\,\Phi_{n,m+2} - 4\,\Phi_{n,m+1} + \Phi_{n,m}. \qquad (3.797)$$

Thus, *a partial differential equation can be approximately discretised (3.790a–c) replacing partial derivatives by partial finite differences of first (3.791a, b; 3.792a, b), second (3.793a–c; 3.794; 3.795a–c), fourth (3.797) or higher (3.796a, b; 3.798) order:*

$$\frac{\partial^{n+m} \Phi}{\partial t^n \partial x^m} \sim h^{-n} k^{-m} \sum_{r=0}^{n} \sum_{s=0}^{m} \frac{n!\,m!\,(-)^{r+s}}{(n-r)!(m-s)!\,r!\,s!} \Phi_{n+r,m+s}, \qquad (3.798)$$

leading in the case of the one-dimensional generalised equation of mathematical physics and engineering (3.767) to (3.799):

$$B_{n,m} = h^{-2} \left(\Phi_{n+2,m} - 2\,\Phi_{n+1,m} + \Phi_{n,m} \right) - c^2 \, k^{-2} \left(\Phi_{n,m+2} - 2\,\Phi_{n,m+1} + \Phi_{n,m} \right)$$

$$+ q^2 \, k^{-4} \left(\Phi_{n,m+4} - 4\,\Phi_{n,m+3} + 6\,\Phi_{n,m+2} - 4\,\Phi_{n,m+1} + \Phi_{n,m} \right)$$

$$+ 2\,v\,h^{-1} \left(\Phi_{n+1,m} - \Phi_{n,m} \right) + \omega_0^2 \, \Phi_{n,m}, \qquad (3.799)$$

where the forcing function on the l.h.s. of (3.799) is also discretised as the dependent variable (3.790a–c) in (3.900a–e):

$$h \equiv \Delta t, \qquad k \equiv \Delta x: \qquad \Phi_{n,m} = \Phi(t + n\,h, \; x + m\,k)$$

$$B_{n,m} = B(t + n\,h, \; x + m\,k) = B(t + n\,\Delta t, \; x + m\,\Delta x). \qquad (3.800a–e)$$

The finite difference equation (3.799) is solved next in the unforced case (note 3.6) followed by forcing by powers (note 3.7).

Note 3.6 – General Solution of Unforced Finite Difference Equation

The partial finite difference equation (3.799) ≡ (3.801) can be re-arranged:

$$h^2 \, B_{n,m} = \Phi_{n+2,m} + A_{1,0} \, \Phi_{n+1,m} + A_{0,0} \, \Phi_{n,m} + A_{0,1} \, \Phi_{n,m+1}$$

$$+ A_{0,2} \, \Phi_{n,m+2} + A_{0,3} \, \Phi_{n,m+3} + A_{0,4} \, \Phi_{n,m+4}, \qquad (3.801)$$

with coefficients (3.802a–f):

$$\left\{ A_{1,0} \; ; \; A_{0,0} \; ; \; A_{0,1} \; ; \; A_{0,2} \; ; \; A_{0,3} \; ; \; A_{0,4} \right\}$$

$$= \{- 2 + 2 \, v \, h \; ; \; 1 - c^2 h^2 / k^2 + q^2 h^2 / k^4 - 2 \, v \, h + \omega_0^2 \, h^2 \; ;$$

$$2 \, c^2 h^2 / k^2 - 4 \, q^2 h^2 / k^4 \; ; \; - c^2 h^2 / k^2 + 6 \, q^2 h^2 / k^4 \; ;$$

$$- 4 \, q^2 h^2 / k^4 \; ; \; q^2 h^2 / k^4 \}. \tag{3.802a–f}$$

Bearing in mind that (3.800b) [(3.800a)] has the dimensions of length L (time T) it follows from (3.769a–e) that all terms in (3.802a–f) are dimensionless. In the unforced case (3.803a), there are power solutions with non-negative integer exponents (3.803b):

$$B_{n,m} = 0 : \qquad\qquad\qquad \Phi_{n,m} = C \, a^n \, b^m, \tag{3.803a, b}$$

leading to a quadratic (quartic) characteristic polynomial (3.804a, b) in a (b)

$$0 = Q_{2,4}(a,b) = a^2 + A_{1,0} \, a + \sum_{r=0}^{4} A_{0,r} \, b^r = \left[a - a_+(b)\right]\left[a - a_-(b)\right]. \tag{3.804a–c}$$

The roots in a are given by (3.804c) $\equiv$ (3.805a–d):

$$\xi = \frac{A_{1,0}}{2}, \qquad \zeta(b) \equiv \sum_{r=0}^{4} A_{0,r} \, b^r :$$

$$a_\pm(b) = - \xi \pm \eta(b); \qquad\qquad \left[\eta(b)\right]^2 = \xi^2 - \zeta(b). \tag{3.805a–d}$$

In the case of distinct roots (3.806a) the natural sequences are (3.806b) and their sum specifies the general solution (3.806c):

$$\eta(b) \neq 0 : \qquad \Phi_{n,m}^\pm = \int b^m \left[a_\pm(b)\right]^n C_\pm(b) \, db, \qquad \Phi_{n,m} = \Phi_{n,m}^+ + \Phi_{n,m}^-. \tag{3.806a–c}$$

In the case (3.807a) of a double root, there are (3.805b, d) eight possible values for b, leading (3.807b–d) to the solutions (3.807e):

$$\eta(b_{1-8}) = 0, \qquad a_\pm(b_{1-8}) = - \xi = - \frac{A_{1,0}}{2} = 1 - v \, h : \tag{3.807a–d}$$

$$\Phi_{n,m}^{1-8} = C_{n,m}^{1-8} (1 - v \, h)^n \, (b_{1-8})^m. \tag{3.807e}$$

Thus, *the partial finite difference equation (3.801) with constant coefficients (3.802a–f) resulting from the discretisation of the one-dimensional generalised equation of mathematical physics and engineering (3.767) has one (eight) solutions (3.806a–c) [(3.807e)] for distinct (3.805a–d) [double (3.807a–d)] roots of the characteristic polynomial (3.804a–c).* Particular solutions of the finite difference equation (3.721; 3.722a–f) are obtained next (note 3.7) for forcing by powers.

Note 3.7 – Particular Solutions for Forcing by Powers

The partial finite difference equation (3.801) forced by powers with non-negative integer exponent:

$$B \, a^n \, b^m = \Phi_{n+2,m} + \sum_{r=0}^{1} A_{r,0} \, \Phi_{n+r,m} + \sum_{s=1}^{4} A_{0,s} \, \Phi_{n,m+s}, \tag{3.808}$$

has particular non-resonant solution (3.809b), using (3.805a, b), provided that the characteristic polynomial (3.804a) is not zero (3.809a):

$$0 \neq Q_{2,4}(a,b) = a^2 + 2 \, \xi \, a + \zeta(b): \qquad \Phi_{n,m} = \frac{B \, a^n \, b^m}{a^2 + 2 \, \xi \, a + \zeta(b)}. \tag{3.809a, b}$$

The distinct roots (3.805a–d) correspond (3.810a–c) to two particular solutions (3.810d) singly resonant in time:

$$E_1(a,b) = \frac{\partial}{\partial a}\left[Q_{2,4}(a,b)\right] = 2 \, a + A_{1,0} = 2 \, a + 2 \, v \, h - 2:$$

$$\Phi_{n,m}^{\pm} = B \, n \, \frac{b^n \left[a_{\pm}(b)\right]^{n-1}}{2 \, v \, h - 2 + 2 \, a_{\pm}(b)}. \tag{3.810a–d}$$

The double roots (3.807a–d) ≡ (3.811c) lead (3.811a, b) to eight particular solutions (3.811d) doubly resonant in time:

$$E_2(a,b) = \frac{\partial^2}{\partial a^2}\left[Q_{2,4}(a,b)\right] = 2, \qquad a_{\pm}(b_{1-8}) = 1 + v \, h:$$

$$\Phi_{n,m}^{1-8} = \frac{B}{2} \, n^2 \, (b_{1-8})^m \, (1 - v \, h)^{n-2}. \tag{3.811a–d}$$

Thus, *the partial finite difference equation (3.808) with constant coefficients (3.802a–f) forced by powers with non-negative integer exponents has one non-resonant solution (3.809a, b) and two (eight) singly (doubly) resonant solutions*

(3.810a–d) [(3.811a–d)]. The exact and approximate solutions of the one-dimensional generalised equation of mathematical physics and engineering (3.767) are indicated in Table 3.17 and apply to all particular cases in Table 3.16. The influence of step size on the convergence of numerical solutions is considered first in the particular case of the wave equation, leading to a causality condition (note 3.8), that is then extended to the general case (notes 3.9–3.12).

Note 3.8 – Causality Condition for the Wave Equation in Space-Time

The generalised equation of mathematical physics and engineering (3.767) reduces to the wave equation (3.812d) in the case of an elastic string, without stiffness (3.812a), damping (3.812b) or spring support (3.812c):

$$q = \lambda = \omega_0 = 0: \qquad \frac{\partial^2 \Phi}{\partial t^2} - c^2 \frac{\partial^2 \Phi}{\partial x^2} = 0. \qquad (3.812a\text{–}d)$$

The discretisation into finite differences (3.793c; 3.794) in space-time (3.800a–e) leads to (3.813):

$$B_{n,m} = \Phi_{n+2,m} - 2\,\Phi_{n+1,m} + \Phi_{n,m} - (ch/k)^2 \left[\Phi_{n,m+2} - 2\,\Phi_{n,m+1} + \Phi_{n,m}\right], \qquad (3.813)$$

corresponding to (3.801; 3.802a–f) with (3.812a–c). The **discretised wave equation** *(3.813) has characteristic polynomial (3.814a, b):*

$$P_{2,2}(a,b) = a^2 - 2\,a + 1 - (ch/k)^2\,(b^2 - 2\,b + 1)$$

$$= (a-1)^2 - (ch/k)^2\,(b-1)^2, \qquad (3.814a, b)$$

with roots (3.815a) for b and (3.815b) for a, that are inverses:

$$b_{\pm}(a) = 1 \pm (a-1)\frac{k}{c\,h} \qquad \Leftrightarrow \qquad a_{\pm}(b) = 1 \pm (b-1)\frac{c\,h}{k}. \qquad (3.815a, b)$$

The roots for a were considered before (3.805a–d; 3.806a–c; 3.807a–e) for the general equation of mathematical physics, with characteristic polynomial (3.804a–c) of second degree in a; the characteristic polynomial (3.804a) is of fourth degree in b and reduces to the second degree (3.814a, b) in the case of the wave equation (3.812a–d) considered next.

In the unforced case (3.816a) the general solution is: (i) for distinct roots (3.816b) there are (3.803b) sequences of powers (3.816c):

$$B_{n,m} = 0, \qquad b_{+}(a) \neq b_{-}(a): \qquad \Phi_{n,m} = a^n \left\{C_{+}\left[b_{+}(a)\right]^m + C_{-}\left[b_{-}(a)\right]^m\right\};$$
$$(3.816a\text{–}c)$$

(ii) for (3.817c) coincident roots (3.817a, b) a constant sequence (3.817d):

$$b_+(a) = b_-(a) = 1, \qquad a = 1: \qquad \Phi_{n,m} = C_{n,m}. \tag{3.817a–d}$$

In the general case (3.816c) convergence as $n \to \infty$ $(m \to \infty)$ requires (3.818a) [(3.818b)] and the extreme values of $a = \pm 1$ lead to (3.818c, d):

$$|a| \le 1 \ge |b_\pm(a)|: \qquad b_\pm(1) = 1, \qquad b_\pm(-1) = 1 \mp \frac{2k}{ch}. \tag{3.818a–d}$$

The condition (3.818b) is met by (3.818c) and is met also by (3.818d) if (3.819a) holds:

$$-1 < 1 \mp \frac{2k}{ch} < +1 \quad \Rightarrow \quad 1 > \frac{k}{ch} = \frac{\Delta x}{c\,\Delta t} \quad \Leftrightarrow \quad v \equiv \frac{\Delta x}{\Delta t} < c. \tag{3.819a–e}$$

*The condition (3.819b) implies (3.819c, e) meaning (Figure 3.7a) that:(i) the distance $c\,\Delta t$ covered at wave speed c in one time step Δt cannot exceed the position step Δx; (ii) the **velocity of discretisation** (3.819d) cannot exceed the wave speed (3.819e). Thus, (i) $\equiv$ (ii) are equivalent to a **causality condition** that signals cannot travel faster than the wave speed; defining the **characteristic lines** in space-time with wave speed (Figure 3.7b) the steps must lie between them. If the roots of (3.814b) had been obtained, instead of for b as $b_\pm(a)$ in (3.815a) for a, as $a_\pm(b)$ in (3.815b), the same reasoning (3.816a–c) to (3.819a–e) would have led to $v > c$; this is of no physical interest since it excludes the case of zero velocity $v = 0$. Thus, the solutions (3.816b, c) [(3.816a, b)] of the unforced (3.816a) discretised (3.813) wave equation (3.812), for a(b) using (3.815a) [(3.815b)] converge in opposite regions $v < c$ $(v > c)$, corresponding to the interchange of (x,t) axis, and is (is not) of physical interest.* The preceding discussion shows that in the discretisation (3.800a–e) the space Δx and time Δt steps cannot be taken arbitrarily. The causality condition for the convergence of the discretisation of the wave equation (note 3.8) is extended next (note 3.9) to the general equation of mathematical physics and engineering.

Note 3.9 – Convergence Conditions for Finite Differences

The discretised solutions (3.790a–c) of the unforced general equation of mathematical physics and engineering (3.766; 3.767; 3.768a–e) in the case (3.806a–c) of distinct roots of the characteristic polynomial (3.804a–c) converge as $n \to \infty$ if the condition (3.820a) is met:

$$1 > |a_\pm(b)| = \left| -\xi \pm \eta(b) \right| = \left| -\xi \pm \sqrt{\xi^2 - \zeta} \right|, \tag{3.820a–c}$$

using (3.805c, d) the condition of convergence (3.820a) is rewritten (3.820b, c), that is equivalent to the square (3.820d):

$$1 > \left[a_\pm(b) \right]^2 > 2\,\xi^2 - \zeta^2 \mp 2\,\xi\,\sqrt{\xi^2 - \zeta}, \tag{3.820d}$$

or alternatively (3.820e):

$$1 + \zeta - 2\,\xi^2 > \mp\, 2\,\xi\,\sqrt{\xi^2 - \zeta}\,. \tag{3.820e}$$

Three cases arise depending on the value of ξ: (i) if it is zero (3.821a), the condition (3.819b) is met by (3.821b); (ii/iii) if it is negative (3.822a) [positive (3.823a)], the square of (3.820b) leads to the inequality (3.822b) [(3.823b)]:

$$\xi = 0: \qquad\qquad\qquad 1 + \zeta > 0; \qquad\qquad\qquad\qquad\quad (3.821a, b)$$

$$\xi < 0: \qquad\qquad \left(1 + \zeta - 2\,\xi^2\right)^2 > 4\,\xi^2\left(\xi^2 - \zeta\right), \qquad\qquad (3.822a, b)$$

$$\xi > 0: \qquad\qquad \left(1 + \zeta - 2\,\xi^2\right)^2 < 4\,\xi^2\left(\xi^2 - \zeta\right). \qquad\qquad (3.823a, b)$$

Thus, *the discretised solutions (3.790a–c) of the unforced general equation of mathematical physics and engineering (3.766; 3.767; 3.768a–e) in the case (3.806a–c) of distinct roots of the characteristic polynomial (3.804a–c) converge as $n \to \infty$ if the equivalent conditions (3.819a–c) $\equiv$ (3.820a, b) are met. The condition is met (3.821b) for (3.805a) zero (3.821a), and if ξ is negative (positive), it leads to opposite inequalities (3.822a, b) [(3.823a, b)]. Since (3.805b) is a polynomial of degree four in b, the inequalities (3.822b) [(3.823b)] involve polynomials of degree eight.* The causality condition for the wave equation (note 3.8) is shown next (note 3.10) to be a particular case of the convergence conditions for the general equation of mathematical physics and engineering (note 3.9). The convergence conditions for the latter have been considered for distinct roots of the characteristic polynomial (note 3.9) and the case of coincident roots leads to an upper bound on temporal step size due to damping.

Note 3.10 – Upper Bound On Temporal Step Size Due To Damping

In the particular case of the wave equation (3.812a–d) $\equiv$ (3.824a–c) of the six coefficients (3.802a–f), only four are non-zero in (3.824d–i):

$$v = q = \omega_0 = 0: \qquad \left\{ A_{1,0}\ ;\ A_{0,0}\ ;\ A_{0,1}\ ;\ A_{0,2}\ ;\ A_{0,3}\ ;\ A_{0,4} \right\}$$

$$= \left\{ -\,2\ ;\ 1 - c^2 h^2/k^2\ ;\ 2\,c^2 h^2/k^2\ ;\ -\,c^2 h^2/k^2;\ 0\ ;\ 0 \right\}. \tag{3.824a–i}$$

Substitution of (3.824d–i) in (3.805b) leads to (3.825a, b):

$$\zeta(b) = 1 - c^2 h^2/k^2 + 2\,b\ c^2 h^2/k^2 - b^2\ c^2 h^2/k^2 = 1 - (b-1)^2\ c^2 h^2/k^2. \tag{3.825a, b}$$

Using (3.824d) in (3.805a) leads to (3.826a), and comparing (3.825b) with (3.805d) implies (3.826b):

$$\xi = -\,1: \qquad \eta(b) = (b-1)\ ch/k; \qquad\qquad a_\pm(b) = 1 \pm \eta, \qquad (3.826a–c)$$

substitution of (3.826a, b) in (3.805a) leads to (3.826c) that agrees with (3.815b); this shows that the causality conditions (3.819a–c) for the wave equation (3.812a–d) are a particular case of the convergence conditions (3.820a, b) for the general equation of mathematical physics and engineering.

The latter have not been considered in the case (3.807a–e) of double root of the characteristic polynomial (3.804a–c) that is addressed next. The solutions (3.807e) converge as $n \to \infty$ if the condition (3.827a) is met, that is equivalent to (3.827b):

$$1 > |1 - v\ h| \quad \Leftrightarrow \quad -1 < 1 - v\ h < +1. \qquad (3.827a, b)$$

Thus, *the discretised solutions (3.790a–e) of the unforced general equation of mathematical physics and engineering (3.766; 3.767; 3.768a–e) in the case (3.807a–e) of double root of the characteristic polynomial (3.804a–c) converge as $n \to \infty$ if the conditions (3.828a, b) are met implying that: (i) there is damping $v > 0$ in (3.827a) and conversely amplification $v < 0$ would lead to divergence; (ii) damping sets an upper limit on temporal step size (3.827b, c):*

$$0 < v < \frac{2}{h} \equiv \frac{2}{\Delta t}; \qquad \Delta t < \frac{2}{v} = \frac{\rho}{\vartheta}, \qquad (3.828a-e)$$

The upper limit on temporal step size (3.828c) ≡ (3.828d) is given (3.768c) by (3.828e) and increases for: (i) increasing mass density; (ii) decreasing friction coefficient. Thus, the smallest temporal step size is needed for small mass and strong friction that cause a faster decay in time. The discretised solution of the general equation of mathematical physics and engineering in the case of double (3.807a–e) [distinct (3.806a–c)] roots of the characteristic polynomial involves one (two) arbitrary constants C $(C_\pm)$, and the convergence conditions as $m \to \infty$ require $|b| < 1$. The two (three) arbitrary constants $(C, b) \left[(C_\pm, b) \right]$ are determined by one starting and one initial (and plus one boundary) conditions [note 3.11 (3.12)].

Note 3.11 – Application of Starting, Boundary, and Initial Conditons

The solutions (3.807e) ≡ (3.829):

$$\Phi_{n,m} = C\ (1 - v\ h)^n\ b^m, \qquad (3.829)$$

require for convergence as $m \to \infty$ $(n \to \infty)$ in (3.830a) [(3.830c)] that the conditions (3.830c) [(3.830d, e) ≡ (3.828a, b)] be met:

$$m \to \infty: \qquad |b| < 1; \qquad n \to \infty: \qquad 0 < v < \frac{2}{h}. \qquad (3.830a-e)$$

The two arbitrary constants (C, b) can be determined from: (i) the **starting condition** (3.831a) for $n = 0 = m$; (ii) the **initial condition** (3.831b) at the first spatial step:

$$\Phi_{0,0} = C, \qquad \Phi_{0,1} = C\ b = \Phi_{0,0}\ b. \qquad (3.831a-c)$$

From (3.831a, b) follows (3.831c), and substitution of (3.831a, c) in (3.829) leads to (3.832b):

$$|\Phi_{0,1}| < |\Phi_{0,0}| : \qquad \Phi_{n,m} = (1 - v\ h)^n\ \left(\Phi_{0,1}\right)^m\ \left(\Phi_{0,0}\right)^{1-m}. \qquad (3.832a, b)$$

Thus, *the discretised solutions (3.790a–c) of the unforced general equation of mathematical physics and engineering (3.766; 3.767; 3.768a–e) in the case (3.829) of double root (3.807a–e) of the characteristic polynomial (3.804a–c) involves two arbitrary constants* (C,b), *that can be determined from one starting (3.831a) and one initial (3.831b) condition, leading to (3.832b) that converges for (3.830a) [(3.830c)] if the condition (3.830b) [(3.832a)] is met*. The condition (3.832a) follows from (3.831c) and (3.830b).

Note 3.12 – Double and Distinct Roots of the Characteristic Polynomial

Considering distinct (3.805a–d) instead of double (3.806a–d) roots of the characteristic polynomial (3.804a–c), the solution (3.807e) is replaced by (3.807b–d) ≡ (3.833):

$$\Phi_{n,m} = b^m \left\{ C_+ \left[a_+(b) \right]^n + C_- \left[a_-(b) \right]^n \right\}, \tag{3.833}$$

involving the distinct roots (3.805b–d) that depend on the arbitrary constant b, and also the arbitrary constant coefficients $C_\pm$. The solution (3.833) converges for (3.834a) [(3.834c)] if the condition (3.834b) [(3.834d, e)] is met:

$$m \to \infty: \qquad |b| < 1; \qquad n \to \infty: \qquad 1 > |a_\pm(b)| = |-\xi \pm \eta(b)|, \tag{3.834a–e}$$

where (3.805c) was used in (3.834e). The three arbitrary constants $(C_\pm, b)$ can be determined from: (i) a starting condition (3.835a); (ii) an initial condition (3.835b) at the first space step; and (iii) the boundary condition at the first time step (3.835c):

$$\Phi_{0,0} = C_+ + C_-, \qquad \Phi_{0,1} = b \left(C_+ + C_- \right), \qquad \Phi_{1,0} = C_+ \, a_+(b) + C_- \, a_-(b). \tag{3.835a–c}$$

The constant b is determined from the ratio (3.836a) of (3.835b, a) and coincides with (3.831c):

$$b = \frac{\Phi_{0,1}}{\Phi_{0,0}}: \qquad\qquad C_\pm = \pm \frac{\Phi_{1,0} - a_\mp(b)\,\Phi_{0,0}}{a_+(b) - a_-(b)}, \tag{3.836a–c}$$

the constants $C_\pm$ are determined (3.836b, c) solving (3.835a, c).

The convergence conditions are (i) for (3.830a) the same (3.830b) ≡ (3.832a) for distinct (3.836a) or double (3.831c) roots; (ii) for (3.830c) ≡ (3.837a) the condition (3.834e) is rewritten (3.837b–d) using (3.807a, b, d) and (3.836a):

$$n \to \infty: \qquad 1 > \bar{a}_\pm \equiv \left| a_\pm \left(\Phi_{0,1} / \Phi_{0,0} \right) \right|$$

$$= \left| -A_{1,0}/2 \pm \left| \left(A_{1,0} \right)^2 / 4 - \sum_{r=1}^{4} A_{0,r} \left(\Phi_{0,1} / \Phi_{0,0} \right)^r \right|^{1/2} \right| \tag{3.837a–d}$$

involving the coefficients (3.802a–f; 3.768a–e). Substitution of (3.836a–c) in (3.833) leads to (3.838):

$$\Phi_{n,m} = \frac{\left(\Phi_{0,1}/\Phi_{0,0}\right)^m}{\bar{a}_+ - \bar{a}_-} \left\{ \left(\Phi_{1,0} - \bar{a}_- \; \Phi_{0,0}\right)\left(\bar{a}_+\right)^n - \left(\Phi_{1,0} - \bar{a}_+ \; \Phi_{0,0}\right)\left(\bar{a}_-\right)^n \right\}. \quad (3.838)$$

Thus, *the discretised solution (3.790a–c) of the unforced general equation of mathematical physics and engineering (3.766) in the case (3.833) of distinct roots (3.805a–d) of the characteristic polynomial (3.804a–c) with starting (3.835a), initial (3.835b), and boundary (3.835c) conditions is (3.838) that converges for (3.834a) [(3.837a)] if the condition (3.834b) [(3.837b)] is met. In the solution (3.838) appears (3.837c, d) that are the double roots (3.805c, d) with (3.805a, b) evaluated for (3.836a), and involving the six constants (3.802a–f) specified by four physical coefficients (3.768a–d).*

TABLE 3.16

Fourth-order P.D.E. of Mathematical Physics and Engineering

$$\rho \frac{\partial^2 \Phi}{\partial t^2} + \vartheta \frac{\partial \Phi}{\partial t} + k\,\Phi - T \frac{\partial^2 \Phi}{\partial x^2} + EI \frac{\partial^4 \Phi}{\partial x^4} = F(x,t)$$

or

$$\frac{\partial^2 \Phi}{\partial t^2} + 2v \frac{\partial \Phi}{\partial t} + \omega_0^2\,\Phi - c^2 \frac{\partial^2 \Phi}{\partial x^2} + q^2 \frac{\partial^4 \Phi}{\partial x^4} = B(x,t)$$

Number	Name	Parameters	Equation
1	Damped oscillator	$T = 0 = EI$	$\rho \dfrac{\partial^2 \Phi}{\partial t^2} + \vartheta \dfrac{\partial \Phi}{\partial t} + k\,\Phi = F$
2	Wave	$\vartheta = k = EI = 0$	$\rho \dfrac{\partial^2 \Phi}{\partial t^2} - T \dfrac{\partial^2 \Phi}{\partial x^2} = F$
3	Diffusion	$k = T = EI = 0$	$\vartheta \dfrac{\partial \Phi}{\partial t} - T \dfrac{\partial^2 \Phi}{\partial x^2} = F$
4	Wave-diffusion	$k = 0 = EI$	$\rho \dfrac{\partial^2 \Phi}{\partial t^2} + \vartheta \dfrac{\partial \Phi}{\partial t} - T \dfrac{\partial^2 \Phi}{\partial x^2} = F$
5	Bar	$\vartheta = k = T = 0$	$\rho \dfrac{\partial^2 \Phi}{\partial t^2} + EI \dfrac{\partial^4 \Phi}{\partial x^4} = F$
6	Beam	$\vartheta = k = 0$	$\rho \dfrac{\partial^2 \Phi}{\partial t^2} - T \dfrac{\partial^2 \Phi}{\partial x^2} + EI \dfrac{\partial^4 \Phi}{\partial x^4} = F$

Wave speed: $c \equiv \sqrt{\dfrac{I}{\rho}}$, damping: $v \equiv \dfrac{\vartheta}{2\rho}$, diffusivity: $\chi \equiv \dfrac{T}{\vartheta}$

Natural frequency: $\omega_0 \equiv \sqrt{\dfrac{k}{\rho}}$, bending stiffness parameter: $q \equiv \sqrt{\dfrac{EI}{\rho}}$

External force per unit mass: $B(x,t) \equiv \dfrac{F(x,t)}{\rho}$

Note: – Combination of the six examples of equations in Tables 3.3 and 3.4 as particular cases of a generalised one-dimensional equation of mathematical physics and engineering.

TABLE 3.17

One-dimensional Equation of Mathematical Physics and Engineering

Type	Partial Differential *	Finite Difference *
Equation	(3.766) ≡ (3.767; 3.768a–e)	(3.799) ≡ (3.801; 3.802a–f)
Characteristic polynomial	(3.771a–d)	(3.804a–c)
Unforced solutions:	(3.773a–d; 3.772a, b)	(3.805a–d)
single roots	(3.776a–f; 3.775a, b)	(3.806a–c)
Double roots	(3.774a–g)	(3.807a–e)
	(3.777a–g)	
Forced solutions	(3.779)	(3.808)
Non-resonant	(3.780a, b)	(3.809a, b)
Singly-resonant	(3.781a–d)	(3.810a–d)
	(3.782a–d)	
Doubly-resonant	(3.783a–e)	(3.811a–d)
	(3.784a–f)	
Power forcing	(3.785; 3.787d)	-
	(3.788; 3.789e)	

* Linear with constant coefficients

Note: – Natural, general, and special unforced solutions of the unforced equations and non-resonant, resonant, and other forced solutions, for the generalised one-dimensional equations of mathematical physics and engineering as a partial differential equation in Table 3.16 and its approximate discretisation into a multiple finite difference equation using a rectangular space-time grid as in Figure 3.5.

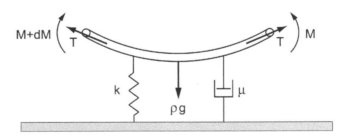

FIGURE 3.5 – The generalised equation of mathematical physics (3.768) is satisfied, for example by the linear transverse deflection ζ of a beam with mass density per unit length ρ, subject to a tangential tension T, supported on a damper with damping μ and a spring with resilience k. The external force could be the weight ρg in the gravity field with acceleration g. The stiffness of the beam resists the bending moment M and equals EI the product of the Young's modulus E of the material by the moment of inertia I of the cross-section.

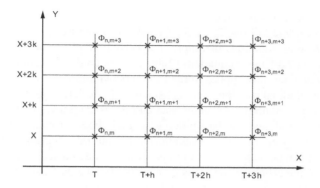

FIGURE 3.6 – Discretisation of a partial differential equation in space-time, respectively vertical–horizontal axis, applied to the generalised equation of mathematical physics and engineering.

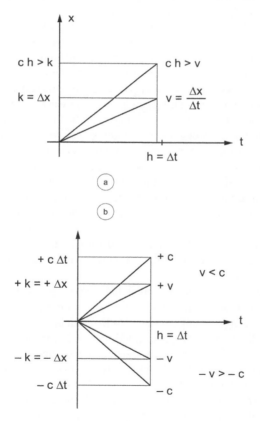

FIGURE 3.7 – In the discretisation of the wave equation (3.812d) in space-time, the space step Δx must be taken at a velocity v relative to the time Δt smaller than the wave speed c to satisfy causality (a), since signals cannot travel faster than the wave speed. This means that the successive grid points (b) must lie within the characteristics that are the curves of propagation at wave speed c in opposite direction $\pm c$.

CONCLUSION 3 – Five methods of solution of linear partial differential equations were considered (Table 3.1 and 3.2) and illustrated by application to the simplest second-order equations of elliptic / hyperbolic / parabolic type, namely the harmonic / wave / diffusion equations (Table 3.3). Not only these, but also the elastic bar (beam) equation, that applies to the transverse vibrations of an elastic rod without (with) longitudinal traction, were considered (Table 3.4) using the method of the characteristic polynomial. The latter was chosen as the most practical of the five methods considered, and can be applied to a single linear partial differential equation with constant coefficients to obtain: (i) the natural integrals of the unforced equation, that can be added to form the general (considered separately as a special) integral involving arbitrary functions (constants); (ii) the particular integrals due to exponential forcing in non-resonant and resonant cases (Table 3.5); (iii) the particular integrals due to forcing by the product of powers, polynomials, exponentials, circular and hyperbolic cosines and sines and other smooth functions (Table 3.6); (iv) the particular integrals due to forcing can also be obtained by the method of the inverse polynomial of partial derivatives (Table 3.7).

The method of the characteristic polynomial applies to six classes of equations (Table 3.8), namely single (simultaneous systems) of linear partial differential equations with constant (homogeneous power) coefficients and linear multiple finite difference equations (simultaneous systems) with constant coefficients (Table 3.10), with several analogies. For example, a linear partial differential equation with all derivatives of the same order and constant (homogeneous power) coefficients has similarity (analogous) solutions (Table 3.9). The analogies in the method of characteristic polynomials applied to these six classes of equations and simultaneous systems include: (i) the general and special integrals in the unforced case (Table 3.11); (ii) the forcing leading to non-resonant and resonant cases (Table 3.12); (iii) the use of the method of the inverse characteristic polynomial of partial ordinary / homogeneous derivatives and finite differences to obtain particular solutions with forcing (Table 3.13); (iv) the examples including analogous equations and simultaneous systems of different classes such as ordinary / homogeneous derivatives and finite differences (Table 3.14); (v) application of boundary and initial (starting) conditions to ensure the unicity of solution of partial differential (finite difference) equations and simultaneous systems (Table 3.15).

The five examples of equations, namely harmonic, wave, diffusion, bar and beam equations can be combined in a generalised one-dimensional equation of mathematical physics and engineering (Table 3.16) that can be discretised in a rectangular grid (Figure 3.4) leading to the corresponding partial or multiple finite difference equation (Table 3.17). To both forms partial differential and finite difference of the one-dimensional equations of mathematical physics, and its particular cases (Diagram 3.2) apply the natural, general, special, non-resonant, resonant and other forced integrals (Diagram 3.1). The physical phenomena involved are combinations and interactions of three types: (i) steady fields specified by the two-dimensional harmonic equation; (ii) similarity solutions representing permanent wave forms propagating without deformation for the wave equation (Figure 3.1); (iii) dissipative phenomena corresponding to decay in space and time (Figure 3.2 – 3.3) for the diffusion equation. An example of combination is the transverse vibrations of an elastic beam, under tangential traction, supported on springs, with friction damping and acted by its own weight (Figure 3.5). In the space-time discretisation (Figure 3.6) the steps may not be arbitrary, for example to satisfy the condition of causality in the case of the wave equation (Figure 3.7).

The Diagram 3.2 indicates the types of solutions of equations and simultaneous systems that appear in the Tables 3.3 – 3.7 and 3.9 – 3.15 and 3.17. The "solutions" are "integrals" ("sequences") for partial differential (multiple finite difference) equations and simultaneous systems. For unforced equations: (i) natural solutions, that are linearly independent particular solutions involving each one arbitrary functions; (ii) the general solution is a sum or linear combination of all natural solutions, and can satisfy general initial and/or boundary conditions; (iii) the special solutions may involve arbitrary constant instead of arbitrary functions, and may exist only for particular boundary and/or initial conditions. For forced equations: (i/ii) non-resonant and resonant solutions exist for equations and simultaneous systems with constant (homogeneous power) coefficients forced by exponentials (powers); (iii) other types of forcing lead to particular solutions that may be obtained by the method of inverse characteristic polynomials of partial derivatives or finite differences.

The preceding methods of solution of linear partial differential equations apply to a wide range of physical and engineering problems, and are applied next (Chapters 4, 6 and 8) to a variety of waves, namely elastic, acoustic and electromagnetic waves: (i/ii) transverse waves in one (two)-dimensional elastic bodies like (i) strings [(ii) membranes)]: (iii/iv) surface water waves in two (three) dimensions, such as in (iii) channels [(iv) basins]; (v-vii) sound waves in one / two / three dimensions, for example in (v) tubes / (vi) cavities / (vii) rooms; (viii-ix) electromagnetic waves in vacuum or matter in (viii) free space or (ix) waveguides; (x) longitudinal and transversal elastic waves in three-dimensional media; (xi-xiii) waves in one-dimensional stiff bodies, such as (xi) longitudinal, (xii) torsional and (xiii) transversal vibrations of rods; (xiv-xv) waves in two-dimensional stiff bodies, such as (xiv) in-plane and (xv) bending vibrations of plates.

References

1744 Euler, L. De curvis elasticis, in Methodus Inveniendi lineas curvas maximi and minimi propritate gaudentes. Lausanne.

1744 Bernoulli, D. Véritable hypothése de la résistance des solides, avec demonstration de la courbure des corps qui font ressort, Geneva, Switzerland (Gesammelte Werke, 2).

1747 D'Alembert, J. R. Essai sur les vibrations des cordes. *Opuscules Mathématiques* 1, 1–47.

1760 Lagrange, J. L. Nouvelles recherches sur la nature et la propagation du son. *Miscellania Taurinensia* 2, 11 (*Ouevres*, Vol. 1, 151–316, Gauthier-Villars, Paris)

1818 Fourier, J. B. J. *Theorie analytique de la chaleur.* Paris, reprinted Dover, New York, 1955.

1820 Laplace, P. S. *Mécanique Celeste.* Vol. 5. Gauthier-Villars, Paris, reprinted Chelsea, New York, 1950.

1821 Cauchy, A. L. *Analyse Algébrique.* Gaulthier-Villars, Paris.

1851 Riemann, B. Grundlagen fur allgemaine Theorie der Functionen einer veränderlichen complexen Grösse. *Inaugunal dissertation.* Göttingen, Werke 1.35–80.

1873 Maxwell, J. C. *Treatise on electricity and magnetism.* Oxford University Press, Oxford, reprinted Clarendon Press, Oxford, 1891, reprinted Dover, New York, 1954.

1892 Heaviside, O. *Electrical Papers.* Vol. 1. Macmillan Co., London and New York.

Bibliography

The bibliography of the series "Mathematics and Physics for Science and Technology" is quite extensive since it covers a variety of subjects. To avoid overlaps, each volume contains only a part of the bibliography on the subjects most closely related to its content. The bibliography of earlier volumes is mostly not repeated, and some of the bibliography may be relevant to earlier and future volumes. The bibliography covered in the four published volumes is:

A. General
- a. Overviews
  - 1. General mathematics: book 1
  - 2. Theoretical physics: book 2
  - 3. Engineering technology: book 3
- b. Reference
  - 4. Collected works: book 6
  - 5. Generic encyclopaedias: book 10
  - 6. Historical accounts: book 12
B. Mathematics
- c. Theory of functions
  - 7. Real functions: books 1 and 2
  - 8. Complex analysis: books 1 and 2
  - 9. Generalized functions: book 3
- d. Differential and integral equations
  - 10. Ordinary differential equations: book 4
  - 11. Partial differential equations: books 10 and 12
  - 12. Non-linear differential and integral equations: book 5
- e. Geometry
  - 13. Tensor calculus: book 3
- f. Higher analysis
  - 14. Special functions: book 8
C. Physics
- g. Classical mechanics
  - 15. Material particles: book 7
- h. Thermodynamics
  - 16. Thermodynamics: books 2 and 11
  - 17. Heat: books 2 and 11
- i. Fluid mechanics
  - 18. Hydrodynamics: book 1
  - 19. Aerodynamics: book 1
- j. Solid mechanics
  - 20. Elasticity: book 2

    21.   Structures: book 2
    22.   Materials: book 3
  k. Oscillations
    23.   Waves: book 12
  l. Electromagnetism
    24.   Optics and Electronics: book 3
This choice of subjects is explained next.

The general bibliography consists of overviews and reference works. The overviews have been completed with mathematics, physics, and engineering, respectively, in volumes I, II, and III corresponding to books 1, 2, and 3. The reference bibliography starts with the collected works of notable authors in book 6 of volume IV, continues with generic encyclopaedias in book 10 of volume V, and is completed with historical accounts in the present book 12 of volume V. Concerning mathematics, the bibliography on the theory of functions has appeared in volumes I, II, and III corresponding to books 1, 2, and 3. The bibliography on ordinary differential equations has appeared in book 4 and on non-linear differential equations in book 5, both in volume IV and partial differential equations in books 10 and 12 of volume V. Volume IV also contains the bibliography on special functions in book 8. The bibliography on tensor calculus appears in book 3 that coincides with volume III. Concerning physics, the bibliography on material particles appears in book 7 of volume IV. The bibliography on thermodynamics and heat appears in book 2 that coincides with volume II and again in book 11 of volume V. The bibliography on solid mechanics, including elasticity and structures, appears in book 2 that coincides with volume II, and materials in book 3 that coincides with volume III. The bibliography on fluid mechanics, including hydrodynamics and aerodynamics, appears in book 1 that coincides with volume I. The bibliography on oscillations starts with waves in the present book 12 of volume V. The bibliography on electromagnetism starts with optics and electronics in book 3 that coincides with volume III. The present book 12 in volume V includes the bibliography on:

  1. Historical accounts
  2. Partial differential equations
  3. Waves

1. HISTORICAL ACCOUNTS

[1] Coolidge, J. L. *A history of geometrical methods.* Oxford University Press 1948, Dover 1963,
[2] Daintith, J., Tootill, E. & Gjertsen, D. *Biographical encyclopedia of scientists.* Institute of Physics Publishing 1981, 3rd edition 1994, 2 vols.,
[3] Descartes. R. *La Géometrie.* Open Court Publishing 1925, Dover 1956,
[4] Heath, T. L. *The Thirteen books of Euclid's elements.* Cambridge University Press 1909, 2nd edition 1925, Dover 1956, 3 vols.,
[5] Livanova, A. *Lev Landau.* Nauka 1978, Editions Mir 1981,

[6] Parke III, N. C. *Guide to the literature of mathematics and physics, works in the engineering science including related subjects.* McGraw-Hill 1947, 2nd edition Dover 1958,
[7] Rouse Ball, W. W. *A short account of the history of mathematics.* 1891, 4th edition 1908, reprinted 1960,
[8] Smith, D. E. *A source book in mathematics.* McGraw-Hill 1929, reprinted Dover 1959,
[9] Struik, D. J. *Concise history of mathematics.* 1948, Dover 1967,
[10] Valson, V. A. *Cauchy: sa vie et ses travaux.* Gaulthier-Villars, 1881.

2. PARTIAL DIFFERENTIAL EQUATIONS

[1] Caratheodory, C. *Calculus of variations and partial differential equations of first order.* Reprinted Chelsea 1982,
[2] Costanda, C. *Solution techniques for elementary partial differential equations.* Chapman & Hall 2020,
[3] Epstein, B. *Partial differential equations: An introduction.* McGraw-Hill 1962,
[4] Garabedian, P. R. *Partial differential equations.* 1964, Chelsea 1986,
[5] Greenspan, D. *Introduction to partial differential equations.* McGraw-Hill 1961,
[6] Godounov, S. *Equations de la physique mathematique.* Editions Mir 1973,
[7] Lie, S. *Differentialgleichungen.* Leipzig 1891, reprinted Chelsea 1967,
[8] Mikhailov, V. *Equations aux derivées partielles.* Nauka 1976, Editions Mir 1980,
[9] Myint-u, T. and Debnath, L. *Partial differential equations for scientists and engineers.* Prentice-Hall 1987,
[10] Sneddon, I. N. *Partial differential equations.* McGraw-Hill 1957,
[11] Sobolev, S. L. *Partial Differential equations of mathematical physics.* Pergamon Press 1964,
[12] Taylor, M. E. *Partial differential equations.* Springer 1996,
[13] Tricomi, F. G. *Equazoni a derivate parziali.* Cremonese 1957,
[14] Webster, A. G. *Partial differential equations of mathematical physics.* 1927, 2nd edition 1933, reprinted Dover 1955,
[15] Young, E. C. *Partial differential equations: An introduction.* Aleyn & Bacon 1972,
[16] Zachmanoglou, E. C. and Theoe, D. N. *Partial differential equations with applications.* Williams & Wilkins 1976, reprinted Dover 1986,
[17] Zauderer, E. *Partial differential equations of applied mathematics.* Wiley 1989.

3. WAVES

[1] Baker, B. B. & Copson, E. T. *Huyghen's principle.* Oxford University Press 1939,
[2] Chernov, L. A. *Wave propagation in a random medium.* McGraw-Hill 1960,
[3] Colton, D. & Kress, R. *Inverse acoustic and electromagnetic scattering theory.* Springer 1992,
[4] Debnath, L. *Non-linear waves.* Cambridge University Press 1983,
[5] Debnath, L. *Non-linear dispersive waves systems.* World Scientific 1992,
[6] Drumheller, D. S. *Wave propagation in non-linear fluids and solids.* Cambridge University Press 1998,
[7] Felsen, L. B. & Marcuwicz, N. *Wave radiation and scattering.* Institute of Electrical and Electronic Engineers 1994,
[8] Friedlander, F. G. *Sound pulses.* Cambridge University Press 1958,
[9] Ishimaru, A. *Wave propagation and scattering in random media.* Academic Press 1978, 2 vols.,
[10] Karpman, V. I. *Non-linear waves in dispersive media.* Pergamon 1975,

[11] Kelbert, M. & Sazonov, I. *Pulses and other wave processes in fluids*. Kluwer 1996,

[12] Lekner, J. *Theory of reflection*. Martinus Nijhoff 1987,

[13] Naugolnykh, K. & Ostrovsky, L. *Non-linear dispersive waves processes in acoustics*. Cambridge University Press 1998,

[14] Novikov, B. K., Rudenko, O. V. & Timoshenko, V. I. *Non-linear underwater acoustics*. Acoustical Society of America 1987,

[15] Ogilvy, J. A. *Wave scattering from random rough surfaces*. Institute of Physics 1992,

[16] Rose, J. K. *Ultrasonic waves in solid media*. Cambridge University Press 1958,

[17] Tatarski, V. I. *Wave propagation in a turbulent medium*. McGraw-Hill 1961,

[18] Uscinski, J. A. *Wave propagation in a random media*. McGraw-Hill 1977,

[19] Whitham, G. B. *Linear and non-linear waves*. Willey 1974.

Index

For Product Safety Concerns and Information please contact our EU
representative GPSR@taylorandfrancis.com
Taylor & Francis Verlag GmbH, Kaufingerstraße 24, 80331 München, Germany

www.ingramcontent.com/pod-product-compliance
Ingram Content Group UK Ltd.
Pitfield, Milton Keynes, MK11 3LW, UK
UKHW021119180425
457613UK00005B/153